Universitext

Universitext

Editors (North America): S. Axler and K.A. Ribet

(continued after index)

Hui-Hsiung Kuo

Introduction to
Stochastic Integration

 Springer

Hui-Hsiung Kuo
Department of Mathematics
Louisiana State University
Baton Rouge, LA 70803-4918
USA
kuo@math.lsu.edu

Mathematics Subject Classification (2000): 60-XX

Library of Congress Control Number: 2005935287

ISBN-10: 0-387-28720-5   Printed on acid-free paper.
ISBN-13: 978-0387-28720-1

Printed in the United States of America.   (EB)

9 8 7 6 5 4 3 2 1

springeronline.com

Dedicated to Kiyosi Itô,
and in memory of
his wife Shizue Itô

Preface

In the Leibniz–Newton calculus, one learns the differentiation and integration of deterministic functions. A basic theorem in differentiation is the chain rule, which gives the derivative of a composite of two differentiable functions. The chain rule, when written in an indefinite integral form, yields the method of substitution. In advanced calculus, the Riemann–Stieltjes integral is defined through the same procedure of "partition-evaluation-summation-limit" as in the Riemann integral.

In dealing with random functions such as functions of a Brownian motion, the chain rule for the Leibniz–Newton calculus breaks down. A Brownian motion moves so rapidly and irregularly that almost all of its sample paths are nowhere differentiable. Thus we cannot differentiate functions of a Brownian motion in the same way as in the Leibniz–Newton calculus.

In 1944 Kiyosi Itô published the celebrated paper "Stochastic Integral" in the Proceedings of the Imperial Academy (Tokyo). It was the beginning of the Itô calculus, the counterpart of the Leibniz–Newton calculus for random functions. In this six-page paper, Itô introduced the stochastic integral and a formula, known since then as Itô's formula.

The Itô formula is the chain rule for the Itô calculus. But it cannot be expressed as in the Leibniz–Newton calculus in terms of derivatives, since a Brownian motion path is nowhere differentiable. The Itô formula can be interpreted only in the integral form. Moreover, there is an additional term in the formula, called the Itô correction term, resulting from the nonzero quadratic variation of a Brownian motion.

Before Itô introduced the stochastic integral in 1944, informal integrals involving white noise (the nonexistent derivative of a Brownian motion) had already been used by applied scientists. It was an innovative idea of Itô to consider the product of white noise and the time differential as a Brownian motion differential, a quantity that can serve as an integrator. The method Itô used to define a stochastic integral is a combination of the techniques in the Riemann–Stieltjes integral (referring to the integrator) and the Lebesgue integral (referring to the integrand).

The Itô calculus was originally motivated by the construction of Markov diffusion processes from infinitesimal generators. The previous construction of such processes had to go through three steps via the Hille–Yosida theory, the Riesz representation theorem, and the Kolmogorov extension theorem. However, Itô constructed these diffusion processes directly in a single step as the solutions of stochastic integral equations associated with the infinitesimal generators. Moreover, the properties of these diffusion processes can be derived from the stochastic integral equations and the Itô formula.

During the last six decades the Itô theory of stochastic integration has been extensively studied and applied in a wide range of scientific fields. Perhaps the most notable application is to the Black–Scholes theory in finance, for which Robert C. Merton and Myron S. Scholes won the 1997 Nobel Prize in Economics. Since the Itô theory is the essential tool for the Black–Scholes theory, many people feel that Itô should have shared the Nobel Prize with Merton and Scholes.

The Itô calculus has a large spectrum of applications in virtually every scientific area involving random functions. But it seems to be a very difficult subject for people without much mathematical background. I have written this introductory book on stochastic integration for anyone who needs or wants to learn the Itô calculus in a short period of time. I assume that the reader has the background of advanced calculus and elementary probability theory. Basic knowledge of measure theory and Hilbert spaces will be helpful. On the other hand, I have written several sections (for example, §2.4 on conditional expectation and §3.2 on the Borel–Cantelli lemma and Chebyshev inequality) to provide background for the sections that follow. I hope the reader will find them helpful. In addition, I have also provided many exercises at the end of each chapter for the reader to further understand the material.

This book is based on the lecture notes of a course I taught at Cheng Kung University in 1998 arranged by Y. J. Lee under an NSC Chair Professorship. I have revised and implemented this set of lecture notes through the courses I have taught at Meijo University arranged by K. Saitô, University of Rome "Tor Vergata" arranged by L. Accardi under a Fulbright Lecturing grant, and Louisiana State University over the past years. The preparation of this book has also benefited greatly from my visits to Hiroshima University, Academic Frontier in Science of Meijo University, University of Madeira, Vito Volterra Center at the University of Rome "Tor Vergata," and the University of Tunis El Manar since 1999.

I am very grateful for financial support to the above-mentioned universities and the following offices: the National Science Council (Taiwan), the Ministry of Education and Science (Japan), the Luso-American Foundation (Portugal), and the Italian Fulbright Commission (Italy). I would like to give my best thanks to Dr. R. W. Pettit, Senior Program Officer of the CIES Fulbright Scholar Program, and Ms. L. Miele, Executive Director of the Italian Fulbright Commission, and the personnel in her office for giving me assistance for my visit to the University of Rome "Tor Vergata."

Many people have helped me to read the manuscript for corrections and improvements. I am especially thankful for comments and suggestions from the following students and colleagues: W. Ayed, J. J. Becnel, J. Esunge, Y. Hara-Mimachi, M. Hitsuda, T. R. Johansen, S. K. Lee, C. Macaro, V. Nigro, H. Ouerdiane, K. Saitô, A. N. Sengupta, H. H. Shih, A. Stan, P. Sundar, H. F. Yang, T. H. Yang, and H. Yin. I would like to give my best thanks to my colleague C. N. Delzell, an amazing T$_E$Xpert, for helping me to resolve many tedious and difficult T$_E$Xnical problems. I am in debt to M. Regoli for drawing the flow chart to outline the chapters on the next page. I thank W. Ayed for her suggestion to include this flow chart. I am grateful to M. Spencer of Springer for his assistance in bringing out this book.

I would like to give my deepest appreciation to L. Accardi, L. Gross, T. Hida, I. Kubo, T. F. Lin, and L. Streit for their encouragement during the preparation of the manuscript. Especially, my Ph. D. advisor, Professor Gross, has been giving me continuous support and encouragement since the first day I met him at Cornell in 1966. I owe him a great deal in my career.

The writing style of this book is very much influenced by Professor K. Itô. I have learned from him that an important mathematical concept always starts with a simple example, followed by the abstract formulation as a definition, then properties as theorems with elaborated examples, and finally extension and concrete applications. He has given me countless lectures in his houses in Ithaca and Kyoto while his wife prepared the most delicious dinners for us. One time, while we were enjoying extremely tasty shrimp-asparagus rolls, he said to me with a proud smile "If one day I am out of a job, my wife can open a restaurant and sell only one item, the shrimp-asparagus rolls." Even today, whenever I am hungry, I think of the shrimp-asparagus rolls invented by Mrs. K. Itô. Another time about 1:30 a.m. in 1991, Professor Itô was still giving me a lecture. His wife came upstairs to urge him to sleep and then said to me, "Kuo san (Japanese for Mr.), don't listen to him." Around 1976, Professor Itô was ranked number 2 table tennis player among the Japanese probabilists. He was so strong that I just could not get any point in a game with him. His wife then said to me, "Kuo san, I will get some points for you." When she succeeded occasionally to win a point, she would joyfully shake hands with me, and Professor Itô would smile very happily.

When I visited Professor Itô in January 2005, my heart was very much touched by the great interest he showed in this book. He read the table of contents and many pages together with his daughter Keiko Kojima and me. It was like the old days when Professor Itô gave me lectures, while I was also thinking about the shrimp-asparagus rolls.

Finally, I must thank my wife, Fukuko, for her patience and understanding through the long hours while I was writing this book.

Hui-Hsiung Kuo
Baton Rouge
September 2005

Outline

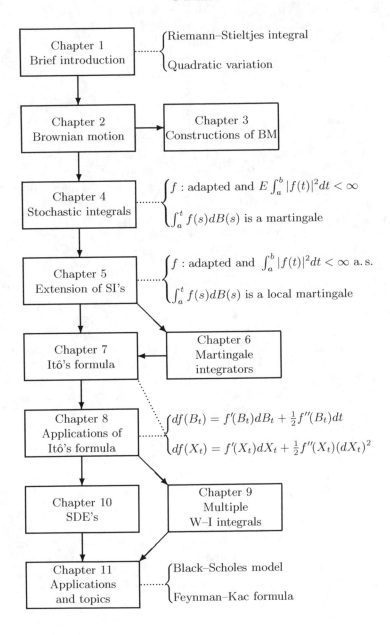

Contents

1

Introduction

1.1 Integrals

First we review the definitions of the Riemann integral in calculus and the Riemann–Stieltjes integral in advanced calculus [74].

(a) **Riemann Integral**

A bounded function f defined on a finite closed interval $[a, b]$ is called *Riemann integrable* if the following limit exists:

$$\int_a^b f(t)\, dt = \lim_{\|\Delta_n\| \to 0} \sum_{i=1}^n f(\tau_i)(t_i - t_{i-1}),$$

where $\Delta_n = \{t_0, t_1, \ldots, t_{n-1}, t_n\}$ is a partition of $[a, b]$ with the convention $a = t_0 < t_1 < \cdots < t_{n-1} < t_n = b$, $\|\Delta_n\| = \max_{1 \le i \le n}(t_i - t_{i-1})$, and τ_i is an evaluation point in the interval $[t_{i-1}, t_i]$. If f is a continuous function on $[a, b]$, then it is Riemann integrable. Moreover, it is well known that a bounded function on $[a, b]$ is Riemann integrable if and only if it is continuous almost everywhere with respect to the Lebesgue measure.

(b) **Riemann–Stieltjes Integral**

Let g be a monotonically increasing function on a finite closed interval $[a, b]$. A bounded function f defined on $[a, b]$ is said to be *Riemann–Stieltjes integrable* with respect to g if the following limit exists:

$$\int_a^b f(t)\, dg(t) = \lim_{\|\Delta_n\| \to 0} \sum_{i=1}^n f(\tau_i)\big(g(t_i) - g(t_{i-1})\big), \tag{1.1.1}$$

where the partition Δ_n and the evaluation points τ_i are given as above. It is a well-known fact that continuous functions on $[a, b]$ are Riemann–Stieltjes integrable with respect to any monotonically increasing function on $[a, b]$.

Suppose f is monotonically increasing and continuous and g is continuous. Then we can use the integration by parts formula to define

$$\int_a^b f(t)\,dg(t) \equiv f(t)g(t)\Big]_a^b - \int_a^b g(t)\,df(t), \tag{1.1.2}$$

where the integral in the right-hand side is defined as in Equation (1.1.1) with f and g interchanged. This leads to the following question.

Question 1.1.1. For any continuous functions f and g on $[a, b]$, can we define the integral $\int_a^b f(t)\,dg(t)$ by Equation (1.1.1)?

Consider the special case $f = g$, namely, the integral

$$\int_a^b f(t)\,df(t).$$

Let $\Delta_n = \{t_0, t_1, \ldots, t_n\}$ be a partition of $[a, b]$. Let L_n and R_n denote the corresponding Riemann sums with the evaluation points $\tau_i = t_{i-1}$ and $\tau_i = t_i$, respectively, namely,

$$L_n = \sum_{i=1}^n f(t_{i-1})\big(f(t_i) - f(t_{i-1})\big), \tag{1.1.3}$$

$$R_n = \sum_{i=1}^n f(t_i)\big(f(t_i) - f(t_{i-1})\big). \tag{1.1.4}$$

Is it true that $\lim L_n = \lim R_n$ as $\|\Delta_n\| \to 0$? Observe that

$$R_n - L_n = \sum_{i=1}^n \big(f(t_i) - f(t_{i-1})\big)^2, \tag{1.1.5}$$

$$R_n + L_n = \sum_{i=1}^n \big(f(t_i)^2 - f(t_{i-1})^2\big) = f(b)^2 - f(a)^2. \tag{1.1.6}$$

Therefore, R_n and L_n are given by

$$R_n = \frac{1}{2}\bigg(f(b)^2 - f(a)^2 + \sum_{i=1}^n \big(f(t_i) - f(t_{i-1})\big)^2\bigg),$$

$$L_n = \frac{1}{2}\bigg(f(b)^2 - f(a)^2 - \sum_{i=1}^n \big(f(t_i) - f(t_{i-1})\big)^2\bigg).$$

The limit of the right-hand side of Equation (1.1.5) as $\|\Delta_n\| \to 0$, if it exists, is called the *quadratic variation* of the function f on $[a, b]$. Obviously, $\lim_{\|\Delta_n\| \to 0} R_n \neq \lim_{\|\Delta_n\| \to 0} L_n$ if and only if the quadratic variation of the function f is nonzero.

Let us consider two simple examples.

Example 1.1.2. Let f be a C^1-function, i.e., $f'(t)$ is a continuous function. Then by the mean value theorem,

$$|R_n - L_n| = \sum_{i=1}^{n} \big(f'(t_i^*)(t_i - t_{i-1})\big)^2 \leq \sum_{i=1}^{n} \|f'\|_\infty^2 (t_i - t_{i-1})^2$$

$$\leq \|f'\|_\infty^2 \|\Delta_n\| \sum_{i=1}^{n}(t_i - t_{i-1}) = \|f'\|_\infty^2 \|\Delta_n\|(b-a)$$

$$\to 0, \qquad \text{as } \|\Delta_n\| \to 0,$$

where $t_{i-1} < t_i^* < t_i$ and $\|\cdot\|_\infty$ is the supremum norm. Thus $\lim L_n = \lim R_n$ as $\|\Delta_n\| \to 0$. Then by Equation (1.1.6) we have

$$\lim_{\|\Delta_n\|\to 0} L_n = \lim_{\|\Delta_n\|\to 0} R_n = \frac{1}{2}\big(f(b)^2 - f(a)^2\big). \tag{1.1.7}$$

On the other hand, for such a C^1-function f, we may simply define the integral $\int_a^b f(t)\,df(t)$ by

$$\int_a^b f(t)\,df(t) = \int_a^b f(t)f'(t)\,dt.$$

Then by the fundamental theorem of calculus,

$$\int_a^b f(t)\,df(t) = \int_a^b f(t)f'(t)\,dt = \frac{1}{2}\big(f(b)^2 - f(a)^2\big),$$

which gives the same value as in Equation (1.1.7).

Example 1.1.3. Suppose f is a continuous function satisfying the condition

$$|f(t) - f(s)| \approx |t - s|^{1/2}.$$

In this case, we have

$$0 \leq R_n - L_n \approx \sum_{i=1}^{n}(t_i - t_{i-1}) = b - a.$$

Hence $\lim R_n \neq \lim L_n$ as $\|\Delta_n\| \to 0$ when $a \neq b$. Consequently, the integral $\int_a^b f(t)\,df(t)$ cannot be defined by Equation (1.1.1) with $f = g$ for such a function f. Observe that the quadratic variation of this function is $b - a$.

We see from the above examples that defining the integral $\int_a^b f(t)\,dg(t)$, even when $f = g$, is a nontrivial problem. In fact, there is no simple definite answer to Question 1.1.1. But then, in view of Example 1.1.3, we can ask another question.

Question 1.1.4. Are there continuous functions f satisfying the condition

$$|f(t) - f(s)| \approx |t - s|^{1/2}?$$

In order to answer this question we consider random walks and take a suitable limit in the next section.

1.2 Random Walks

Consider a random walk starting at 0 with jumps h and $-h$ equally likely at times $\delta, 2\delta, \ldots$, where h and δ are positive numbers. More precisely, let $\{X_n\}_{n=1}^{\infty}$ be a sequence of independent and identically distributed random variables with

$$P\{X_j = h\} = P\{X_j = -h\} = \frac{1}{2}.$$

Let $Y_{\delta,h}(0) = 0$ and put

$$Y_{\delta,h}(n\delta) = X_1 + X_2 + \cdots + X_n.$$

For $t > 0$, define $Y_{\delta,h}(t)$ by linearization, i.e., for $n\delta < t < (n+1)\delta$, define

$$Y_{\delta,h}(t) = \frac{(n+1)\delta - t}{\delta} Y_{\delta,h}(n\delta) + \frac{t - n\delta}{\delta} Y_{\delta,h}((n+1)\delta).$$

We can think of $Y_{\delta,h}(t)$ as the position of the random walk at time t. In particular, $X_1 + X_2 + \cdots + X_n$ is the position of this random walk at time $n\delta$.

Question 1.2.1. What is the limit of the random walk $Y_{\delta,h}$ as $\delta, h \to 0$?

In order to find out the answer, let us compute the following limit of the characteristic function of $Y_{\delta,h}(t)$:

$$\lim_{\delta,h \to 0} E \exp\left[i\lambda Y_{\delta,h}(t)\right],$$

where $\lambda \in \mathbb{R}$ is fixed. For the heuristic derivation, let $t = n\delta$ and so $n = t/\delta$. Then we have

$$
\begin{aligned}
E \exp\left[i\lambda Y_{\delta,h}(t)\right] &= \prod_{j=1}^{n} E e^{i\lambda X_j} \\
&= \left(E e^{i\lambda X_1}\right)^n \\
&= \left(e^{i\lambda h \frac{1}{2}} + e^{-i\lambda h \frac{1}{2}}\right)^n \\
&= \left(\cos(\lambda h)\right)^n \\
&= \left(\cos(\lambda h)\right)^{t/\delta}. \qquad (1.2.1)
\end{aligned}
$$

Obviously, for fixed λ and t, the limit of $\exp\left[i\lambda Y_{\delta,h}(t)\right]$ does not exist when δ and h tend to zero independently. Thus in order for the limit to exist we must impose a certain relationship between δ and h. However, depending on this relationship, we may obtain different limits, as shown in the exercise problems at the end of this chapter.

Let $u = \left(\cos(\lambda h)\right)^{1/\delta}$. Then $\ln u = \frac{1}{\delta} \ln \cos(\lambda h)$. Note that

$$\cos(\lambda h) \approx 1 - \frac{1}{2}\lambda^2 h^2, \quad \text{for small } h.$$

But $\ln(1 + x) \approx x$ for small x. Hence

$$\ln \cos(\lambda h) \approx \ln \left(1 - \frac{1}{2}\lambda^2 h^2\right) \approx -\frac{1}{2}\lambda^2 h^2.$$

Therefore, for small δ and h, we have $\ln u \approx -\frac{1}{2\delta}\lambda^2 h^2$ and so

$$u \approx \exp\left[-\frac{1}{2\delta}\lambda^2 h^2\right].$$

Then by Equation (1.2.1),

$$E \exp\left[i\lambda Y_{\delta,h}(t)\right] \approx \exp\left[-\frac{1}{2\delta}t\lambda^2 h^2\right]. \tag{1.2.2}$$

In particular, if δ and h are related by $h^2 = \delta$, then

$$\lim_{\delta \to 0} E \exp\left[i\lambda Y_{\delta,h}(t)\right] = e^{-\frac{1}{2}t\lambda^2}, \quad \lambda \subset \mathbb{R}.$$

Thus we have derived the following theorem about the limit of the random walk $Y_{\delta,h}$ as $\delta, h \to 0$ in such a way that $h^2 = \delta$.

Theorem 1.2.2. *Let $Y_{\delta,h}(t)$ be the random walk starting at 0 with jumps h and $-h$ equally likely at times $\delta, 2\delta, 3\delta, \ldots$. Assume that $h^2 = \delta$. Then for each $t \geq 0$, the limit*

$$B(t) = \lim_{\delta \to 0} Y_{\delta,h}(t)$$

exists in distribution. Moreover, we have

$$E e^{i\lambda B(t)} = e^{-\frac{1}{2}t\lambda^2}, \quad \lambda \in \mathbb{R}. \tag{1.2.3}$$

Remark 1.2.3. On the basis of the above discussion, we would expect the stochastic process $B(t)$ to have the following properties:

(1) The absolute value of the slope of $Y_{\delta,h}$ in each step is $h/\delta = 1/\sqrt{\delta} \to \infty$ as $\delta \to 0$. Thus it is plausible that every Brownian path $B(t)$ is nowhere differentiable. In fact, if we let $\delta = |t - s|$, then

$$|B(t) - B(s)| \approx \frac{1}{\sqrt{\delta}}|t - s| = |t - s|^{1/2}. \tag{1.2.4}$$

Thus almost all sample paths of $B(t)$ have the property in Question 1.1.4.
(2) Almost all sample paths of $B(t)$ are continuous.
(3) For each t, $B(t)$ is a Gaussian random variable with mean 0 and variance t. This is a straightforward consequence of Equation (1.2.3).
(4) The stochastic process $B(t)$ has independent increments, namely, for any $0 \leq t_1 < t_2 < \cdots < t_n$, the random variables

$$B(t_1), \, B(t_2) - B(t_1), \, \ldots, B(t_n) - B(t_{n-1}),$$

are independent.

The above properties (2), (3), and (4) specify a fundamental stochastic process called Brownian motion, which we will study in the next chapter.

Exercises

1. Let g be a monotone function on a finite closed interval $[a, b]$. Show that a bounded function f defined on $[a, b]$ is Riemann–Stieltjes integrable with respect to g if and only if f is continuous almost everywhere with respect to the measure induced by g.

2. Let $Y_{\delta,h}(t)$ be the random walk described in Section 1.2. Assume that $h^2 = o(\delta)$, i.e., $h^2/\delta \to 0$ as $\delta \to 0$. Show that $X(t) = \lim_{\delta \to 0} Y_{\delta,h}(t)$ exists and that $X(t) \equiv 0$.

3. Let $Y_{\delta,h}(t)$ be the random walk described in Section 1.2. Show that for small δ and h we have

$$E \exp\left[i\lambda Y_{\delta,h}(t)\right] \approx \exp\left[-\frac{t\lambda^2 h^2}{2\delta} - \frac{t\lambda^4 h^4}{12\delta}\right].$$

Assume that $\delta \to 0, h \to 0$, but $h^2/\delta \to \infty$. Then $\lim_{\delta \to 0} Y_{\delta,h}(t)$ does not exist. However, consider the following renormalization:

$$\exp\left[i\lambda Y_{\delta,h}(t) + \frac{t\lambda^2 h^2}{2\delta}\right]. \tag{1.2.5}$$

Then we have

$$E \exp\left[i\lambda Y_{\delta,h}(t) + \frac{t\lambda^2 h^2}{2\delta}\right] \approx \exp\left[-\frac{t\lambda^4 h^4}{12\delta}\right].$$

Thus, if $\delta, h \to 0$ in such a way that $h^2/\delta \to \infty$ and $h^4/\delta \to 0$, then

$$\lim_{\delta \to 0} E \exp\left[i\lambda Y_{\delta,h}(t) + \frac{t\lambda^2 h^2}{2\delta}\right] = 1, \quad \text{for all } \lambda \in \mathbb{R}.$$

Hence for this choice of δ and h, the limit $\lim_{\delta \to 0} Y_{\delta,h}(t)$ does not exist. However, the existence of the limit of the renormalization in Equation (1.2.5) indicates that the limit $\lim_{\delta \to 0} Y_{\delta,h}(t)$ exists in some generalized sense. This limit is called a *white noise*. It is informally regarded as the derivative $\dot{B}(t)$ of the Brownian motion $B(t)$. Note that $h = \delta^\alpha$ satisfies the conditions $h^2/\delta \to \infty$ and $h^4/\delta \to 0$ if and only if $\frac{1}{4} < \alpha < \frac{1}{2}$.

2

Brownian Motion

2.1 Definition of Brownian Motion

Let $(\Omega, \mathcal{F}, P)$ be a probability space. A stochastic process is a measurable function $X(t, \omega)$ defined on the product space $[0, \infty) \times \Omega$. In particular,

(a) for each t, $X(t, \cdot)$ is a random variable,
(b) for each ω, $X(\cdot, \omega)$ is a measurable function (called a *sample path*).

For convenience, the random variable $X(t, \cdot)$ will be written as $X(t)$ or X_t. Thus a stochastic process $X(t, \omega)$ can also be expressed as $X(t)(\omega)$ or simply as $X(t)$ or X_t.

Definition 2.1.1. *A stochastic process $B(t, \omega)$ is called a* Brownian motion *if it satisfies the following conditions:*

(1) $P\{\omega \,;\, B(0, \omega) = 0\} = 1$.
(2) *For any $0 \le s < t$, the random variable $B(t) - B(s)$ is normally distributed with mean 0 and variance $t - s$, i.e., for any $a < b$,*

$$P\{a \le B(t) - B(s) \le b\} = \frac{1}{\sqrt{2\pi(t-s)}} \int_a^b e^{-x^2/2(t-s)} \, dx.$$

(3) *$B(t, \omega)$ has independent increments, i.e., for any $0 \le t_1 < t_2 < \cdots < t_n$, the random variables*

$$B(t_1), \; B(t_2) - B(t_1), \; \ldots, B(t_n) - B(t_{n-1}),$$

are independent.

(4) *Almost all sample paths of $B(t, \omega)$ are continuous functions, i.e.,*

$$P\{\omega \,;\, B(\cdot, \omega) \text{ is continuous}\} = 1.$$

In Remark 1.2.3 we mentioned that the limit

$$B(t) = \lim_{\delta \to 0} Y_{\delta, \sqrt{\delta}}(t)$$

is a Brownian motion. However, this fact comes only as a consequence of an intuitive observation. In the next chapter we will give several constructions of Brownian motion. But before these constructions we shall give some simple properties of a Brownian motion and define the Wiener integral.

A Brownian motion is sometimes defined as a stochastic process $B(t, \omega)$ satisfying conditions (1), (2), (3) in Definition 2.1.1. Such a stochastic process always has a continuous realization, i.e., there exists Ω_0 such that $P(\Omega_0) = 1$ and for any $\omega \in \Omega_0$, $B(t, \omega)$ is a continuous function of t. This fact can be easily checked by applying the Kolmogorov continuity theorem in Section 3.3. Thus condition (4) is automatically satisfied.

The Brownian motion $B(t)$ in the above definition starts at 0. Sometimes we will need a Brownian motion starting at x. Such a process is given by $x + B(t)$. If the starting point is not 0, we will explicitly mention the starting point x.

2.2 Simple Properties of Brownian Motion

Let $B(t)$ be a fixed Brownian motion. We give below some simple properties that follow directly from the definition of Brownian motion.

Proposition 2.2.1. *For any $t > 0$, $B(t)$ is normally distributed with mean 0 and variance t. For any $s, t \geq 0$, we have $E[B(s)B(t)] = \min\{s, t\}$.*

Remark 2.2.2. Regarding Definition 2.1.1, it can be proved that condition (2) and $E[B(s)B(t)] = \min\{s, t\}$ imply condition (3).

Proof. By condition (1), we have $B(t) = B(t) - B(0)$ and so the first assertion follows from condition (2). To show that $EB(s)B(t) = \min\{s, t\}$ we may assume that $s < t$. Then by conditions (2) and (3),

$$E\big[B(s)B(t)\big] = E\big[B(s)\big(B(t) - B(s)\big) + B(s)^2\big] = 0 + s = s,$$

which is equal to $\min\{s, t\}$. $\qquad\qquad\qquad\qquad\qquad\qquad\qquad\square$

Proposition 2.2.3. (Translation invariance) *For fixed $t_0 \geq 0$, the stochastic process $\widetilde{B}(t) = B(t + t_0) - B(t_0)$ is also a Brownian motion.*

Proof. The stochastic process $\widetilde{B}(t)$ obviously satisfies conditions (1) and (4) of a Brownian motion. For any $s < t$,

$$\widetilde{B}(t) - \widetilde{B}(s) = B(t + t_0) - B(s + t_0). \qquad\qquad (2.2.1)$$

By condition (2) of $B(t)$, we see that $\widetilde{B}(t) - \widetilde{B}(s)$ is normally distributed with mean 0 and variance $(t + t_0) - (s + t_0) = t - s$. Thus $\widetilde{B}(t)$ satisfies condition (2). To check condition (3) for $\widetilde{B}(t)$, we may assume that $t_0 > 0$. Then for any $0 \le t_1 < t_2 < \cdots < t_n$, we have $0 < t_0 \le t_1 + t_0 < \cdots < t_n + t_0$. Hence by condition (3) of $B(t)$, $B(t_k + t_0) - B(t_{k-1} + t_0)$, $k = 1, 2, \ldots, n$, are independent random variables. Thus by Equation (2.2.1), the random variables $\widetilde{B}(t_k) - \widetilde{B}(t_{k-1})$, $k = 1, 2, \ldots, n$, are independent and so $\widetilde{B}(t)$ satisfies condition (3) of a Brownian motion. $\qquad\square$

The above translation invariance property says that a Brownian motion starts afresh at any moment as a new Brownian motion.

Proposition 2.2.4. (Scaling invariance) *For any real number $\lambda > 0$, the stochastic process $\widetilde{B}(t) = B(\lambda t)/\sqrt{\lambda}$ is also a Brownian motion.*

Proof. Conditions (1), (3), and (4) of a Brownian motion can be readily checked for the stochastic process $\widetilde{B}(t)$. To check condition (2), note that for any $s < t$,

$$\widetilde{B}(t) - \widetilde{B}(s) = \frac{1}{\sqrt{\lambda}} \big(B(\lambda t) - B(\lambda s) \big),$$

which shows that $\widetilde{B}(t) - \widetilde{B}(s)$ is normally distributed with mean 0 and variance $\frac{1}{\lambda}(\lambda t - \lambda s) = t - s$. Hence $\widetilde{B}(t)$ satisfies condition (2). $\qquad\square$

It follows from the scaling invariance property that for any $\lambda > 0$ and $0 \le t_1 < t_2 < \cdots < t_n$ the random vectors

$$\big(B(\lambda t_1), B(\lambda t_2), \ldots, B(\lambda t_n) \big), \quad \big(\sqrt{\lambda} B(t_1), \sqrt{\lambda} B(t_2), \ldots, \sqrt{\lambda} B(t_n) \big)$$

have the same distribution.

2.3 Wiener Integral

In Section 1.1 we raised the question of defining the integral $\int_a^b f(t)\, dg(t)$. We see from Example 1.1.3 that in general this integral cannot be defined as a Riemann–Stieltjes integral.

Now let us consider the following integral:

$$\int_a^b f(t)\, dB(t, \omega),$$

where f is a deterministic function (i.e., it does not depend on ω) and $B(t, \omega)$ is a Brownian motion. Suppose for each $\omega \in \Omega$ we want to use Equation (1.1.2) to define this integral in the Riemann–Stieltjes sense by

$$(RS) \int_a^b f(t)\, dB(t, \omega) = f(t)B(t, \omega) \Big]_a^b - (RS) \int_a^b B(t, \omega)\, df(t). \qquad (2.3.1)$$

Then the class of functions $f(t)$ for which the integral $(RS)\int_a^b f(t)\,dB(t,\omega)$ is defined for each $\omega \in \Omega$ is rather limited, i.e., $f(t)$ needs to be a continuous function of bounded variation. Hence for a continuous function of unbounded variation such as $f(t) = t\sin\frac{1}{t}$, $0 < t \leq 1$, and $f(0) = 0$, we cannot use Equation (2.3.1) to define the integral $\int_0^1 f(t)\,dB(t,\omega)$ for each $\omega \in \Omega$.

We need a different idea in order to define the integral $\int_a^b f(t)\,dB(t,\omega)$ for a wider class of functions $f(t)$. This new integral, called the Wiener integral of f, is defined for all functions $f \in L^2[a,b]$. Here $L^2[a,b]$ denotes the Hilbert space of all real-valued square integrable functions on $[a,b]$. For example, $\int_0^1 t\sin\frac{1}{t}\,dB(t)$ is a Wiener integral.

Now we define the Wiener integral in two steps:

Step 1. Suppose f is a step function given by $f = \sum_{i=1}^{n} a_i\,1_{[t_{i-1},t_i)}$, where $t_0 = a$ and $t_n = b$. In this case, define

$$I(f) = \sum_{i=1}^{n} a_i\big(B(t_i) - B(t_{i-1})\big). \tag{2.3.2}$$

Obviously, $I(af + bg) = aI(f) + bI(g)$ for any $a,b \in \mathbb{R}$ and step functions f and g. Moreover, we have the following lemma.

Lemma 2.3.1. *For a step function f, the random variable $I(f)$ is Gaussian with mean 0 and variance*

$$E\big(I(f)^2\big) = \int_a^b f(t)^2\,dt. \tag{2.3.3}$$

Proof. It is well known that a linear combination of independent Gaussian random variables is also a Gaussian random variable. Hence by conditions (2) and (3) of Brownian motion, the random variable $I(f)$ defined by Equation (2.3.2) is Gaussian with mean 0. To check Equation (2.3.3), note that

$$E\big(I(f)^2\big) = E\sum_{i,j=1}^{n} a_i\,a_j\big(B(t_i) - B(t_{i-1})\big)\big(B(t_j) - B(t_{j-1})\big).$$

By conditions (2) and (3) of Brownian motion,

$$E\big(B(t_i) - B(t_{i-1})\big)^2 = t_i - t_{i-1},$$

and for $i \neq j$,

$$E\big(B(t_i) - B(t_{i-1})\big)\big(B(t_j) - B(t_{j-1})\big) = 0.$$

Therefore,

$$E\big(I(f)^2\big) = \sum_{i=1}^{n} a_i^2\,(t_i - t_{i-1}) = \int_a^b f(t)^2\,dt. \qquad \square$$

Step 2. We will use $L^2(\Omega)$ to denote the Hilbert space of square integrable real-valued random variables on Ω with inner product $\langle X, Y \rangle = E(XY)$. Let $f \in L^2[a, b]$. Choose a sequence $\{f_n\}_{n=1}^\infty$ of step functions such that $f_n \to f$ in $L^2[a, b]$. By Lemma 2.3.1 the sequence $\{I(f_n)\}_{n=1}^\infty$ is Cauchy in $L^2(\Omega)$. Hence it converges in $L^2(\Omega)$. Define

$$I(f) = \lim_{n\to\infty} I(f_n), \quad \text{in } L^2(\Omega). \tag{2.3.4}$$

Question 2.3.2. Is $I(f)$ well-defined?

In order for $I(f)$ to be well-defined, we need to show that the limit in Equation (2.3.4) is independent of the choice of the sequence $\{f_n\}$. Suppose $\{g_m\}$ is another such sequence, i.e., the g_m's are step functions and $g_m \to f$ in $L^2[a, b]$. Then by the linearity of the mapping I and Equation (2.3.3),

$$E\big(|I(f_n) - I(g_m)|^2\big) = E\big(|I(f_n - g_m)|^2\big) = \int_a^b \big(f_n(t) - g_m(t)\big)^2 \, dt.$$

Write $f_n(t) - g_m(t) = [f_n(t) - f(t)] - [g_m(t) - f(t)]$ and then use the inequality $(x - y)^2 \le 2(x^2 + y^2)$ to get

$$\int_a^b \big(f_n(t) - g_m(t)\big)^2 \, dt \le 2 \int_a^b \Big([f_n(t) - f(t)]^2 + [g_m(t) - f(t)]^2\Big) \, dt$$

$$\to 0, \quad \text{as } n, m \to \infty.$$

It follows that $\lim_{n\to\infty} I(f_n) = \lim_{m\to\infty} I(g_m)$ in $L^2(\Omega)$. This shows that $I(f)$ is well-defined.

Definition 2.3.3. *Let $f \in L^2[a, b]$. The limit $I(f)$ defined in Equation (2.3.4) is called the* Wiener integral *of f.*

The Wiener integral $I(f)$ of f will be denoted by

$$I(f)(\omega) = \left(\int_a^b f(t) \, dB(t)\right)(\omega), \quad \omega \in \Omega, \text{ almost surely.}$$

For simplicity, it will be denoted by $\int_a^b f(t) \, dB(t)$ or $\int_a^b f(t) \, dB(t, \omega)$. Note that the mapping I is linear on $L^2[a, b]$.

Theorem 2.3.4. *For each $f \in L^2[a, b]$, the Wiener integral $\int_a^b f(t) \, dB(t)$ is a Gaussian random variable with mean 0 and variance $\|f\|^2 = \int_a^b f(t)^2 \, dt$.*

Proof. By Lemma 2.3.1, the assertion is true when f is a step function. For a general $f \in L^2[a, b]$, the assertion follows from the following well-known fact: If X_n is Gaussian with mean μ_n and variance σ_n^2 and X_n converges to X in $L^2(\Omega)$, then X is Gaussian with mean $\mu = \lim_{n\to\infty} \mu_n$ and variance $\sigma^2 = \lim_{n\to\infty} \sigma_n^2$. $\qquad \square$

Thus the Wiener integral $I : L^2[a,b] \to L^2(\Omega)$ is an isometry. In fact, it preserves the inner product, as shown by the next corollary.

Corollary 2.3.5. *If $f, g \in L^2[a,b]$, then*

$$E\big(I(f)\,I(g)\big) = \int_a^b f(t)g(t)\,dt. \qquad (2.3.5)$$

In particular, if f and g are orthogonal, then the Gaussian random variables $I(f)$ and $I(g)$ are independent.

Proof. By the linearity of I and Theorem 2.3.4 we have

$$E\big[(I(f)+I(g))^2\big] = E\big[(I(f+g))^2\big]$$
$$= \int_a^b \big(f(t)+g(t)\big)^2\,dt$$
$$= \int_a^b f(t)^2\,dt + 2\int_a^b f(t)g(t)\,dt + \int_a^b g(t)^2\,dt. \quad (2.3.6)$$

On the other hand, we can also use Theorem 2.3.4 to obtain

$$E\big[(I(f)+I(g))^2\big]$$
$$= E\big[I(f)^2 + 2I(f)I(g) + I(g)^2\big]$$
$$= \int_a^b f(t)^2\,dt + 2E\big[I(f)I(g)\big] + \int_a^b g(t)^2\,dt. \qquad (2.3.7)$$

Obviously, Equation (2.3.5) follows from Equations (2.3.6) and (2.3.7). □

Example 2.3.6. The Wiener integral $\int_0^1 s\,dB(s)$ is a Gaussian random variable with mean 0 and variance $\int_0^1 s^2\,ds = \frac{1}{3}$.

Theorem 2.3.7. *Let f be a continuous function of bounded variation. Then for almost all $\omega \in \Omega$,*

$$\left(\int_a^b f(t)\,dB(t)\right)(\omega) = (RS)\int_a^b f(t)\,dB(t,\omega),$$

where the left-hand side is the Wiener integral of f and the right-hand side is the Riemann–Stieltjes integral of f defined by Equation (2.3.1).

Proof. For each partition $\Delta_n = \{t_0, t_1, \ldots, t_{n-1}, t_n\}$ of $[a,b]$, we define a step function f_n by

$$f_n = \sum_{i=1}^n f(t_{i-1})\mathbf{1}_{[t_{i-1},t_i)}.$$

Note that f_n converges to f in $L^2[a,b]$ as $n \to \infty$, i.e., as $\|\Delta_n\| \to 0$. Hence by the definition of the Wiener integral in Equation (2.3.4),

$$\int_a^b f(t)\,dB(t) = \lim_{n\to\infty}\sum_{i=1}^n f(t_{i-1})\big(B(t_i) - B(t_{i-1})\big), \quad \text{in } L^2(\Omega). \qquad (2.3.8)$$

On the other hand, by Equation (2.3.1), the following limit holds for each $\omega \in \Omega_0$ for some Ω_0 with $P(\Omega_0) = 1$,

$$(RS)\int_a^b f(t)\,dB(t,\omega)$$

$$= f(b)B(b,\omega) - f(a)B(a,\omega) - \lim_{n\to\infty}\sum_{i=1}^n B(t_i,\omega)\big(f(t_i) - f(t_{i-1})\big)$$

$$= \lim_{n\to\infty}\Big(f(b)B(b,\omega) - f(a)B(a,\omega) - \sum_{i=1}^n B(t_i,\omega)\big(f(t_i) - f(t_{i-1})\big)\Big),$$

which, after regrouping the terms, yields the following equality for each ω in Ω_0:

$$(RS)\int_a^b f(t)\,dB(t,\omega) = \lim_{n\to\infty}\sum_{i=1}^n f(t_{i-1})\big(B(t_i) - B(t_{i-1})\big). \qquad (2.3.9)$$

Since $L^2(\Omega)$-convergence implies the existence of a subsequence converging almost surely, we can pick such a subsequence of $\{f_n\}$ to get the conclusion of the theorem from Equations (2.3.8) and (2.3.9).   □

Example 2.3.8. Consider the Riemann integral $\int_0^1 B(t,\omega)\,dt$ defined for each $\omega \in \Omega_0$ for some Ω_0 with $P(\Omega_0) = 1$. Let us find the distribution of this random variable. Use the integration by parts formula to get

$$\int_0^1 B(t,\omega)\,dt = B(t,\omega)(t-1)\Big]_0^1 - \int_0^1 (t-1)\,dB(t,\omega)$$

$$= (RS)\int_0^1 (1-t)\,dB(t,\omega).$$

Hence by Theorem 2.3.7 we see that for almost all $\omega \in \Omega$,

$$\int_0^1 B(t,\omega)\,dt = \Big(\int_0^1 (1-t)\,dB(t)\Big)(\omega),$$

where the right-hand side is a Wiener integral. Thus $\int_0^1 B(t)\,dt$ and the Wiener integral $\int_0^1 (1-t)\,dB(t)$ have the same distribution, which is easily seen to be Gaussian with mean 0 and variance

$$E\Big(\int_0^1 (1-t)\,dB(t)\Big)^2 = \int_0^1 (1-t)^2\,dt = \frac{1}{3}.$$

2.4 Conditional Expectation

In this section we explain the concept of conditional expectation, which will be needed in the next section and other places. Let $(\Omega, \mathcal{F}, P)$ be a fixed probability space. For $1 \leq p < \infty$, we will use $L^p(\Omega)$ to denote the space of all random variables X with $E(|X|^p) < \infty$. It is a Banach space with norm

$$\|X\|_p = \left(E\left(|X|^p \right) \right)^{1/p}.$$

In particular, $L^2(\Omega)$ is the Hilbert space used in Section 2.3. In this section we use the space $L^1(\Omega)$ with norm given by $\|X\|_1 = E|X|$. Sometimes we will write $L^1(\Omega, \mathcal{F})$ when we want to emphasize the σ-field $\mathcal{F}$.

Suppose we have another σ-field $\mathcal{G} \subset \mathcal{F}$. Let X be a random variable with $E|X| < \infty$, i.e., $X \in L^1(\Omega)$. Define a real-valued function μ on $\mathcal{G}$ by

$$\mu(A) = \int_A X(\omega) \, dP(\omega), \quad A \in \mathcal{G}. \tag{2.4.1}$$

Note that $|\mu(A)| \leq \int_A |X| \, dP \leq \int_\Omega |X| \, dP = E|X|$ for all $A \in \mathcal{G}$. Moreover, the function μ satisfies the following conditions:

(a) $\mu(\emptyset) = 0$;
(b) $\mu\left(\cup_{n \geq 1} A_n \right) = \sum_{n \geq 1} \mu(A_n)$ for any disjoint sets $A_n \in \mathcal{G}$, $n = 1, 2, \ldots$;
(c) If $P(A) = 0$ and $A \in \mathcal{G}$, then $\mu(A) = 0$.

A function $\mu \colon \mathcal{G} \to \mathbb{R}$ satisfying conditions (a) and (b) is called a *signed measure* on $(\Omega, \mathcal{G})$. A signed measure μ is said to be *absolutely continuous* with respect to P if it satisfies condition (c). Therefore, the function μ defined in Equation (2.4.1) is a signed measure on $(\Omega, \mathcal{G})$ and is absolutely continuous with respect to P.

Apply the Radon–Nikodym theorem (see, e.g., the book by Royden [73]) to the signed measure μ defined in Equation (2.4.1) to get a $\mathcal{G}$-measurable random variable Y with $E|Y| < \infty$ such that

$$\mu(A) = \int_A Y(\omega) \, dP(\omega), \quad \forall A \in \mathcal{G}. \tag{2.4.2}$$

Suppose $\widetilde{Y}$ is another such random variable, namely, it is $\mathcal{G}$-measurable with $E|\widetilde{Y}| < \infty$ and satisfies

$$\mu(A) = \int_A \widetilde{Y}(\omega) \, dP(\omega), \quad \forall A \in \mathcal{G}. \tag{2.4.3}$$

Then by Equations (2.4.2) and (2.4.3), we have $\int_A (Y - \widetilde{Y}) \, dP = 0$ for all $A \in \mathcal{G}$. This implies that $Y = \widetilde{Y}$ almost surely.

The above discussion shows the existence and uniqueness of the conditional expectation in the next definition.

Definition 2.4.1. *Let $X \in L^1(\Omega, \mathcal{F})$. Suppose $\mathcal{G}$ is a σ-field and $\mathcal{G} \subset \mathcal{F}$. The* conditional expectation *of X given $\mathcal{G}$ is defined to be the unique random variable Y (up to P-measure 1) satisfying the following conditions:*

(1) *Y is $\mathcal{G}$-measurable;*

(2) *$\int_A X \, dP = \int_A Y \, dP$ for all $A \in \mathcal{G}$.*

We will freely use $E[X|\mathcal{G}]$, $E(X|\mathcal{G})$, or $E\{X|\mathcal{G}\}$ to denote the conditional expectation of X given $\mathcal{G}$. Notice that the $\mathcal{G}$-measurability in condition (1) is a crucial requirement. Otherwise, we could take $Y = X$ to satisfy condition (2), and the above definition would not be so meaningful. The conditional expectation $E[X|\mathcal{G}]$ can be interpreted as the best guess of the value of X based on the information provided by $\mathcal{G}$.

Example 2.4.2. Suppose $\mathcal{G} = \{\emptyset, \Omega\}$. Let X be a random variable in $L^1(\Omega)$ and let $Y = E[X|\mathcal{G}]$. Since Y is $\mathcal{G}$-measurable, it must be a constant, say $Y = c$. Then use condition (2) in Definition 2.4.1 with $A = \Omega$ to get

$$\int_\Omega X \, dP = \int_\Omega Y \, dP = c.$$

Hence $c = EX$ and we have $E[X|\mathcal{G}] = EX$. This conclusion is intuitively obvious. Since the σ-field $\mathcal{G} = \{\emptyset, \Omega\}$ provides no information, the best guess of the value of X is its expectation.

Example 2.4.3. Suppose $\Omega = \cup_n A_n$ is a disjoint union (finite or countable) with $P(A_n) > 0$ for each n. Let $\mathcal{G} = \sigma\{A_1, A_2, \ldots\}$, the σ-field generated by the A_n's. Let $X \in L^1(\Omega)$ and $Y = E[X|\mathcal{G}]$. Since Y is $\mathcal{G}$-measurable, it must be constant, say c_n, on A_n for each n. Use condition (2) in Definition 2.4.1 with $A = A_n$ to show that $c_n = P(A_n)^{-1} \int_{A_n} X \, dP$. Therefore, $E[X|\mathcal{G}]$ is given by

$$E[X|\mathcal{G}] = \sum_n \left(\frac{1}{P(A_n)} \int_{A_n} X \, dP \right) 1_{A_n},$$

where 1_{A_n} denotes the characteristic function of A_n.

Example 2.4.4. Let Z be a discrete random variable taking values $a_1, a_2, \ldots$ (finite or countable). Let $\sigma\{Z\}$ be the σ-field generated by Z. Then

$$\sigma\{Z\} = \sigma\{A_1, A_2, \ldots\},$$

where $A_n = \{Z = a_n\}$. Let $X \in L^1(\Omega)$. We can use Example 2.4.3 to obtain

$$E[X|\sigma\{Z\}] = \sum_n \left(\frac{1}{P(A_n)} \int_{A_n} X \, dP \right) 1_{A_n},$$

which can be rewritten as $E[X|\sigma\{Z\}] = \theta(Z)$ with the function θ defined by

$$\theta(x) = \begin{cases} \dfrac{1}{P(Z = a_n)} \displaystyle\int_{Z=a_n} X \, dP, & \text{if } x = a_n, \ n \geq 1; \\ 0, & \text{if } x \notin \{a_1, a_2, \ldots\}. \end{cases}$$

Note that the conditional expectation $E[X|\mathcal{G}]$ is a random variable, while the expectation EX is a real number. Below we list several properties of conditional expectation and leave most of the proofs as exercises at the end of this chapter.

Recall that $(\Omega, \mathcal{F}, P)$ is a fixed probability space. The random variable X below is assumed to be in $L^1(\Omega, \mathcal{F})$ and $\mathcal{G}$ is a sub-σ-field of $\mathcal{F}$, namely, $\mathcal{G}$ is a σ-field and $\mathcal{G} \subset \mathcal{F}$. All equalities and inequalities below hold almost surely.

1. $E\big(E[X|\mathcal{G}]\big) = EX$.

   Remark: Hence the conditional expectation $E[X|\mathcal{G}]$ and X have the same expectation. When written in the form $EX = E\big(E[X|\mathcal{G}]\big)$, the equality is often referred to as *computing expectation by conditioning*. To prove this equality, simply put $A = \Omega$ in condition (2) of Definition 2.4.1.

2. *If X is $\mathcal{G}$-measurable, then $E[X|\mathcal{G}] = X$.*

3. *If X and $\mathcal{G}$ are independent, then $E[X|\mathcal{G}] = EX$.*

   Remark: Here X and $\mathcal{G}$ being independent means that $\{X \in U\}$ and A are independent events for any Borel subset U of $\mathbb{R}$ and $A \in \mathcal{G}$, or equivalently, the events $\{X \leq x\}$ and A are independent for any $x \in \mathbb{R}$ and $A \in \mathcal{G}$.

4. *If Y is $\mathcal{G}$-measurable and $E|XY| < \infty$, then $E[XY|\mathcal{G}] = Y E[X|\mathcal{G}]$.*

5. *If $\mathcal{H}$ is a sub-σ-field of $\mathcal{G}$, then $E[X|\mathcal{H}] = E\big[E[X|\mathcal{G}]\,|\,\mathcal{H}\big]$.*

   Remark: This property is useful when X is a product of random variables. In that case, in order to find $E[X|\mathcal{H}]$, we can use some factors in X to choose a suitable σ-field $\mathcal{G}$ between $\mathcal{H}$ and $\mathcal{F}$ and then apply this property.

6. *If $X, Y \in L^1(\Omega)$ and $X \leq Y$, then $E[X|\mathcal{G}] \leq E[Y|\mathcal{G}]$.*

7. $\big|E[X|\mathcal{G}]\big| \leq E[|X|\,|\,\mathcal{G}]$.

   Remark: For the proof, let $X^+ = \max\{X, 0\}$ and $X^- = -\min\{X, 0\}$ be the positive and negative parts of X, respectively. Then apply Property 6 to X^+ and X^-.

8. $E[aX + bY|\mathcal{G}] = aE[X|\mathcal{G}] + bE[Y|\mathcal{G}]$, $\forall a, b \in \mathbb{R}$ *and* $X, Y \in L^1(\Omega)$.

   Remark: By Properties 7 and 8, the conditional expectation $E[\,\cdot\,|\mathcal{G}]$ is a bounded linear operator from $L^1(\Omega, \mathcal{F})$ into $L^1(\Omega, \mathcal{G})$

9. (Conditional Fatou's lemma) *Let $X_n \geq 0$, $X_n \in L^1(\Omega)$, $n = 1, 2, \ldots$, and assume that $\liminf_{n \to \infty} X_n \in L^1(\Omega)$. Then*

$$E\Big[\liminf_{n \to \infty} X_n \,\Big|\, \mathcal{G}\Big] \leq \liminf_{n \to \infty} E[X_n|\mathcal{G}].$$

10. (Conditional monotone convergence theorem) *Let $0 \leq X_1 \leq X_2 \leq \cdots \leq X_n \leq \cdots$ and assume that $X = \lim_{n \to \infty} X_n \in L^1(\Omega)$. Then*

$$E[X|\mathcal{G}] = \lim_{n \to \infty} E[X_n|\mathcal{G}].$$

11. (Conditional Lebesgue dominated convergence theorem) *Assume that* $|X_n| \le Y$, $Y \in L^1(\Omega)$, *and* $X = \lim_{n \to \infty} X_n$ *exists almost surely. Then*

$$E[X|\mathcal{G}] = \lim_{n \to \infty} E[X_n|\mathcal{G}].$$

12. (Conditional Jensen's inequality) *Let* $X \in L^1(\Omega)$. *Suppose* ϕ *is a convex function on* $\mathbb{R}$ *and* $\phi(X) \in L^1(\Omega)$. *Then*

$$\phi(E[X|\mathcal{G}]) \le E[\phi(X)|\mathcal{G}].$$

2.5 Martingales

Let $f \in L^2[a, b]$ and consider the stochastic process defined by

$$M_t = \int_a^t f(s)\, dB(s), \quad a \le t \le b. \tag{2.5.1}$$

We will show that M_t is a martingale. But first we review the concept of the martingale. Let T be either an interval in $\mathbb{R}$ or the set of positive integers.

Definition 2.5.1. *A* filtration *on* T *is an increasing family* $\{\mathcal{F}_t | t \in T\}$ *of σ-fields. A stochastic process $X_t, t \in T$, is said to be* adapted *to* $\{\mathcal{F}_t | t \in T\}$ *if for each t, the random variable X_t is $\mathcal{F}_t$-measurable.*

Remark 2.5.2. A σ-field $\mathcal{F}$ is called *complete* if $A \in \mathcal{F}$ and $P(A) = 0$ imply that $B \in \mathcal{F}$ for any subset B of A. We will always assume that all σ-fields $\mathcal{F}_t$ are complete.

Definition 2.5.3. *Let X_t be a stochastic process adapted to a filtration $\{\mathcal{F}_t\}$ and $E|X_t| < \infty$ for all $t \in T$. Then X_t is called a* martingale *with respect to* $\{\mathcal{F}_t\}$ *if for any $s \le t$ in T,*

$$E\{X_t | \mathcal{F}_s\} = X_s, \quad \text{a.s.} \ (almost\ surely). \tag{2.5.2}$$

In case the filtration is not explicitly specified, then the filtration $\{\mathcal{F}_t\}$ is understood to be the one given by $\mathcal{F}_t = \sigma\{X_s\,;\, s \le t\}$.

The concept of the martingale is a generalization of the sequence of partial sums arising from a sequence $\{X_n\}$ of independent and identically distributed random variables with mean 0. Let $S_n = X_1 + \cdots + X_n$. Then the sequence $\{S_n\}$ is a martingale.

Submartingale and supermartingale are defined by replacing the equality in Equation (2.5.2) with $\ge$ and $\le$, respectively, i.e., for any $s \le t$ in T,

$$E\{X_t | \mathcal{F}_s\} \ge X_s, \quad \text{a.s.} \quad \text{(submartingale)},$$

$$E\{X_t | \mathcal{F}_s\} \le X_s, \quad \text{a.s.} \quad \text{(supermartingale)}.$$

Let $\{X_n\}$ be a sequence of independent and identically distributed random variables with finite expectation and let $S_n = X_1 + \cdots + X_n$. Then $\{S_n\}$ is a submartingale if $EX_1 \geq 0$ and a supermartingale if $EX_1 \leq 0$.

A Brownian motion $B(t)$ is a martingale. To see this fact, let

$$\mathcal{F}_t = \sigma\{B(s)\,;\, s \leq t\}.$$

Then for any $s \leq t$,

$$E\{B(t)|\,\mathcal{F}_s\} = E\{B(t) - B(s)|\,\mathcal{F}_s\} + E\{B(s)|\,\mathcal{F}_s\}.$$

Since $B(t) - B(s)$ is independent of $\mathcal{F}_s$, we have $E\{B(t) - B(s)|\,\mathcal{F}_s\} = E\{B(t) - B(s)\}$. But $EB(t) = 0$ for any t. Hence $E\{B(t) - B(s)|\,\mathcal{F}_s\} = 0$. On the other hand, $E\{B(s)|\,\mathcal{F}_s\} = B(s)$ because $B(s)$ is $\mathcal{F}_s$-measurable. Thus $E\{B(t)|\,\mathcal{F}_s\} = B(s)$ for any $s \leq t$ and this shows that $B(t)$ is a martingale. In fact, it is the most basic martingale stochastic process with time parameter in an interval.

Now we return to the stochastic process M_t defined in Equation (2.5.1) and show that it is a martingale in the next theorem.

Theorem 2.5.4. *Let $f \in L^2[a, b]$. Then the stochastic process*

$$M_t = \int_a^t f(s)\, dB(s), \quad a \leq t \leq b,$$

is a martingale with respect to $\mathcal{F}_t = \sigma\{B(s)\,;\, s \leq t\}$.

Proof. First we need to show that $E|M_t| < \infty$ for all $t \in [a, b]$ in order to take the conditional expectation of M_t. Apply Theorem 2.3.4 to get

$$E\big(|M_t|^2\big) = \int_a^t |f(s)|^2\, ds \leq \int_a^b |f(s)|^2\, ds.$$

Hence $E|M_t| \leq \big\{E\big(|M_t|^2\big)\big\}^{1/2} < \infty$. Next we need to prove that $E\{M_t|\,\mathcal{F}_s\} = M_s$ a.s. for any $s \leq t$. But

$$M_t = M_s + \int_s^t f(u)\, dB(u)$$

and M_s is $\mathcal{F}_s$-measurable. Hence

$$E\{M_t|\,\mathcal{F}_s\} = M_s + E\bigg\{\int_s^t f(u)\, dB(u)\,\Big|\, \mathcal{F}_s\bigg\}.$$

Thus it suffices to show that for any $s \leq t$,

$$E\bigg\{\int_s^t f(u)\, dB(u)\,\Big|\, \mathcal{F}_s\bigg\} = 0. \tag{2.5.3}$$

First suppose f is a step function $f = \sum_{i=1}^{n} a_i 1_{[t_{i-1}, t_i)}$, where $t_0 = s$ and $t_n = t$. In this case, we have

$$\int_s^t f(u) \, dB(u) = \sum_{i=1}^{n} a_i \left(B(t_i) - B(t_{i-1}) \right).$$

But $B(t_i) - B(t_{i-1})$, $i = 1, \ldots, n$, are all independent of the σ-field $\mathcal{F}_s$. Hence $E\{B(t_i) - B(t_{i-1}) | \mathcal{F}_s\} = 0$ for all i and so Equation (2.5.3) holds.

Next suppose $f \in L^2[a, b]$. Choose a sequence $\{f_n\}_{n=1}^{\infty}$ of step functions converging to f in $L^2[a, b]$. Then by the conditional Jensen's inequality with $\phi(x) = x^2$ in Section 2.4 we have the inequality

$$|E\{X | \mathcal{F}\}|^2 \le E\{X^2 | \mathcal{F}\},$$

which implies that

$$\left| E\left\{ \int_s^t \left(f_n(u) - f(u) \right) dB(u) \Big| \mathcal{F}_s \right\} \right|^2$$
$$\le E\left\{ \left(\int_s^t \left(f_n(u) - f(u) \right) dB(u) \right)^2 \Big| \mathcal{F}_s \right\}.$$

Next we use the property $E(E\{X | \mathcal{F}\}) = EX$ of conditional expectation and then apply Theorem 2.3.4 to get

$$E\left| E\left\{ \int_s^t \left(f_n(u) - f(u) \right) dB(u) \Big| \mathcal{F}_s \right\} \right|^2 \le \int_s^t \left(f_n(u) - f(u) \right)^2 du$$
$$\le \int_a^b \left(f_n(u) - f(u) \right)^2 du$$
$$\to 0,$$

as $n \to \infty$. Hence the sequence $E\{\int_s^t f_n(u) \, dB(u) | \mathcal{F}_s\}$ of random variables converges to $E\{\int_s^t f(u) \, dB(u) | \mathcal{F}_s\}$ in $L^2(\Omega)$. Note that the convergence of a sequence in $L^2(\Omega)$ implies convergence in probability, which implies the existence of a subsequence converging almost surely. Hence by choosing a subsequence if necessary, we can conclude that with probability 1,

$$\lim_{n \to \infty} E\left\{ \int_s^t f_n(u) \, dB(u) \Big| \mathcal{F}_s \right\} = E\left\{ \int_s^t f(u) \, dB(u) \Big| \mathcal{F}_s \right\}. \qquad (2.5.4)$$

Now $E\{\int_s^t f_n(u) \, dB(u) | \mathcal{F}_s\} = 0$ since we have already shown that Equation (2.5.3) holds for step functions. Hence by Equation (2.5.4),

$$E\left\{ \int_s^t f(u) \, dB(u) \Big| \mathcal{F}_s \right\} = 0,$$

and so Equation (2.5.3) holds for any $f \in L^2[a, b]$. □

2.6 Series Expansion of Wiener Integrals

Let $\{\phi_n\}_{n=1}^{\infty}$ be an orthonormal basis for the Hilbert space $L^2[a, b]$. Each $f \in L^2[a, b]$ has the following expansion:

$$f = \sum_{n=1}^{\infty} \langle f, \phi_n \rangle \phi_n, \tag{2.6.1}$$

where $\langle \cdot, \cdot \rangle$ is the inner product on $L^2[a, b]$ given by $\langle f, g \rangle = \int_a^b f(t)g(t)\, dt$. Moreover, we have the Parseval identity

$$\|f\|^2 = \sum_{n=1}^{\infty} \langle f, \phi_n \rangle^2. \tag{2.6.2}$$

Take the Wiener integral in both sides of Equation (2.6.1) and informally interchange the order of integration and summation to get

$$\int_a^b f(t)\, dB(t) = \sum_{n=1}^{\infty} \langle f, \phi_n \rangle \int_a^b \phi_n(t)\, dB(t). \tag{2.6.3}$$

Question 2.6.1. Does the random series in the right-hand side converge to the left-hand side and in what sense?

First observe that by Theorem 2.3.4 and the remark following Equation (2.3.5), the random variables $\int_a^b \phi_n(t)\, dB(t), n \geq 1$, are independent and have the Gaussian distribution with mean 0 and variance 1. Thus the right-hand side of Equation (2.6.3) is a random series of independent and identically distributed random variables. By the Lévy equivalence theorem [10] [37] this random series converges almost surely if and only if it converges in probability and, in turn, if and only if it converges in distribution. On the other hand, we can easily check the $L^2(\Omega)$ convergence of this random series as follows. Apply Equations (2.3.5) and (2.6.2) to show that

$$E\left(\int_a^b f(t)\, dB(t) - \sum_{n=1}^{N} \langle f, \phi_n \rangle \int_a^b \phi_n(t)\, dB(t) \right)^2$$

$$= \int_a^b f(t)^2\, dt - 2\sum_{n=1}^{N} \langle f, \phi_n \rangle^2 + \sum_{n=1}^{N} \langle f, \phi_n \rangle^2$$

$$= \int_a^b f(t)^2\, dt - \sum_{n=1}^{N} \langle f, \phi_n \rangle^2$$

$$\to 0,$$

as $N \to \infty$. Hence the random series in Equation (2.6.3) converges in $L^2(\Omega)$ to the random variable in the left-hand side of Equation (2.6.3). But the $L^2(\Omega)$ convergence implies convergence in probability. Therefore we have proved the next theorem for the series expansion of the Wiener integral.

Theorem 2.6.2. *Let $\{\phi_n\}_{n=1}^{\infty}$ be an orthonormal basis for $L^2[a, b]$. Then for each $f \in L^2[a, b]$, the Wiener integral of f has the series expansion*

$$\int_a^b f(t)\, dB(t) = \sum_{n=1}^{\infty} \langle f, \phi_n \rangle \int_a^b \phi_n(t)\, dB(t),$$

with probability 1, where the random series converges almost surely.

In particular, apply the theorem to $a = 0$, $b = 1$, and $f = 1_{[0,t)}$, $0 \le t \le 1$. Then $\int_0^1 f(s)\, dB(s) = B(t)$ and we have the random series expansion,

$$B(t, \omega) = \sum_{n=1}^{\infty} \left(\int_0^t \phi_n(s)\, ds \right) \left(\int_0^1 \phi_n(s)\, dB(s, \omega) \right).$$

Note that the variables t and ω are separated in the right-hand side. In view of this expansion, we expect that $B(t)$ can be represented by

$$B(t, \omega) = \sum_{n=1}^{\infty} \xi_n(\omega) \int_0^t \phi_n(s)\, ds,$$

where $\{\xi_n\}_{n=1}^{\infty}$ is a sequence of independent random variables having the same Gaussian distribution with mean 0 and variance 1. This method of defining a Brownian motion has been studied in [29] [41] [67].

Exercises

1. Let $B(t)$ be a Brownian motion. Show that $E|B(s) - B(t)|^4 = 3|s - t|^2$.

2. Show that the marginal distribution of a Brownian motion $B(t)$ at times $0 < t_1 < t_2 < \cdots t_n$ is given by

   $P\{B(t_1) \le a_1, B(t_2) \le a_2, \ldots, B(t_n) \le a_n\}$

   $$= \frac{1}{\sqrt{(2\pi)^n t_1 (t_2 - t_1) \cdots (t_n - t_{n-1})}} \int_{-\infty}^{a_n} \cdots \int_{-\infty}^{a_1}$$

   $$\exp\left[-\frac{1}{2}\left(\frac{x_1^2}{t_1} + \frac{(x_2 - x_1)^2}{t_2 - t_1} + \cdots + \frac{(x_n - x_{n-1})^2}{t_n - t_{n-1}} \right) \right] dx_1 dx_2 \cdots dx_n.$$

3. Let $B(t)$ be a Brownian motion. For fixed t and s, find the distribution function of the random variable $X = B(t) + B(s)$.

4. Let $B(t)$ be a Brownian motion and let $0 < s \le t \le u \le v$. Show that the random variables $\frac{1}{t}B(t) - \frac{1}{s}B(s)$ and $aB(u) + bB(v)$ are independent for any $a, b \in \mathbb{R}$.

5. Let $B(t)$ be a Brownian motion and let $0 < s \le t \le u \le v$. Show that the random variables $aB(s) + bB(t)$ and $\frac{1}{v}B(v) - \frac{1}{u}B(u)$ are independent for any $a, b \in \mathbb{R}$ satisfying the condition $as + bt = 0$.

6. Let $B(t)$ be a Brownian motion. Show that $\lim_{t \to 0+} tB(1/t) = 0$ almost surely. Define $W(0) = 0$ and $W(t) = tB(1/t)$ for $t > 0$. Prove that $W(t)$ is a Brownian motion.

7. Let $B(t)$ be a Brownian motion. Find all constants a and b such that $X(t) = \int_0^t \left(a + b\frac{u}{t}\right) dB(u)$ is also a Brownian motion.

8. Let $B(t)$ be a Brownian motion. Find all constants a, b, and c such that $X(t) = \int_0^t \left(a + b\frac{u}{t} + c\frac{u^2}{t^2}\right) dB(u)$ is also a Brownian motion.

9. Let $B(t)$ be a Brownian motion. Show that for any integer $n \ge 1$, there exist nonzero constants $a_0, a_1, \ldots, a_n$ such that $X(t) = \int_0^t \left(a_0 + a_1\frac{u}{t} + a_2\frac{u^2}{t^2} + \cdots + a_n\frac{u^n}{t^n}\right) dB(u)$ is also a Brownian motion.

10. Let $B(t)$ be a Brownian motion. Show that both $X(t) = \int_0^t (2t - u) dB(u)$ and $Y(t) = \int_0^t (3t - 4u) dB(u)$ are Gaussian processes with mean function 0 and the same covariance function $3s^2t - \frac{2}{3}s^3$ for $s \le t$.

11. Let $B(t) = (B_1(t), \ldots, B_n(t))$ be an $\mathbb{R}^n$-valued Brownian motion. Find the density functions of $R(t) = |B(t)|$ and $S(t) = |B(t)|^2$.

12. For each $n \ge 1$, let X_n be a Gaussian random variable with mean μ_n and variance σ_n^2. Suppose the sequence X_n converges to X in $L^2(\Omega)$. Show that the limits $\mu = \lim_{n \to \infty} \mu_n$ and $\sigma^2 = \lim_{n \to \infty} \sigma_n^2$ exist and that X is a Gaussian random variable with mean μ and variance σ^2.

13. Let $f(x, y)$ be the joint density function of random variables X and Y. The *marginal density function* of Y is given by $f_Y(y) = \int_{-\infty}^{\infty} f(x, y) dx$. The *conditional density function* of X given $Y = y$ is defined by $f_{X|Y}(x|y) = f(x, y)/f_Y(y)$. The *conditional expectation* of X given $Y = y$ is defined by $E[X|Y = y] = \int_{-\infty}^{\infty} x f_{X|Y}(x|y) dx$. Let $\sigma(Y)$ be the σ-field generated by Y. Prove that
$$E[X|\sigma(Y)] = \theta(Y),$$
where θ is the function $\theta(y) = E[X|Y = y]$.

14. Prove the properties of conditional expectation listed in Section 2.4.

15. Let $B(t)$ be a Brownian motion. Find the distribution of $\int_0^t e^{t-s} dB(s)$. Check whether $X_t = \int_0^t e^{t-s} dB(s)$ is a martingale.

16. Let $B(t)$ be a Brownian motion. Find the distribution of $\int_0^t B(s) ds$. Check whether $Y_t = \int_0^t B(s) ds$ is a martingale.

17. Let $B(t)$ be a Brownian motion. Find the distribution of the integral $\int_0^t B(s) \cos(t - s) ds$.

18. Let $B(t)$ be a Brownian motion. Show that $X_t = \frac{1}{3}B(t)^3 - \int_0^t B(s) ds$ is a martingale.

3

Constructions of Brownian Motion

In Section 1.2 we gave an intuitive description of Brownian motion as the limit of a random walk $Y_{\delta,\sqrt{\delta}}$ as $\delta \to 0$. The purpose of this chapter is to give three constructions of Brownian motion. The first construction, due to N. Wiener in 1923, will be explained in Section 3.1 without proof. The second construction, using Kolmogorov's extension and continuity theorems, will be discussed in detail in Section 3.3. We will briefly explain Lévy's interpolation method for constructing a Brownian motion in Section 3.4.

3.1 Wiener Space

Let C be the Banach space of real-valued continuous functions ω on $[0,1]$ with $\omega(0) = 0$. The norm on C is $\|\omega\|_\infty = \sup_{t \in [0,1]} |\omega(t)|$.

A cylindrical subset A of C is a set of the form

$$A = \{\omega \in C\,;\, (\omega(t_1), \omega(t_2), \ldots, \omega(t_n)) \in U\}, \qquad (3.1.1)$$

where $0 < t_1 < t_2 < \cdots < t_n \leq 1$ and $U \in \mathcal{B}(\mathbb{R}^n)$, the Borel σ-field of $\mathbb{R}^n$. Let $\mathcal{R}$ be the collection of all cylindrical subsets of C. Obviously, $\mathcal{R}$ is a field. However, it is not a σ-field.

Suppose $A \in \mathcal{R}$ is given by Equation (3.1.1). Define $\mu(A)$ by

$$\mu(A) = \int_U \prod_{i=1}^{n} \left(\frac{1}{\sqrt{2\pi(t_i - t_{i-1})}} \exp\left[-\frac{(u_i - u_{i-1})^2}{2(t_i - t_{i-1})} \right] \right) du_1 \cdots du_n, \quad (3.1.2)$$

where $t_0 = u_0 = 0$. Observe that a cylindrical set A can be expressed in many different ways as in Equation (3.1.1). For instance,

$$A = \{\omega \in C\,;\, \omega(1/2) \in [a,b]\} = \{\omega \in C\,;\, (\omega(1/2), \omega(2/3)) \in [a,b] \times \mathbb{R}\}.$$

However, $\mu(A)$ as defined by Equation (3.1.2) is independent of the choice of different expressions in the right-hand side of Equation (3.1.1). This means that $\mu(A)$ is well-defined. Hence μ is a mapping from $\mathcal{R}$ into $[0,1]$. It can be easily checked that $\mu : \mathcal{R} \to [0,1]$ is finitely additive, i.e., for any disjoint $A_1, A_2, \ldots, A_m \in \mathcal{R}$, we have

$$\mu\left(\bigcup_{i=1}^{m} A_i\right) = \sum_{i=1}^{m} \mu(A_i).$$

Theorem 3.1.1. (Wiener 1923 [82]) *The mapping μ on $\mathcal{R}$ is σ-additive, i.e., for any disjoint $A_1, A_2, \ldots \in \mathcal{R}$ with $\cup_n A_n \in \mathcal{R}$, the following equality holds:*

$$\mu\left(\bigcup_{i=1}^{\infty} A_i\right) = \sum_{i=1}^{\infty} \mu(A_i).$$

The proof of this theorem can be found in [55]. Thus μ has a unique σ-additive extension to $\sigma(\mathcal{R})$, the σ-field generated by $\mathcal{R}$. The σ-field $\sigma(\mathcal{R})$ turns out to be the same as the Borel σ-field $\mathcal{B}(C)$ of C. To check this fact, we essentially need to show that the closed unit ball $\{\omega \in C \,;\, \|\omega\|_\infty \leq 1\}$ belongs to $\sigma(\mathcal{R})$. But this is so in view of the following equality:

$$\{\omega \in C \,;\, \|\omega\|_\infty \leq 1\} = \bigcap_{n=1}^{\infty} \{\omega \in C \,;\, |\omega(k/n)| \leq 1, \, \forall k = 1, 2, \ldots, n\}.$$

We will use the same notation μ to denote the extension of μ to $\mathcal{B}(C)$. Hence (C, μ) is a probability space. It is called the *Wiener space*. The measure μ is called the *Wiener measure*. Wiener himself called (C, μ) the *differential space* [82]. In [24], Gross generalized the theorem to abstract Wiener spaces. See also the book [55].

Theorem 3.1.2. *The stochastic process $B(t, \omega) = \omega(t)$, $0 \leq t \leq 1$, $\omega \in C$, is a Brownian motion.*

Proof. We need to check that the conditions in Definition 2.1.1 are satisfied. Conditions (1) and (4) are obviously satisfied. To check condition (2), let $0 < s < t \leq 1$. Then

$$\mu\{B(t) - B(s) \leq a\} = \mu\{\omega(t) - \omega(s) \leq a\}$$
$$= \mu\{\omega \in C \,;\, (\omega(s), \omega(t)) \in U\},$$

where $U = \{(u_1, u_2) \in \mathbb{R}^2 \,;\, u_2 - u_1 \leq a\}$. Hence by the definition of the Wiener measure μ we have

$$\mu\{B(t) - B(s) \leq a\}$$
$$= \int_U \frac{1}{\sqrt{(2\pi)^2 \, s(t-s)}} \exp\left[-\frac{1}{2}\left(\frac{u_1^2}{s} + \frac{(u_2 - u_1)^2}{t-s}\right)\right] du_1 du_2.$$

Make a change of variables $u_1 = x, u_2 - u_1 = y$ to get

$$\mu\{B(t) - B(s) \le a\}$$

$$= \int_{-\infty}^{a} \int_{-\infty}^{\infty} \frac{1}{\sqrt{(2\pi)^2 \, s(t-s)}} \exp\left[-\frac{1}{2}\left(\frac{x^2}{s} + \frac{y^2}{t-s}\right)\right] dx \, dy$$

$$= \int_{-\infty}^{a} \frac{1}{\sqrt{2\pi(t-s)}} \, e^{-\frac{y^2}{2(t-s)}} \, dy.$$

Thus $B(t) - B(s)$ is normally distributed with mean 0 and variance $t - s$, so condition (2) is satisfied.

To check condition (3), let $0 < t_1 < t_2 < \cdots < t_n$. By arguments similar to those above, we can show that

$$\mu\{B(t_1) \le a_1, B(t_2) - B(t_1) \le a_2, \ldots, B(t_n) - B(t_{n-1}) \le a_n\}$$
$$= \mu\{B(t_1) \le a_1\}\mu\{B(t_2) - B(t_1) \le a_2\} \cdots \mu\{B(t_n) - B(t_{n-1}) \le a_n\}.$$

This implies that the random variables

$$B(t_1), \ B(t_2) - B(t_1), \ \ldots, \ B(t_n) - B(t_{n-1})$$

are independent. Hence condition (3) is satisfied.   □

The above theorem provides a Brownian motion $B(t)$ for $0 \le t < 1$. We can define a Brownian motion $B(t)$ for $t \ge 0$ as follows. Take a sequence of independent Brownian motions $B_1(t), B_2(t), \ldots, B_n(t), \ldots$ for $0 \le t \le 1$. Then define

$$B(t) = \begin{cases} B_1(t), & \text{if } 0 \le t < 1; \\ B_1(1) + B_2(t-1), & \text{if } 1 \le t < 2; \\ \cdots\cdots\cdots & \cdots\cdots \\ B_1(1) + \cdots + B_n(1) + B_{n+1}(t-n), & \text{if } n \le t < n+1; \\ \cdots\cdots\cdots & \cdots\cdots. \end{cases}$$

It is easy to check that $B(t)$, $t \ge 0$, is a Brownian motion.

3.2 Borel–Cantelli Lemma and Chebyshev Inequality

In this section we will explain the Borel–Cantelli lemma and the Chebyshev inequality, which will be needed in the next section and elsewhere.

Let $\{A_n\}_{n=1}^{\infty}$ be a sequence of events in some probability space. Consider the event A given by $A = \cap_{n=1}^{\infty} \cup_{k=n}^{\infty} A_k$. It is easy to see that $\omega \in A$ if and only if $\omega \in A_n$ for infinitely many n's. Thus we can think of the event A as the event that A_n's occur infinitely often. We will use the notation introduced in [5]:

$$\{A_n \text{ i.o.}\} = \{A_n \text{ infinitely often}\} = \bigcap_{n=1}^{\infty} \bigcup_{k=n}^{\infty} A_k.$$

Theorem 3.2.1. (Borel–Cantelli lemma) *Let $\{A_n\}_{n=1}^\infty$ be a sequence of events such that $\sum_{n=1}^\infty P(A_n) < \infty$. Then $P\{A_n \text{ i.o.}\} = 0$.*

This theorem is often called the first part of the Borel–Cantelli lemma. The second part states that if $\sum_{n=1}^\infty P(A_n) = \infty$ and the events A_n are independent, then $P\{A_n \text{ i.o.}\} = 1$. We will need only the first part, which can be proved rather easily as follows:

$$P\{A_n \text{ i.o.}\} = \lim_{n\to\infty} P\left(\cup_{k=n}^\infty A_k \right) \leq \lim_{n\to\infty} \sum_{k=n}^\infty P(A_k) = 0.$$

The complement of $\{A_n \text{ i.o.}\}$ is the event $\{A_n \text{ f.o.}\} = \{A_n \text{ finitely often}\}$ that A_n's occur finitely often. Thus whenever we have a situation like

$$P\{A_n \text{ i.o.}\} = 0,$$

then, by taking the complement, we get $P\{A_n \text{ f.o.}\} = 1$. Hence there exists an event $\widetilde{\Omega}$ such that $P(\widetilde{\Omega}) = 1$ and for each $\omega \in \widetilde{\Omega}$, $\omega \in A_n$ for only finitely many n's, namely, for each $\omega \in \widetilde{\Omega}$, there exists a positive integer $N(\omega)$ such that $\omega \in A_n^c$ for all $n \geq N(\omega)$. Here A_n^c is the complement of A_n. Then we can use this fact to conclude useful information.

Example 3.2.2. Let $\{X_n\}_{n=1}^\infty$ be a sequence of random variables. Suppose we can find positive numbers α_n and β_n such that

$$P(|X_n| \geq \alpha_n\} \leq \beta_n, \quad \forall n \geq 1.$$

If $\sum_n \beta_n < \infty$, then we can apply the Borel–Cantelli lemma to see that

$$P\{|X_n| \geq \alpha_n \text{ i.o.}\} = 0,$$

or equivalently,

$$P\{|X_n| \geq \alpha_n \text{ f.o.}\} = 1.$$

If $\alpha_n \to 0$ as $n \to \infty$, then we can conclude that $X_n \to 0$ almost surely. On the other hand, if $\sum_n \alpha_n < \infty$, then we can conclude that the random series $\sum_n X_n$ is absolutely convergent almost surely.

The Borel–Cantelli lemma is often used with the Chebyshev inequality. Let X be a random variable with $E|X| < \infty$. Then for any $a > 0$,

$$E|X| \geq \int_{|X|\geq a} |X|\, dP \geq a \int_{|X|\geq a} dP = aP(|X| \geq a).$$

Divide both sides by a to get a very useful inequality in the next theorem.

Theorem 3.2.3. (Chebyshev inequality) *Let X be a random variable with $E|X| < \infty$. Then for any $a > 0$,*

$$P(|X| \geq a) \leq \frac{1}{a} E|X|, \quad \forall a > 0.$$

Example 3.2.4. Let $\{X_n\}_{n=1}^{\infty}$ be a sequence of random variables such that $E|X_n| \leq 1$ for all n. Let $c > 2$ be a fixed number. Choose $\alpha = c/2$. By the Chebyshev inequality, we have

$$P\{|X_n| \geq n^{\alpha}\} \leq \frac{1}{n^{\alpha}} E|X_n| \leq \frac{1}{n^{\alpha}}.$$

Since $\sum_n \frac{1}{n^{\alpha}} < \infty$, we can apply the Borel–Cantelli lemma to show that

$$P\{|X_n| \geq n^{\alpha} \text{ i.o.}\} = 0.$$

Hence

$$P\{n^{-c}|X_n| \geq n^{\alpha-c} \text{ i.o.}\} = 0,$$

or equivalently,

$$P\{n^{-c}|X_n| \geq n^{\alpha-c} \text{ f.o.}\} = 1.$$

Note that $c - \alpha = c/2 > 1$, so that $\sum_n n^{\alpha-c} < \infty$. It follows that the random series $\sum_n n^{-c} X_n$ is absolutely convergent almost surely.

3.3 Kolmogorov's Extension and Continuity Theorems

In this section we will explain Kolmogorov's extension theorem, give a detailed proof of Kolmogorov's continuity theorem, and then apply these theorems to construct a Brownian motion.

Let $X(t)$, $t \geq 0$, be a stochastic process. For any $0 \leq t_1 < t_2 < \cdots < t_n$, define

$$\mu_{t_1, t_2, \ldots, t_n}(A) = P\{(X(t_1), X(t_2), \ldots, X(t_n)) \in A\}, \quad A \in \mathcal{B}(\mathbb{R}^n).$$

The probability measure $\mu_{t_1, t_2, \ldots, t_n}$ on $\mathbb{R}^n$ is called a *marginal distribution* of the stochastic process $X(t)$. Observe that for any $0 \leq t_1 < \cdots < t_n$ and any $1 \leq i \leq n$,

$$P\{(X(t_1), \ldots, X(t_{i-1}), X(t_{i+1}), \ldots, X(t_n)) \in A_1 \times A_2\}$$
$$= P\{(X(t_1), \ldots, X(t_n)) \in A_1 \times \mathbb{R} \times A_2\}, \tag{3.3.1}$$

where $A_1 \in \mathcal{B}(\mathbb{R}^{i-1})$ and $A_2 \in \mathcal{B}(\mathbb{R}^{n-i})$. When $i = 1$, Equation (3.3.1) is understood to be

$$P\{(X(t_2), \ldots, X(t_n)) \in A_2\} = P\{(X(t_1), \ldots, X(t_n)) \in \mathbb{R} \times A_2\}.$$

Similarly, when $i = n$, Equation (3.3.1) is understood to be

$$P\{(X(t_1), \ldots, X(t_{n-1})) \in A_1\} = P\{(X(t_1), \ldots, X(t_n)) \in A_1 \times \mathbb{R}\}.$$

Therefore, we have the equality

$$\mu_{t_1,\ldots,t_{i-1},\widehat{t_i},t_{i+1},\ldots,t_n}(A_1 \times A_2) = \mu_{t_1,\ldots,t_n}(A_1 \times \mathbb{R} \times A_2), \qquad (3.3.2)$$

where $1 \le i \le n$ and $\widehat{t_i}$ means that t_i is deleted. Hence the family of marginal distributions of a stochastic process $X(t)$ satisfies Equation (3.3.2).

Conversely, suppose that for any $0 \le t_1 < t_2 < \cdots < t_n$, there is a probability measure $\mu_{t_1,t_2,\ldots,t_n}$ on $\mathbb{R}^n$. The family of probability measures

$$\left\{ \mu_{t_1,t_2,\ldots,t_n} \,;\, 0 \le t_1 < t_2 < \cdots < t_n, \, n = 1, 2, \ldots \right\}$$

is said to satisfy the *consistency condition* if Equation (3.3.2) holds for any $0 \le t_1 < \cdots < t_n$ and $A_1 \in \mathcal{B}(\mathbb{R}^{i-1})$, $A_2 \in \mathcal{B}(\mathbb{R}^{n-i})$ with $1 \le i \le n$, $n \ge 1$.

Question 3.3.1. Given a family of probability measures satisfying the above consistency condition, does there exist a stochastic process whose marginal distributions are the given family of probability measures?

The answer to this question is given by Kolmogorov's extension theorem. Let $\mathbb{R}^{[0,\infty)}$ denote the space of all real-valued functions ω on the interval $[0,\infty)$. Let $\mathcal{F}$ be the σ-field generated by *cylindrical sets*, i.e., sets of the form

$$\left\{ \omega \in \mathbb{R}^{[0,\infty)} \,;\, (\omega(t_1), \omega(t_2), \ldots, \omega(t_n)) \in A \right\},$$

where $0 \le t_1 < t_2 < \cdots < t_n$ and $A \in \mathcal{B}(\mathbb{R}^n)$. The next theorem ensures the existence of a probability measure on the function space $\mathbb{R}^{[0,\infty)}$ corresponding to a consistent family of probability measures. Equivalently, the theorem says that a stochastic process can be specified by its marginal distributions.

Theorem 3.3.2. (Kolmogorov's extension theorem) *Suppose that associated with each $0 \le t_1 < t_2 < \cdots < t_n$, $n \ge 1$, there is a probability measure $\mu_{t_1,t_2,\ldots,t_n}$ on $\mathbb{R}^n$. Assume that the family*

$$\left\{ \mu_{t_1,t_2,\ldots,t_n} \,;\, 0 \le t_1 < t_2 < \cdots < t_n, \, n = 1, 2, \ldots \right\} \qquad (3.3.3)$$

satisfies the consistency condition in Equation (3.3.2). Then there exists a unique probability measure P on the space $(\mathbb{R}^{[0,\infty)}, \mathcal{F})$ such that

$$P\left\{ \omega \in \mathbb{R}^{[0,\infty)} \,;\, (\omega(t_1), \omega(t_2), \ldots, \omega(t_n)) \in A \right\} = \mu_{t_1,t_2,\ldots,t_n}(A)$$

for any $0 \le t_1 < t_2 < \cdots < t_n$, $A \in \mathcal{B}(\mathbb{R}^n)$, and $n \ge 1$.

The proof of this theorem can be found, e.g., in the book by Lamperti [57]. On the probability space $(\mathbb{R}^{[0,\infty)}, \mathcal{F}, P)$ we define a stochastic process

$$X(t,\omega) = \omega(t), \quad t \ge 0, \ \omega \in \mathbb{R}^{[0,\infty)}.$$

Then $X(t)$ is a stochastic process with marginal distributions specified by the probability measures given by Equation (3.3.3), i.e.,

$$P\{(X(t_1), X(t_2), \ldots, X(t_n)) \in A\} = \mu_{t_1,t_2,\ldots,t_n}(A).$$

Example 3.3.3. For $0 \leq t_1 < t_2 < \cdots < t_n$, let $\mu_{t_1,t_2,\ldots,t_n}$ be the probability measure on $\mathbb{R}^n$ defined by

$$\mu_{t_1,t_2,\ldots,t_n}(U) = \int_{\mathbb{R}} \int_U$$

$$\prod_{i=1}^n \left(\frac{1}{\sqrt{2\pi(t_i - t_{i-1})}} \exp\left[-\frac{(u_i - u_{i-1})^2}{2(t_i - t_{i-1})} \right] \right) du_1 \cdots du_n d\nu(u_0), \quad (3.3.4)$$

where $U \in \mathcal{B}(\mathbb{R}^n)$, $t_0 = 0$, ν is a probability measure on $\mathbb{R}$, and we use the following convention for the integrand:

$$\frac{1}{\sqrt{2\pi t_1}} e^{-\frac{(u_1 - u_0)^2}{2t_1}} du_1 \bigg|_{t_1=0} = d\delta_{u_0}(u_1), \qquad (3.3.5)$$

where δ_{u_0} is the Dirac delta measure at u_0.

Observe that the integral in the right-hand side of Equation (3.3.4) with $\nu = \delta_0$ is exactly the same as the one in the right-hand side of Equation (3.1.2) for the Wiener measure μ.

The family of probability measures defined by Equation (3.3.4) can be easily verified to satisfy the consistency condition.

Now apply Kolmogorov's extension theorem to the family of probability measures in Example 3.3.3 to get a probability measure P on $\mathbb{R}^{[0,\infty)}$. Define a stochastic process $Y(t)$ by

$$Y(t,\omega) = \omega(t), \quad \omega \in \mathbb{R}^{[0,\infty)}. \qquad (3.3.6)$$

Question 3.3.4. Is this stochastic process $Y(t)$ a Brownian motion?

First we notice some properties of the stochastic process $Y(t)$:

(a) For any $0 \leq s < t$, the random variable $Y(t) - Y(s)$ is normally distributed with mean 0 and variance $t - s$.

(b) $Y(t)$ has independent increments, i.e., for any $0 = t_0 \leq t_1 < t_2 < \cdots < t_n$, the random variables $Y(t_i) - Y(t_{i-1})$, $i = 1, 2, \ldots, n$, are independent.

Moreover, by Equations (3.3.4) and (3.3.5) with $n = 1$, $t_1 = 0$, we have

$$P\{Y(0) \in U\} = \int_{\mathbb{R}} \int_U \frac{1}{\sqrt{2\pi t_1}} e^{-\frac{(u_1 - u_0)^2}{2t_1}} du_1 \bigg|_{t_1=0} d\nu(u_0)$$

$$= \int_{\mathbb{R}} \left(\int_U d\delta_{u_0}(u_1) \right) d\nu(u_0)$$

$$= \int_{\mathbb{R}} \delta_{u_0}(U) \, d\nu(u_0).$$

$$= \nu(U), \quad U \in \mathcal{B}(\mathbb{R}).$$

Hence ν is the initial distribution of the stochastic process $Y(t)$.

In particular, the stochastic process $Y(t)$ with $\nu = \delta_0$ satisfies conditions (1), (2), and (3) in Definition 2.1.1. How about condition (4)? Before we can answer this question, we need to introduce some concepts concerning stochastic processes.

Definition 3.3.5. *A stochastic process* $\widetilde{X}(t)$ *is called a* version *of* $X(t)$ *if* $P\{\widetilde{X}(t) = X(t)\} = 1$ *for each* t.

Definition 3.3.6. *A stochastic process* $\widetilde{X}(t)$ *is called a* realization *of* $X(t)$ *if there exists* Ω_0 *such that* $P(\Omega_0) = 1$ *and for each* $\omega \in \Omega_0$, $\widetilde{X}(t,\omega) = X(t,\omega)$ *for all* t.

Obviously, a realization is also a version, but not vice versa. But if the time parameter set is countable, then these two concepts are equivalent.

Let $X(t)$, $0 \le t \le 1$, be a stochastic process. In general $\sup_{0 \le t \le 1} X(t)$ does not define a random variable. For example, take the probability space $\Omega = [0,1]$ with the Lebesgue measure. Let $S \subset [0,1]$ be a nonmeasurable set and define a stochastic process by

$$X(t,\omega) = \begin{cases} 1, & \text{if } t \in S \text{ and } \omega = t; \\ 0, & \text{otherwise.} \end{cases}$$

Then the function $\sup_{0 \le t \le 1} X(t)$ is given by

$$\left(\sup_{0 \le t \le 1} X(t)\right)(\omega) = \begin{cases} 1, & \text{if } \omega \in S; \\ 0, & \text{if } \omega \notin S. \end{cases}$$

This shows that $\sup_{0 \le t \le 1} X(t)$ is not measurable. Hence it does not define a random variable. In order to overcome the difficulty involving supremum and infimum, we impose the condition of the separability of stochastic processes.

Definition 3.3.7. *A stochastic process* $X(t,\omega)$, $0 \le t \le 1$, $\omega \in \Omega$, *is called* separable *if there exist* Ω_0 *with* $P(\Omega_0) = 1$ *and a countable dense subset* S *of* $[0,1]$ *such that for any closed set* $F \subset \mathbb{R}$ *and any open interval* $I \subset [0,1]$, *the set difference*

$$\left\{\omega \in \Omega \,;\, X(t,\omega) \in F,\, \forall t \in I \cap S\right\} \setminus \left\{\omega \in \Omega \,;\, X(t,\omega) \in F,\, \forall t \in I\right\}$$

is a subset of the complement Ω_0^c *of* Ω_0. *The set* S *is called a* separating set.

It turns out that a stochastic process always possesses a separable version. Moreover, if a separable stochastic process $X(t)$ is continuous in probability, i.e., for each t and any $\epsilon > 0$,

$$P\{|X(t) - X(s)| \ge \epsilon\} \to 0, \quad \text{as } s \to t,$$

then any countable dense subset of $[0,1]$ can be taken as a separating set. For details, see the book by Doob [9].

The next theorem gives a very simple sufficient condition for a separable stochastic process to have a continuous realization.

Theorem 3.3.8. (Kolmogorov's continuity theorem) *Let $X(t)$, $0 \leq t \leq 1$, be a separable stochastic process. Assume that there exist constants $\alpha, \beta, K > 0$ satisfying the inequality*

$$E|X(t) - X(s)|^\alpha \leq K|t - s|^{1+\beta}, \quad \forall 0 \leq t, s \leq 1. \tag{3.3.7}$$

Then $X(t)$ has a continuous realization, namely, there exists Ω_0 such that $P(\Omega_0) = 1$ and for each $\omega \in \Omega_0$, $X(t, \omega)$ is a continuous function of t.

Now, before we proceed to prove this theorem, we return to the stochastic process $Y(t)$ defined by Equation (3.3.6) with initial distribution $\nu = \delta_0$. By the above discussion we may assume that $Y(t)$ is separable. Moreover, for any $0 \leq s < t$, $Y(t) - Y(s)$ is normally distributed with mean 0 and variance $t - s$, and so we have

$$E|Y(t) - Y(s)|^4 = 3|t - s|^2.$$

This shows that the stochastic process $Y(t)$ satisfies the condition in Equation (3.3.7). Hence by Kolmogorov's continuity theorem, $Y(t)$ satisfies condition (4) in Definition 2.1.1. Therefore, $Y(t)$ is a Brownian motion.

The rest of this section is devoted to the proof of Kolmogorov's continuity theorem. The proof is based on the book by Wong and Hajek [86].

By Equation (3.3.7) and the Chebyshev inequality,

$$P\{|X(t) - X(s)| \geq \epsilon\} \leq \frac{1}{\epsilon^\alpha} E|X(t) - X(s)|^\alpha \leq \frac{K}{\epsilon^\alpha}|t - s|^{1+\beta} \to 0,$$

as $s \to t$. Hence $X(t)$ is continuous in probability. Thus we can take the following set of dyadic rational numbers as a separating set:

$$D = \left\{ \frac{k}{2^n} ; k = 0, 1, 2, \ldots, 2^n, n = 1, 2, \ldots \right\}. \tag{3.3.8}$$

Hence there exists Ω_1 with $P\{\Omega_1\} = 1$ such that for any $h > 0$ and all $\omega \in \Omega_1$,

$$\sup_{\substack{0 \leq t, s \leq 1 \\ |t-s| < h}} |X(t, \omega) - X(s, \omega)| = \sup_{\substack{t, s \in D \\ |t-s| < h}} |X(t, \omega) - X(s, \omega)|. \tag{3.3.9}$$

Next, we prepare a key lemma.

Lemma 3.3.9. *Let D be the set of numbers defined by Equation (3.3.8) and let f be a function on $[0, 1]$. For each integer $n \geq 1$, let*

$$g_n = \sup_{1 \leq k \leq 2^n} \left| f\left(\frac{k}{2^n}\right) - f\left(\frac{k-1}{2^n}\right) \right|.$$

Then for any integer $m \geq 0$, the following inequality holds:

$$\sup_{\substack{t, s \in D \\ |t-s| < 2^{-m}}} |f(t) - f(s)| \leq 2 \sum_{n=m+1}^{\infty} g_n. \tag{3.3.10}$$

Proof. Suppose $t, s \in D$ and $|t - s| < 2^{-m}$. Then there exists some integer $k \in \{0, 1, 2, \ldots, 2^m\}$ such that

$$\left| t - \frac{k}{2^m} \right| < \frac{1}{2^m}, \quad \left| s - \frac{k}{2^m} \right| < \frac{1}{2^m}.$$

Therefore, there exists some integer $j \geq 1$ such that

$$t = \frac{k}{2^m} \pm \sum_{\ell=1}^{j} \frac{a_\ell}{2^{m+\ell}}, \quad a_\ell = 0, 1.$$

This implies that

$$\left| f(t) - f\left(\frac{k}{2^m}\right) \right| \leq \sum_{n=m+1}^{m+j} g_n \leq \sum_{n=m+1}^{\infty} g_n. \tag{3.3.11}$$

Similarly, we have

$$\left| f(s) - f\left(\frac{k}{2^m}\right) \right| \leq \sum_{n=m+1}^{\infty} g_n. \tag{3.3.12}$$

Hence Equation (3.3.10) follows from Equations (3.3.11) and (3.3.12).   □

Now we proceed to prove Kolmogorov's continuity theorem.

Step 1: Let $c > 0$. By the Chebyshev inequality

$$P\{|X(t) - X(s)| \geq c\} \leq \frac{1}{c^\alpha} E|X(t) - X(s)|^\alpha \leq \frac{K}{c^\alpha} |t - s|^{1+\beta}.$$

Choose a constant γ such that $0 < \gamma < \beta/\alpha$. Let $\delta = \beta - \alpha\gamma$ and put $c = |t-s|^\gamma$ to get

$$P\{|X(t) - X(s)| \geq |t - s|^\gamma\} \leq K|t - s|^{1+\delta}. \tag{3.3.13}$$

Step 2: Define a sequence of random variables by

$$Z_n = \sup_{1 \leq k \leq 2^n} \left| X\left(\frac{k}{2^n}\right) - X\left(\frac{k-1}{2^n}\right) \right|.$$

Note that for any constant $C \geq 0$,

$$\{Z_n \geq C\} = \bigcup_{k=1}^{2^n} \left\{ \left| X\left(\frac{k}{2^n}\right) - X\left(\frac{k-1}{2^n}\right) \right| \geq C \right\}.$$

Hence we have

$$P\{Z_n \geq C\} \leq \sum_{k=1}^{2^n} P\left\{ \left| X\left(\frac{k}{2^n}\right) - X\left(\frac{k-1}{2^n}\right) \right| \geq C \right\}.$$

Take $C = 2^{-n\gamma}$ and then use Equation (3.3.13) to get

$$P\{Z_n \geq 2^{-n\gamma}\} \leq \sum_{k=1}^{2^n} K2^{-n(1+\delta)} = K2^{-n\delta}. \tag{3.3.14}$$

Step 3: In this step we will show that $\sum_{n=1}^{\infty} Z_n$ converges almost surely. The above Equation (3.3.14) implies that

$$\sum_{n=1}^{\infty} P\{Z_n \geq 2^{-n\gamma}\} < \infty.$$

Hence by the Borel–Cantelli lemma in Theorem 3.2.1, we have

$$P\{Z_n \geq 2^{-n\gamma} \text{ infinitely often}\} = 0,$$

or equivalently,

$$P\{Z_n \geq 2^{-n\gamma} \text{ finitely often}\} = 1.$$

Hence there exists Ω_2 with $P\{\Omega_2\} = 1$ and for any $\omega \in \Omega_2$ there is a positive integer $N(\omega)$ such that

$$Z_n(\omega) < 2^{-n\gamma}, \quad \forall n \geq N(\omega).$$

It follows that $\sum_{n=1}^{\infty} Z_n(\omega)$ converges for all $\omega \in \Omega_2$.

Step 4: For each $\omega \in \Omega_2$ in Step 3, apply Lemma 3.3.9 to $f(t) = X(t, \omega)$ to conclude that

$$\sup_{\substack{t,s \in D \\ |t-s| < 2^{-m}}} |X(t, \omega) - X(s, \omega)| \leq 2 \sum_{n=m+1}^{\infty} Z_n(\omega).$$

But $\sum_{n=m+1}^{\infty} Z_n(\omega) \to 0$ as $m \to \infty$, since the series $\sum_{n=1}^{\infty} Z_n(\omega)$ converges as shown in Step 3. Hence for each $\omega \in \Omega_2$,

$$\sup_{\substack{t,s \in D \\ |t-s| < 2^{-m}}} |X(t, \omega) - X(s, \omega)| \longrightarrow 0, \quad \text{as } m \to \infty.$$

On the other hand, recall from Equation (3.3.9) with $h = 2^{-m}$ that for all $\omega \in \Omega_1$,

$$\sup_{\substack{0 \leq t,s < 1 \\ |t-s| < 2^{-m}}} |X(t, \omega) - X(s, \omega)| = \sup_{\substack{t,s \in D \\ |t-s| < 2^{-m}}} |X(t, \omega) - X(s, \omega)|.$$

Finally, let $\Omega_0 = \Omega_1 \cap \Omega_2$. Then $P\{\Omega_0\} = 1$ and for any $\omega \in \Omega_0$,

$$\sup_{\substack{0 \leq t,s \leq 1 \\ |t-s| < 2^{-m}}} |X(t, \omega) - X(s, \omega)| \longrightarrow 0, \quad \text{as } m \to \infty,$$

which implies that the sample function $X(t, \omega)$ is uniformly continuous on $[0, 1]$. This completes the proof of Kolmogorov's continuity theorem. □

3.4 Lévy's Interpolation Method

In [58] P. Lévy used the interpolation method to construct a Brownian motion. Below we outline this method. For details, see the book by Hida [25].

Let $\{\phi_n \, ; \, n \geq 1\}$ be a sequence of independent Gaussian random variables with mean 0 and variance 1. Define a sequence $\{X_n(t) \, ; \, n \geq 1\}$ of stochastic processes for $t \in [0, 1]$ by interpolation as follows. For $n = 1$, define

$$X_1(t) = \begin{cases} 0, & \text{if } t = 0; \\ \phi_1, & \text{if } t = 1, \end{cases}$$

and interpolate $X_1(t)$ linearly for other values of t. From $X_1(t)$, define

$$X_2(t) = \begin{cases} X_1(t), & \text{if } t = 0, 1; \\ X_1(t) + 2^{-1} \phi_2, & \text{if } t = 1/2, \end{cases}$$

and interpolate $X_2(t)$ linearly for other values of t. Note that $X_2(t)$ keeps the values of $X_1(t)$ at $t = 0, 1$. At $t = \frac{1}{2}$, we use ϕ_2 and $X_1(t)$ to define $X_2(t)$ in order to get the correct variance $\frac{1}{2}$, i.e., the variance of a Brownian motion at time $\frac{1}{2}$. We go one more step in order to see the situation more clearly. Define $X_3(t)$ from $X_2(t)$ as follows:

$$X_3(t) = \begin{cases} X_2(t), & \text{if } t = 0, 1/2, 1; \\ X_2(t) + 2^{-3/2} \phi_3, & \text{if } t = 1/4; \\ X_2(t) + 2^{-3/2} \phi_4, & \text{if } t = 3/4, \end{cases}$$

and interpolate $X_3(t)$ linearly for other values of t. Again note that $X_3(t)$ keeps the values of $X_2(t)$ at $t = 0, \frac{1}{2}, 1$. At $t = \frac{1}{4}$ and $\frac{3}{4}$, we use ϕ_3 and ϕ_4, respectively, and $X_2(t)$ to define $X_3(t)$ in order to get the correct variances $\frac{1}{4}$ and $\frac{3}{4}$ of a Brownian motion at times $\frac{1}{4}$ and $\frac{3}{4}$, respectively. The verification of this fact is tedious, but straightforward.

Now we use induction to define $X_{n+1}(t)$ from $X_n(t)$ for $n \geq 1$ by

$$X_{n+1}(t) = \begin{cases} X_n(t), & \text{if } t = \dfrac{k}{2^n}, \; k = 0, 2, \ldots, 2^n; \\ X_n(t) + 2^{-\frac{n+1}{2}} \phi_{2^{n-1} + \frac{k+1}{2}}, & \text{if } t = \dfrac{k}{2^n}, \; k = 1, 3, \ldots, 2^n - 1, \end{cases}$$

and interpolate $X_{n+1}(t)$ linearly for other values of t.

Recall the set D of dyadic rational numbers defined in Equation (3.3.8). For each fixed $t \in D$, it follows from the definition of the $X_n(t)$'s that there exists a number $N = N(t, \omega)$ such that $X_n(t, \omega) = X_N(t, \omega)$ for all $n \geq N$. Hence for each $t \in D$, the following limit exists:

$$\lim_{n \to \infty} X_n(t) = X(t), \quad \text{in } L^2(\Omega).$$

It can be easily verified from this limit and the definition of the $X_n(t)$'s that $X(t)$, $t \in D$, is a Gaussian process and for any $s, t \in D$,

$$E\{X(t)\} = 0,$$

$$E\{|X(t) - X(s)|^2\} = |t - s|. \tag{3.4.1}$$

In particular, Equation (3.4.1) implies that the mapping $X : D \to L^2(\Omega)$ is continuous. Thus we can extend $X(t)$ by continuity to all $t \in [0, 1]$, and then for any $t \in [0, 1]$, we have

$$X(t) = \lim_{n \to \infty} X_n(t), \quad \text{in } L^2(\Omega).$$

The stochastic process $X(t)$, $t \in [0, 1]$, is a Gaussian process. Moreover, for any $s, t \in [0, 1]$,

$$E\{X(t)\} = 0, \quad E\{X(s)X(t)\} = \min\{s, t\}.$$

It follows from Remark 2.2.2 that $X(t)$ has independent increments. By the above construction $X(0) = 0$ almost surely and for $0 \le s < t$, $X(t) - X(s)$ is normally distributed with mean 0 and variance $t - s$. Therefore, $X(t)$ satisfies conditions (1), (2), and (3) in Definition 2.1.1.

On the other hand, by Proposition 2.3 of the book by Hida [25], there exists $\Omega_0 \subset \Omega$ with $P(\Omega_0) = 1$ and for each $\omega \in \Omega_0$, the limit

$$\lim_{n \to \infty} X_n(t, \omega) = \widetilde{X}(t, \omega)$$

exists and $\widetilde{X}(t, \omega)$ is a continuous function of t. Observe that $\widetilde{X}(t)$ is a version of $X(t)$, i.e., for each $t \in [0, 1]$,

$$\widetilde{X}(t) = X(t), \quad \text{almost surely}.$$

Therefore, the stochastic process $\widetilde{X}(t)$ satisfies conditions (1), (2), (3), and (4) in Definition 2.1.1, namely, $\widetilde{X}(t)$ is a Brownian motion.

Exercises

1. Show that the collection $\mathcal{R}$ given in Section 3.1 is not a σ-field.
2. Let $0 \le t_1 < \cdots < t_n$ and $0 \le t_{m_1} < \cdots < t_{m_k}$ be related by the inclusion $\{t_{m_1}, \ldots, t_{m_k}\} \subset \{t_1, \ldots, t_n\}$. Define a projection $\theta : \mathbb{R}^n \to \mathbb{R}^k$ by $\theta(x_1, \ldots, x_n) = (x_{m_1}, \ldots, x_{m_k})$. Show that the family of probability measures given by Equation (3.3.3) satisfies the consistency condition if and only if the equality

$$\mu_{t_{m_1}, \ldots, t_{m_k}}(A) = \mu_{t_1, \ldots, t_n}(\theta^{-1}A), \quad A \in \mathcal{B}(\mathbb{R}^k),$$

holds for all such choices of $0 \le t_1 < \cdots < t_n$ and $0 \le t_{m_1} < \cdots < t_{m_k}$.

3. Prove that the family of probability measures given in Example 3.3.3 satisfies the consistency condition.

4. (Kolmogorov's extension theorem for sequences of probability measures) *For each $n \geq 1$, let μ_n be a probability measure on $\mathbb{R}^n$. Assume that the sequence $\{\mu_n\}$ satisfies the following consistency condition: for any $n \geq 1$,*

$$\mu_n(A) = \mu_{n+1}(A \times \mathbb{R}), \quad A \in \mathcal{B}(\mathbb{R}^n).$$

Then there exists a unique probability measure P on $\mathbb{R}^\infty = \mathbb{R} \times \mathbb{R} \times \cdots$ with the σ-field $\mathcal{F}$ generated by cylindrical sets such that for any $n \geq 1$,

$$P\{(\omega_1, \omega_2, \ldots) \, ; \, (\omega_1, \omega_2, \ldots, \omega_n) \in A\} = \mu_n(A), \quad A \in \mathcal{B}(\mathbb{R}^n). \quad (3.4.2)$$

For the proof, see the book by Itô [37]. Below are some examples.

(a) Let ν be the probability measure on $\mathbb{R}$ given by $\nu(\{0\}) = \nu(\{1\}) = \frac{1}{2}$. For $n \geq 1$, define $\mu_n = \nu \times \nu \times \cdots \times \nu$ (n factors). Show that the sequence $\{\mu_n\}_{n=1}^\infty$ satisfies the consistency condition in Equation (3.4.2). Let P be the resulting probability measure on $\mathbb{R}^\infty$ from the above Kolmogorov's extension theorem. Consider the set

$$A = \left\{(\omega_1, \omega_2, \ldots, \omega_n, \ldots) \, ; \, \omega_n = 0 \text{ for infinitely many } n\text{'s}\right\}.$$

Show that $A \in \mathcal{F}$ and find $P(A)$.

(b) Let ν_n be the probability measure on $\mathbb{R}$ given by $\nu_n(\{0\}) = 1/n^2$ and $\nu_n(\{1\}) = 1 - 1/n^2$. For $n \geq 1$, define $\mu_n = \nu_1 \times \nu_2 \times \cdots \times \nu_n$. Let P be the resulting probability measure on $\mathbb{R}^\infty$. Find $P(A)$ for the same set A given in part (a).

(c) Let μ_n be the probability measure on $\mathbb{R}^n$ supported by the "diagonal line" $\{(t, t, \ldots, t) \, ; \, t \in \mathbb{R}\}$. Assume that μ_n has normal distribution with mean 0 and variance σ_n^2. Suppose $\sigma_1 = 1$. Find σ_n for $n \geq 2$ such that the sequence $\{\mu_n\}_{n=1}^\infty$ satisfies the consistency condition in Equation (3.4.2). Describe the resulting probability measure P on $\mathbb{R}^\infty$.

5. Consider the sequence $\{X_n(t) \, ; \, n \geq 1\}$ of stochastic processes defined in Section 3.4. Let D be the set of dyadic rational numbers given by Equation (3.3.8). Show that for each $t \in D$ there exists $N = N(t, \omega)$ such that $X_n(t, \omega) = X_N(t, \omega)$ for all $n \geq N$. Moreover, check that $X_N(t)$ is a Gaussian random variable with mean 0 and variance t.

4

Stochastic Integrals

Let $B(t, \omega)$ be a Brownian motion. In this chapter we will study the very first stochastic integral $\int_a^b f(t, \omega)\, dB(t, \omega)$ defined by K. Itô in his 1944 paper [31]. The integrand $f(t, \omega)$ is a nonanticipating stochastic process with respect to the filtration $\mathcal{F}_t = \sigma\{B(s); s \leq t\}$ and $\int_a^b E\big(|f(t)|^2\big)\, dt < \infty$. The term "nonanticipating" used by Itô is nowadays commonly called "adapted," which we have defined in Definition 2.5.1. When the integrand is a deterministic function $f(t)$, the Itô integral $\int_a^b f(t)\, dB(t, \omega)$ reduces to the Wiener integral defined in Section 2.3.

4.1 Background and Motivation

Itô's theory of stochastic integration was originally motivated as a direct method to construct diffusion processes (a subclass of Markov processes) as solutions of stochastic differential equations. It can also be motivated from the viewpoint of martingales. Let $B(t)$ be a Brownian motion. Suppose $f(t)$ is a deterministic function in $L^2[a, b]$. We showed in Theorem 2.5.4 that the stochastic process

$$M_t = \int_a^t f(s)\, dB(s), \quad a \leq t \leq b,$$

is a martingale. Now we pose a natural question.

Question 4.1.1. How can one define a stochastic integral $\int_a^b f(t, \omega)\, dB(t, \omega)$ for a stochastic process $f(t, \omega)$ in such a way that the stochastic process

$$M_t = \int_a^t f(s, \omega)\, dB(s, \omega), \quad a \leq t \leq b,$$

is a martingale?

In order to get key ideas to answer this question, let us consider a simple example with $f(t) = B(t)$ so that the integral in question is $\int_a^b B(t)\,dB(t)$. From Equations (1.1.3) and (1.1.4), we have

$$L_n = \sum_{i=1}^n B(t_{i-1})\big(B(t_i) - B(t_{i-1})\big), \tag{4.1.1}$$

$$R_n = \sum_{i=1}^n B(t_i)\big(B(t_i) - B(t_{i-1})\big), \tag{4.1.2}$$

where the evaluation points for L_n and R_n are the left endpoint t_{i-1} and right endpoint t_i of $[t_{i-1}, t_i]$, respectively. As in Equation (1.1.5), we have

$$R_n - L_n = \sum_{i=1}^n \big(B(t_i) - B(t_{i-1})\big)^2. \tag{4.1.3}$$

Here the limit $\lim_{n\to\infty}(R_n - L_n)$, if it exists, is the quadratic variation of the Brownian motion $B(t)$. The next fundamental theorem shows that $B(t)$ fluctuates so much that its quadratic variation is nonzero.

Theorem 4.1.2. *Let $\Delta_n = \{a = t_0, t_1, \ldots, t_{n-1}, t_n = b\}$ be a partition of a finite interval $[a, b]$. Then*

$$\sum_{i=1}^n \big(B(t_i) - B(t_{i-1})\big)^2 \longrightarrow b - a \tag{4.1.4}$$

in $L^2(\Omega)$ as $\|\Delta_n\| = \max_{1\le i \le n}(t_i - t_{i-1})$ tends to 0.

Remark 4.1.3. Recall two facts: (1) $L^2(\Omega)$-convergence implies convergence in probability; (2) convergence in probability of a sequence implies almost sure convergence of some subsequence. Hence there exists a subsequence $\{\widetilde{\Delta}_n\}$ of $\{\Delta_n\}$ such that the convergence in Equation (4.1.4) is almost sure convergence as $\|\widetilde{\Delta}_n\|$ tends to 0. In fact, almost sure convergence in Equation (4.1.4) is guaranteed if the sequence $\{\Delta_n\}$ satisfies the condition:

$$\Delta_1 \subset \Delta_2 \subset \cdots \subset \Delta_n \subset \cdots.$$

For the proof, see the book by Hida [25]. Almost sure convergence is also guaranteed when $\{\Delta_n\}$ satisfies the condition $\sum_{n=1}^\infty \|\Delta_n\|^2 < \infty$.

Proof. Note that $b - a = \sum_{i=1}^n (t_i - t_{i-1})$ and so let

$$\Phi_n = \sum_{i=1}^n \Big[\big(B(t_i) - B(t_{i-1})\big)^2 - (t_i - t_{i-1})\Big] = \sum_{i=1}^n X_i, \tag{4.1.5}$$

where $X_i = \big(B(t_i) - B(t_{i-1})\big)^2 - (t_i - t_{i-1})$. Then

$$\Phi_n^2 = \sum_{i,j=1}^n X_i X_j. \qquad (4.1.6)$$

For $i \neq j$, we have $E(X_i X_j) = 0$ since $B(t)$ has independent increments and $E(B(t) - B(s))^2 = |t - s|$. On the other hand, $E\big[(B(t) - B(s))^4\big] = 3(t - s)^2$ and so for $i = j$ in Equation (4.1.6), we have

$$\begin{aligned}
E(X_i^2) &= E\big\{ \big(B(t_i) - B(t_{i-1})\big)^4 - 2(t_i - t_{i-1})\big(B(t_i) - B(t_{i-1})\big)^2 \\
&\quad + (t_i - t_{i-1})^2 \big\} \\
&= 3(t_i - t_{i-1})^2 - 2(t_i - t_{i-1})^2 + (t_i - t_{i-1})^2 \\
&= 2(t_i - t_{i-1})^2. \qquad (4.1.7)
\end{aligned}$$

Therefore, from Equation (4.1.6), we get

$$\begin{aligned}
E\Phi_n^2 &= \sum_{i=1}^n 2(t_i - t_{i-1})^2 \leq 2\|\Delta_n\| \sum_{i=1}^n (t_i - t_{i-1}) \\
&= 2(b - a)\|\Delta_n\| \to 0, \quad \text{as } \|\Delta_n\| \to 0.
\end{aligned}$$

This shows that Φ_n converges to 0 in $L^2(\Omega)$. Hence from Equation (4.1.5), we see that Equation (4.1.4) holds. □

Now apply Theorem 4.1.2 to Equation (4.1.3) to conclude that

$$\lim_{\|\Delta_n\| \to 0} (R_n - L_n) = b - a, \quad \text{in } L^2(\Omega).$$

Hence $\lim_{\|\Delta_n\| \to 0} R_n \neq \lim_{\|\Delta_n\| \to 0} L_n$. But what are these limits? In order to find out the answer, note that

$$\begin{aligned}
R_n + L_n &= \sum_{i=1}^n \big(B(t_i) + B(t_{i-1})\big)\big(B(t_i) - B(t_{i-1})\big) \\
&= \sum_{i=1}^n \big(B(t_i)^2 - B(t_{i-1})^2\big) \\
&= B(t_n)^2 - B(t_0)^2 \\
&= B(b)^2 - B(a)^2. \qquad (4.1.8)
\end{aligned}$$

Obviously, it follows from Equations (4.1.3) and (4.1.8) that

$$R_n = \frac{1}{2}\left(B(b)^2 - B(a)^2 + \sum_{i=1}^n \big(B(t_i) - B(t_{i-1})\big)^2 \right),$$

$$L_n = \frac{1}{2}\left(B(b)^2 - B(a)^2 - \sum_{i=1}^n \big(B(t_i) - B(t_{i-1})\big)^2 \right).$$

We can use Theorem 4.1.2 to take the $L^2(\Omega)$ limits of R_n and L_n to get

$$\lim_{\|\Delta_n\|\to 0} R_n = \frac{1}{2}\left(B(b)^2 - B(a)^2 + (b-a)\right), \qquad (4.1.9)$$

$$\lim_{\|\Delta_n\|\to 0} L_n = \frac{1}{2}\left(B(b)^2 - B(a)^2 - (b-a)\right). \qquad (4.1.10)$$

Question 4.1.4. Which one of Equations (4.1.9) and (4.1.10) should we take to be the integral $\int_a^b B(t)\,dB(t)$? Namely, which endpoint (left or right) should we use for the evaluation of the integrand?

To answer this question, let us take $a = 0$ and $b = t$ in Equations (4.1.9) and (4.1.10) to define the stochastic processes

$$R(t) = \frac{1}{2}\left(B(t)^2 + t\right), \quad L(t) = \frac{1}{2}\left(B(t)^2 - t\right).$$

Note that $ER(t) = t$. Hence $R(t)$ is not a martingale, since $EM(t)$ must be a constant for any martingale $M(t)$. On the other hand, $L(t)$ is a martingale. This can be verified as follows. Let $\mathcal{F}_t = \sigma\{B(s); s \le t\}$. Then for any $s \le t$,

$$E\left(L(t)|\mathcal{F}_s\right) = \frac{1}{2}E\left(B(t)^2|\mathcal{F}_s\right) - \frac{1}{2}t. \qquad (4.1.11)$$

Recall that the conditional expectation (see Section 2.6) has the following properties:

(a) If X and $\mathcal{F}$ are independent, then $E(X|\mathcal{F}) = EX$.
(b) If X is $\mathcal{F}$-measurable, then $E(XY|\mathcal{F}) = XE(Y|\mathcal{F})$, in particular, we have $E(X|\mathcal{F}) = X$.

Since $B(t) - B(s)$ and $B(u)$ are independent for all $u \le s$, it follows that $B(t) - B(s)$ and $\mathcal{F}_s$ are independent. Therefore,

$$\begin{aligned}
E\left(B(t)^2|\mathcal{F}_s\right) &= E\left((B(t) - B(s) + B(s))^2|\mathcal{F}_s\right) \\
&= E\left((B(t) - B(s))^2 + 2B(s)(B(t) - B(s)) + B(s)^2|\mathcal{F}_s\right) \\
&= E(B(t) - B(s))^2 + 2B(s)E(B(t) - B(s)) + B(s)^2 \\
&= t - s + B(s)^2.
\end{aligned}$$

Thus $E(B(t)^2|\mathcal{F}_s) = t - s + B(s)^2$, which we can put into Equation (4.1.11) to get

$$E\left(L(t)|\mathcal{F}_s\right) = L(s), \quad \forall s \le t.$$

This shows that $L(t)$ is a martingale. From this simple example we can draw the following crucial conclusion:

If we want to have the martingale property for this yet to be defined stochastic integral $\int_a^t f(s)\,dB(s)$, we should take the left endpoint of each subinterval as the evaluation point.

Furthermore, consider another simple example:

$$X(t) = \int_0^t B(1)\, dB(s), \quad 0 \le t \le 1.$$

Intuitively, we would expect that $X(t) = B(1)B(t)$. But the stochastic process $X(t)$ is not a martingale since $E[B(1)B(t)] = \min\{1, t\} = t$, which is not a constant. Thus the integral $\int_0^t B(1)\, dB(s)$ is not what we expect to define if we want to obtain martingale processes. The reason for such a simple integral to be undefined (when we want to obtain martingales) is because the integrand $B(1)$ is not adapted to the filtration $\sigma\{B(s); s \le t\}$, $0 \le t \le 1$. So here is an important requirement for the integrand:

If we want to have the martingale property for this yet to be defined stochastic integral $\int_a^t f(s)\, dB(s)$, we need to assume that the integrand is adapted to a filtration $\{\mathcal{F}_t\}$.

In general we will allow $\{\mathcal{F}_t\}$ to be a larger filtration than the one given by a Brownian motion, i.e., $\mathcal{F}_t \supset \sigma\{B(s); s \le t\}$ for all t; see the next section.

4.2 Filtrations for a Brownian Motion

As pointed out in the previous section, the yet to be defined stochastic integral $\int_a^b f(t)\, dB(t)$ should have the property that when the upper limit b is replaced by t, the resulting stochastic process $X_t = \int_a^t f(s)\, dB(s)$, $a \le t \le b$, is a martingale with respect to the filtration $\mathcal{F}_t^B = \sigma\{B(s); s \le t\}$.

Recall that $B(t)$ has independent increments. This property implies that $B(t)$ is a martingale with respect to the filtration $\{\mathcal{F}_t^B\}$. In fact, the property of independent increments is related to the independence of increments with respect to a filtration $\{\mathcal{F}_t; t \ge 0\}$ as shown in the next proposition.

Proposition 4.2.1. *If $W(t), t \ge 0$, is an $\{\mathcal{F}_t\}$-adapted stochastic process such that $W(t) - W(s)$ is independent of $\mathcal{F}_s$ for any $s \le t$, then the stochastic process $W(t)$ has independent increments.*

Proof. Let $0 \le t_1 < t_2 < \cdots < t_n$. For any real numbers $\lambda_k, 1 \le k \le n$, we use the assumption and the first three properties of conditional expectation in Section 2.4 to show that

$$E e^{i \sum_{k=1}^n \lambda_k (W(t_k) - W(t_{k-1}))}$$

$$= E\left\{ E\left[e^{i \sum_{k=1}^n \lambda_k (W(t_k) - W(t_{k-1}))} \,\big|\, \mathcal{F}_{t_{n-1}} \right] \right\}$$

$$= E\left\{ e^{i \sum_{k=1}^{n-1} \lambda_k (W(t_k) - W(t_{k-1}))} E\left[e^{i \lambda_n (W(t_n) - W(t_{n-1}))} \,\big|\, \mathcal{F}_{t_{n-1}} \right] \right\}$$

$$= E e^{i \lambda_n (W(t_n) - W(t_{n-1}))} E e^{i \sum_{k=1}^{n-1} \lambda_k (W(t_k) - W(t_{k-1}))},$$

where $t_0 = 0$ by convention. Then repeat the above arguments inductively to conclude that the equality

$$Ee^{i\sum_{k=1}^{n} \lambda_k(W(t_k)-W(t_{k-1}))} = \prod_{k=1}^{n} Ee^{i\lambda_k(W(t_k)-W(t_{k-1}))} \qquad (4.2.1)$$

holds for all real numbers $\lambda_k, 1 \le k \le n$. It is well known (e.g., see Theorem 6.6.1 in [5]) that Equation (4.2.1) holds if and only if the random variables $W(t_k) - W(t_{k-1}), 1 \le k \le n$, are independent. Hence the stochastic process $W(t)$ has independent increments.   □

Now suppose $B(t)$ is a stochastic process satisfying conditions (1), (2), and (4) of a Brownian motion in Definition 2.1.1. Moreover, suppose there is a filtration $\{\mathcal{F}_t; t \ge 0\}$ such that $B(t)$ satisfies the assumption in Proposition 4.2.1, namely, $B(t)$ is $\{\mathcal{F}_t\}$-adapted and $B(t) - B(s)$ is independent of $\mathcal{F}_s$ for any $s \le t$. Then by Proposition 4.2.1 the stochastic process $B(t)$ has independent increments, i.e., it satisfies condition (3) of a Brownian motion in Definition 2.1.1. Therefore, $B(t)$ is a Brownian motion according to Definition 2.1.1. Hence we may say that $B(t)$ is a *Brownian motion with respect to a filtration* $\{\mathcal{F}_t; t \ge 0\}$ if it satisfies conditions (1), (2), and (4) in Definition 2.1.1 and the assumption in Proposition 4.2.1 with respect to $\{\mathcal{F}_t; t \ge 0\}$.

Suppose $B(t)$ is a Brownian motion with respect to a filtration $\{\mathcal{F}_t; t \ge 0\}$. Let $\{\mathcal{G}_t; t \ge 0\}$ be another filtration such that $\mathcal{F}_t \subset \mathcal{G}_t$ for all $t \ge 0$. In general, it is not true that $B(t)$ is still a Brownian motion with respect to $\{\mathcal{G}_t; t \ge 0\}$, as shown in the next example.

Example 4.2.2. Let $B(t)$ be a Brownian motion. Then it is a Brownian motion with respect to the filtration $\mathcal{F}_t^B = \sigma\{B(s); s \le t\}$. Consider the filtration $\{\mathcal{G}_t; t \ge 0\}$ given by

$$\mathcal{G}_t = \text{the } \sigma\text{-field generated by } B(1) \text{ and } \mathcal{F}_t^B, \quad t \ge 0.$$

Then $B(t)$ is not a Brownian motion with respect to the filtration $\{\mathcal{G}_t; t \ge 0\}$. To see this fact, simply note that for any $0 < t < 1$,

$$E\big[B(1) \,\big|\, \mathcal{G}_t\big] = B(1) \ne B(t).$$

Hence $B(t)$ is not a martingale with respect to $\{\mathcal{G}_t; t \ge 0\}$. It follows that $B(t)$ is not a Brownian motion with respect to $\{\mathcal{G}_t; t \ge 0\}$.

On the other hand, suppose a filtration $\{\mathcal{H}_t; t \ge 0\}$ is independent of $\{\mathcal{F}_t^B; t \ge 0\}$, i.e., $\mathcal{H}_t$ and $\mathcal{F}_s^B$ are independent for all t and s. Define

$$\mathcal{F}_t = \text{the } \sigma\text{-field generated by } \mathcal{H}_t \text{ and } \mathcal{F}_t^B, \quad t \ge 0.$$

It is obvious that $B(t)$ satisfies the assumption in Proposition 4.2.1. Hence $B(t)$ is still a Brownian motion with respect to the filtration $\{\mathcal{F}_t; t \ge 0\}$, which is a larger filtration than $\{\mathcal{F}_t^B; t \ge 0\}$.

In fact, there is another reason why we need to consider a filtration with respect to which $B(t)$ is a Brownian motion. We want to allow the integrand $f(t)$ in the yet to be defined stochastic integral $\int_a^b f(t)\, dB(t)$ to be from a large class of stochastic processes. In particular, the integrand $f(t)$ is not required to be adapted with respect to the filtration $\{\mathcal{F}_t^B; t \geq 0\}$. This will be discussed in the next section.

4.3 Stochastic Integrals

Being motivated by the discussion in the previous section, we will from now on fix a Brownian motion $B(t)$ and a filtration $\{\mathcal{F}_t; a \leq t \leq b\}$ satisfying the following conditions:

(a) For each t, $B(t)$ is $\mathcal{F}_t$-measurable;
(b) For any $s \leq t$, the random variable $B(t) - B(s)$ is independent of the σ-field $\mathcal{F}_s$.

Notation 4.3.1. For convenience, we will use $L_{\mathrm{ad}}^2([a,b] \times \Omega)$ to denote the space of all stochastic processes $f(t,\omega)$, $a \leq t \leq b$, $\omega \in \Omega$, satisfying the following conditions:

(1) $f(t,\omega)$ is adapted to the filtration $\{\mathcal{F}_t\}$;
(2) $\int_a^b E(|f(t)|^2)\, dt < \infty$.

In this section we will use Itô's original ideas in his paper [31] to define the stochastic integral

$$\int_a^b f(t)\, dB(t) \tag{4.3.1}$$

for $f \in L_{\mathrm{ad}}^2([a,b] \times \Omega)$. For clarity, we divide the discussion into three steps. In Step 1 we define the stochastic integral for step stochastic processes in $L_{\mathrm{ad}}^2([a,b] \times \Omega)$. In Step 2 we prove a crucial approximation lemma. In Step 3 we define the stochastic integral for general stochastic processes in $L_{\mathrm{ad}}^2([a,b] \times \Omega)$.

Step 1. *f is a step stochastic process in $L_{\mathrm{ad}}^2([a,b] \times \Omega)$.*

Suppose f is a step stochastic process[1] given by

$$f(t,\omega) = \sum_{i=1}^n \xi_{i-1}(\omega)\, 1_{[t_{i-1}, t_i)}(t),$$

---

[1] In [31], K. Itô used $f(t,\omega) = \sum_{i=1}^n \xi_{i-1}(\omega)\, 1_{(t_{i-1}, t_i]}(t)$. This makes no difference for a stochastic integral with respect to the integrator $dB(t)$. However, we will see in Chapter 6 that the left continuity of an integrand must be assumed when the integrator is a certain martingale. For notational consistence, here we also use the left continuity of an integrand.

where ξ_{i-1} is $\mathcal{F}_{t_{i-1}}$-measurable and $E(\xi_{i-1}^2) < \infty$. In this case we define

$$I(f) = \sum_{i=1}^{n} \xi_{i-1} \left(B(t_i) - B(t_{i-1}) \right). \tag{4.3.2}$$

Obviously, $I(af + bg) = aI(f) + bI(g)$ for any $a, b \in \mathbb{R}$ and any such step stochastic processes f and g. Moreover, we have the next lemma.

Lemma 4.3.2. *Let* $I(f)$ *be defined by Equation* (4.3.2). *Then* $EI(f) = 0$ *and*

$$E\left(|I(f)|^2 \right) = \int_a^b E\left(|f(t)|^2 \right) dt. \tag{4.3.3}$$

Proof. For each $1 \le i \le n$ in Equation (4.3.2),

$$
\begin{aligned}
E\big\{ \xi_{i-1} \left(B(t_i) - B(t_{i-1}) \right) \big\} &= E\big\{ E\big[\xi_{i-1} \left(B(t_i) - B(t_{i-1}) \right) \big| \mathcal{F}_{t_{i-1}} \big] \big\} \\
&= E\big\{ \xi_{i-1} E\big[B(t_i) - B(t_{i-1}) \big| \mathcal{F}_{t_{i-1}} \big] \big\} \\
&= E\big\{ \xi_{i-1} E\big(B(t_i) - B(t_{i-1}) \big) \big\} \\
&= 0.
\end{aligned}
$$

Hence $EI(f) = 0$. Moreover, we have

$$|I(f)|^2 = \sum_{i,j=1}^{n} \xi_{i-1}\xi_{j-1} \left(B(t_i) - B(t_{i-1}) \right)\left(B(t_j) - B(t_{j-1}) \right).$$

Note that for $i \ne j$, say $i < j$,

$$
\begin{aligned}
&E\big\{ \xi_{i-1}\xi_{j-1} \left(B(t_i) - B(t_{i-1}) \right)\left(B(t_j) - B(t_{j-1}) \right) \big\} \\
&= E\big\{ E\big[\cdots\cdots \big| \mathcal{F}_{t_{j-1}} \big] \big\} \\
&= E\big\{ \xi_{i-1}\xi_{j-1} \left(B(t_i) - B(t_{i-1}) \right) E\big[B(t_j) - B(t_{j-1}) \big| \mathcal{F}_{t_{j-1}} \big] \big\} \\
&= 0, \tag{4.3.4}
\end{aligned}
$$

since $E\big[B(t_j) - B(t_{j-1}) \big| \mathcal{F}_{t_{j-1}} \big] = E\big(B(t_j) - B(t_{j-1}) \big) = 0$ as before. On the other hand, for $i = j$ we have

$$
\begin{aligned}
E\big\{ \xi_{i-1}^2 \big(B(t_i) - B(t_{i-1}) \big)^2 \big\} &= E\big\{ E\big[\cdots\cdots \big| \mathcal{F}_{t_{i-1}} \big] \big\} \\
&= E\big\{ \xi_{i-1}^2 E\big[\big(B(t_i) - B(t_{i-1}) \big)^2 \big] \big\} \\
&= E\big\{ \xi_{i-1}^2 (t_i - t_{i-1}) \big\} \\
&= (t_i - t_{i-1}) E(\xi_{i-1}^2). \tag{4.3.5}
\end{aligned}
$$

Equation (4.3.3) follows from Equations (4.3.4) and (4.3.5).   □

Step 2. *An approximation lemma.*

We need to prove an approximation lemma in this step in order to be able to define the stochastic integral $\int_a^b f(t)\, dB(t)$ for general stochastic processes $f \in L^2_{\mathrm{ad}}([a,b] \times \Omega)$.

Lemma 4.3.3. *Suppose $f \in L^2_{\mathrm{ad}}([a,b] \times \Omega)$. Then there exists a sequence $\{f_n(t)\,;\, n \geq 1\}$ of step stochastic processes in $L^2_{\mathrm{ad}}([a,b] \times \Omega)$ such that*

$$\lim_{n \to \infty} \int_a^b E\big\{|f(t) - f_n(t)|^2\big\}\, dt = 0. \tag{4.3.6}$$

Proof. We divide the proof into special cases and the general case.

Case 1: $E\big(f(t)f(s)\big)$ *is a continuous function of $(t,s) \in [a,b]^2$.*

In this case, let $\Delta_n = \{t_0, t_1, \ldots, t_{n-1}, t_n\}$ be a partition of $[a,b]$ and define a stochastic process $f_n(t,\omega)$ by

$$f_n(t,\omega) = f(t_{i-1}, \omega), \quad t_{i-1} < t \leq t_i. \tag{4.3.7}$$

Then $\{f_n(t,\omega)\}$ is a sequence of adapted step stochastic processes. By the continuity of $E\big(f(t)f(s)\big)$ on $[a,b]^2$ we have

$$\lim_{s \to t} E\big\{|f(t) - f(s)|^2\big\} = 0,$$

which implies that for each $t \in [a,b]$,

$$\lim_{n \to \infty} E\big\{|f(t) - f_n(t)|^2\big\} = 0. \tag{4.3.8}$$

Moreover, use the inequality $|\alpha - \beta|^2 \leq 2(|\alpha|^2 + |\beta|^2)$ to get

$$|f(t) - f_n(t)|^2 \leq 2\big(|f(t)|^2 + |f_n(t)|^2\big).$$

Hence for all $a \leq t \leq b$,

$$E\big\{|f(t) - f_n(t)|^2\big\} \leq 2\Big(E\big\{|f(t)|^2\big\} + E\big\{|f_n(t)|^2\big\}\Big)$$
$$\leq 4 \sup_{a \leq s \leq b} E\big\{|f(s)|^2\big\}. \tag{4.3.9}$$

Therefore, from Equations (4.3.8) and (4.3.9), we can apply the Lebesgue dominated convergence theorem to conclude that

$$\lim_{n \to \infty} \int_a^b E\big\{|f(t) - f_n(t)|^2\big\}\, dt = 0.$$

Case 2: f *is bounded.*

In this case, define a stochastic process g_n by

$$g_n(t,\omega) = \int_0^{n(t-a)} e^{-\tau} f(t - n^{-1}\tau, \omega)\, d\tau.$$

Note that g_n is adapted to $\mathcal{F}_t$ and $\int_a^b E\big(|g_n(t)|^2\big)\, dt < \infty$.

○ Claim (a): *For each n, $E\big(g_n(t)g_n(s)\big)$ is a continuous function of (t, s).*

To prove this claim, let $u = t - n^{-1}\tau$ to rewrite $g_n(t, \omega)$ as

$$g_n(t, \omega) = \int_a^t n e^{-n(t-u)} f(u, \omega)\, du,$$

which can be used to verify that

$$\lim_{t \to s} E\big(|g_n(t) - g_n(s)|^2\big) = 0,$$

and the claim follows.

○ Claim (b): $\int_a^b E\big(|f(t) - g_n(t)|^2\big)\, dt \to 0$ as $n \to \infty$.

To prove this claim, note that

$$f(t) - g_n(t) = \int_0^\infty e^{-\tau}\big(f(t) - f(t - n^{-1}\tau)\big)\, d\tau,$$

where $f(t)$ is understood to be zero for $t < a$. Since $e^{-\tau}\, d\tau$ is a probability measure on $[0, \infty)$, we can apply the Schwarz inequality to get

$$|f(t) - g_n(t)|^2 \le \int_0^\infty |f(t) - f(t - n^{-1}\tau)|^2 e^{-\tau}\, d\tau.$$

Therefore,

$$\int_a^b E\big(|f(t) - g_n(t)|^2\big)\, dt$$

$$\le \int_a^b \int_0^\infty e^{-\tau} E\big(|f(t) - f(t - n^{-1}\tau)|^2\big)\, d\tau dt$$

$$= \int_0^\infty e^{-\tau} \left(\int_a^b E\big(|f(t) - f(t - n^{-1}\tau)|^2\big)\, dt \right) d\tau$$

$$= \int_0^\infty e^{-\tau} E\left(\int_a^b |f(t) - f(t - n^{-1}\tau)|^2\, dt \right) d\tau. \qquad (4.3.10)$$

Since f is assumed to be bounded, we have

$$\int_a^b |f(t, \cdot) - f(t - n^{-1}\tau, \cdot)|^2\, dt \to 0, \quad \text{almost surely}, \qquad (4.3.11)$$

as $n \to \infty$. Then claim (b) follows from Equations (4.3.10) and (4.3.11).

Now, by claim (a) we can apply case 1 to g_n for each n to pick up an adapted step stochastic process $f_n(t, \omega)$ such that

$$\int_a^b E\big(|g_n(t) - f_n(t)|^2\big)\, dt \le \frac{1}{n}. \qquad (4.3.12)$$

Hence by claim (b) and Equation (4.3.12) we have

$$\lim_{n \to \infty} \int_a^b E\{|f(t) - f_n(t)|^2\} \, dt = 0,$$

which completes the proof for the second case.

Case 3: *The general case for* $f \in L^2_{ad}([a, b] \times \Omega)$.

Let $f \in L^2_{ad}([a, b] \times \Omega)$. For each n, define

$$g_n(t, \omega) = \begin{cases} f(t, \omega), & \text{if } |f(t, \omega)| \leq n; \\ 0, & \text{if } |f(t, \omega)| > n. \end{cases}$$

Then by the Lebesgue dominated convergence theorem,

$$\int_a^b E(|f(t) - g_n(t)|^2) \, dt \to 0, \quad \text{as } n \to \infty. \tag{4.3.13}$$

Now, for each n we apply case 2 to g_n to pick up an adapted step stochastic process $f_n(t, \omega)$ such that

$$\int_a^b E(|g_n(t) - f_n(t)|^2) \, dt \leq \frac{1}{n}. \tag{4.3.14}$$

Hence Equation (4.3.6) follows from Equations (4.3.13) and (4.3.14), and we have completed the proof of the lemma. ☐

Step 3. *Stochastic integral* $\int_a^b f(t) \, dB(t)$ *for* $f \in L^2_{ad}([a, b] \times \Omega)$.

Now we can use what we proved in Steps 1 and 2 to define the stochastic integral

$$\int_a^b f(t) \, dB(t), \quad f \in L^2_{ad}([a, b] \times \Omega).$$

Apply Lemma 4.3.3 to get a sequence $\{f_n(t, \omega); n \geq 1\}$ of adapted step stochastic processes such that Equation (4.3.6) holds. For each n, $I(f_n)$ is defined by Step 1. By Lemma 4.3.2 we have

$$E(|I(f_n) - I(f_m)|^2) = \int_a^b E(|f_n(t) - f_m(t)|^2) \, dt$$
$$\to 0, \quad \text{as } n, m \to \infty.$$

Hence the sequence $\{I(f_n)\}$ is a Cauchy sequence in $L^2(\Omega)$. Define

$$I(f) = \lim_{n \to \infty} I(f_n), \quad \text{in } L^2(\Omega). \tag{4.3.15}$$

We can use arguments similar to those in Section 2.3 for the Wiener integral to show that the above $I(f)$ is well-defined.

Definition 4.3.4. *The limit $I(f)$ defined in Equation (4.3.15) is called the* Itô integral *of f and is denoted by $\int_a^b f(t)\, dB(t)$.*

Thus the Itô integral $I(f)$ is defined for $f \in L_{\mathrm{ad}}^2([a,b] \times \Omega)$ and the mapping I is linear, namely, for any $a, b \in \mathbb{R}$ and $f, g \in L_{\mathrm{ad}}^2([a,b] \times \Omega)$,

$$I(af + bg) = aI(f) + bI(g).$$

From the discussion in Step 2 we clearly see that Lemma 4.3.2 remains valid for $f \in L_{\mathrm{ad}}^2([a,b] \times \Omega)$. We state this fact as the next theorem.

Theorem 4.3.5. *Suppose $f \in L_{\mathrm{ad}}^2([a,b] \times \Omega)$. Then the Itô integral $I(f) = \int_a^b f(t)\, dB(t)$ is a random variable with $E\{I(f)\} = 0$ and*

$$E\big(|I(f)|^2\big) = \int_a^b E\big(|f(t)|^2\big)\, dt. \qquad (4.3.16)$$

By this theorem, the Itô integral $I : L_{\mathrm{ad}}^2([a,b] \times \Omega) \to L^2(\Omega)$ is an isometry. Since I is also linear, we have the following corollary.

Corollary 4.3.6. *For any $f, g \in L_{\mathrm{ad}}^2([a,b] \times \Omega)$, the following equality holds:*

$$E\left(\int_a^b f(t)\, dB(t) \int_a^b g(t)\, dB(t) \right) = \int_a^b E\big(f(t)g(t)\big)\, dt.$$

4.4 Simple Examples of Stochastic Integrals

Example 4.4.1. $\int_a^b B(t)\, dB(t) = \frac{1}{2}\big\{B(b)^2 - B(a)^2 - (b-a)\big\}$.

In Section 4.1 we tried to define the integral $\int_a^b B(t)\, dB(t)$. When we use the left endpoint of each subinterval in a partition of $[a,b]$ to evaluate the integrand, we get the sum L_n in Equation (4.1.1). If we take as the integral the limit of L_n as $n \to \infty$, then from Equation (4.1.10) we have

$$\int_a^b B(t)\, dB(t) = \frac{1}{2}\big\{B(b)^2 - B(a)^2 - (b-a)\big\}. \qquad (4.4.1)$$

Is this value equal to the integral $\int_a^b B(t)\, dB(t)$ as defined in Section 4.3? First note that $E\big(B(t)B(s)\big) = \min\{t, s\}$, which is a continuous function of t and s. Hence we can apply Case 1 in the proof of Lemma 4.3.3 to the integrand $f(t) = B(t)$, namely, for a partition $\Delta_n = \{t_0, t_1, \ldots, t_{n-1}, t_n\}$ of $[a,b]$, define a stochastic process $f_n(t, \omega)$ by

$$f_n(t, \omega) = B(t_{i-1}, \omega), \quad t_{i-1} < t \leq t_i.$$

Then by Step 2 of defining the stochastic integral we see that the stochastic integral $\int_a^b B(t)\, dB(t)$ as defined in Section 4.3 is given by

$$\int_a^b B(t)\,dB(t) = \lim_{n\to\infty} I(f_n), \quad \text{in } L^2(\Omega).$$

Now, by Equation (4.3.2) in Step 1 of defining the stochastic integral, $I(f_n)$ is given by

$$I(f_n) = \sum_{i=1}^n B(t_{i-1})\big(B(t_i) - B(t_{i-1})\big),$$

which is L_n in Equation (4.1.1). Thus the stochastic integral $\int_a^b B(t)\,dB(t)$ as defined in Section 4.3 has the same value as the one in Equation (4.4.1).

Example 4.4.2. We use the same idea as in Example 4.4.1 to show that

$$\int_a^b B(t)^2\,dB(t) = \frac{1}{3}\big(B(b)^3 - B(a)^3\big) - \int_a^b B(t)\,dt, \qquad (4.4.2)$$

where the integral in the right-hand side is the Riemann integral of $B(t,\omega)$ for almost all ω in Ω. Note that

$$
\begin{aligned}
E[B(t)^2 B(s)^2] \\
&= E[\big((B(t) - B(s)) + B(s)\big)^2 B(s)^2] \\
&= E[\{\big(B(t) - B(s)\big)^2 + 2B(s)\big(B(t) - B(s)\big) + B(s)^2\}B(s)^2] \\
&= (t - s)s + 3s^2 \\
&= ts + 2s^2,
\end{aligned}
$$

which shows that $E[B(t)^2 B(s)^2]$ is a continuous function of t and s. Hence we can apply Case 1 in the proof of Lemma 4.3.3 to the integrand $f(t) = B(t)^2$. For a partition $\Delta_n = \{t_0, t_1, \dots, t_{n-1}, t_n\}$ of $[a, b]$, define a stochastic process $f_n(t, \omega)$ by

$$f_n(t, \omega) = B(t_{i-1}, \omega), \quad t_{i-1} < t \le t_i.$$

Then the stochastic integral $\int_a^b B(t)^2\,dB(t)$ is given by

$$\int_a^b B(t)^2\,dB(t) = \lim_{n\to\infty} \sum_{i=1}^n B(t_{i-1})^2\big(B(t_i) - B(t_{i-1})\big), \qquad (4.4.3)$$

where the series converges in $L^2(\Omega)$. It can be directly checked that

$$
3\sum_{i=1}^n B(t_{i-1})^2\big(B(t_i) - B(t_{i-1})\big)
$$

$$
= B(b)^3 - B(a)^3 - \sum_{i=1}^n \big(B(t_i) - B(t_{i-1})\big)^3
$$

$$
- 3\sum_{i=1}^n B(t_{i-1})\big(B(t_i) - B(t_{i-1})\big)^2. \qquad (4.4.4)
$$

To take care of the first summation in the right-hand side of this equation, we use routine arguments and the fact that $E|B(t) - B(s)|^6 = 15|t - s|^3$ to show that

$$E\left|\sum_{i=1}^{n}(B(t_i) - B(t_{i-1}))^3\right|^2 = 15\sum_{i=1}^{n}(t_i - t_{i-1})^3$$

$$\leq 15\|\Delta_n\|^2(b - a) \to 0. \qquad (4.4.5)$$

On the other hand, for the second summation in Equation (4.4.4), we can modify the arguments in the proof of Theorem 4.1.2 by the same idea of taking expectation by conditioning as in the proof of Lemma 4.3.2 to derive the inequality

$$E\left|\sum_{i=1}^{n}B(t_{i-1})(B(t_i) - B(t_{i-1}))^2 - \sum_{i=1}^{n}B(t_{i-1})(t_i - t_{i-1})\right|^2$$

$$= \sum_{i=1}^{n}2t_{i-1}(t_i - t_{i-1})^2 \leq 2b(b - a)\|\Delta_n\| \to 0. \qquad (4.4.6)$$

Equation (4.4.5) means that the first summation in the right-hand side of Equation (4.4.4) converges to 0 in $L^2(\Omega)$, while Equation (4.4.6) means that the second summation in the right-hand side of Equation (4.4.4) converges to $\int_a^b B(t)\, dt$. Therefore, we conclude from Equations (4.4.3) and (4.4.4) that the equality in Equation (4.4.2) holds.

Example 4.4.3. Put $a = 0$ and $b = t$ in Equation (4.4.2). Then we have a stochastic process

$$X_t = \int_0^t B(u)^2\, dB(u) = \frac{1}{3}B(t)^3 - \int_0^t B(u)\, du, \quad t \geq 0.$$

It will follow from Theorem 4.6.1 in the next section that the stochastic process X_t is a martingale. However, this fact can also be directly checked as follows. Let $0 \leq s \leq t$. First write $B(t)^3$ as

$$(B(t) - B(s))^3 + 3(B(t) - B(s))^2 B(s) + 3(B(t) - B(s))B(s)^2 + B(s)^3$$

and then take conditional expectation to get

$$E[B(t)^3 \mid \mathcal{F}_s] = 3(t - s)B(s) + B(s)^3. \qquad (4.4.7)$$

Moreover, we have

$$E\left[\int_0^t B(u)\, du \,\Big|\, \mathcal{F}_s\right] = \int_0^s B(u)\, du + \int_s^t E[B(u) \mid \mathcal{F}_s]\, du$$

$$= \int_0^s B(u)\, du + B(s)(t - s). \qquad (4.4.8)$$

It follows from Equations (4.4.7) and (4.4.8) that $E[X_t \mid \mathcal{F}_s] = X_s$. Hence X_t is a martingale.

4.5 Doob Submartingale Inequality

In this section we explain the Doob submartingale inequality, which will be needed for the proof of Theorem 4.6.2 in the next section.

Let X_t, $a \leq t \leq b$, be a martingale and φ a convex function such that $\varphi(X_t)$ is integrable for each $t \in [a, b]$. Then $\varphi(X_t)$ is a submartingale by the conditional Jensen's inequality. For example, with the function $\varphi(x) = |x|$, we get a submartingale $|X_t|$.

A stochastic process $Y(t)$, $a \leq t \leq b$, is called *right continuous* if almost all sample paths are right continuous functions on $[a, b]$.

Theorem 4.5.1. (Doob submartingale inequality) *Let $Y(t)$, $a \leq t \leq b$, be a right continuous submartingale. Then for any $\epsilon > 0$,*

$$P\left\{ \sup_{a \leq t \leq b} Y(t) \geq \epsilon \right\} \leq \frac{1}{\epsilon} E[Y(b)^+], \qquad (4.5.1)$$

where $Y(b)^+$ is the positive part of $Y(b)$, namely, $Y(b)^+ = \max\{Y(b), 0\}$. In particular, if X_t is a right continuous martingale, then for any $\epsilon > 0$,

$$P\left\{ \sup_{a \leq t \leq b} |X_t| \geq \epsilon \right\} \leq \frac{1}{\epsilon} E|X_b|. \qquad (4.5.2)$$

Proof. We adopt the proof from the book by Hida and Hitsuda [26]. Let $Q = \{r_1, r_2, \ldots\}$ be an enumeration of all rational numbers in $[a, b]$. Then by the right continuity of $Y(t)$ we have

$$\sup_{a \leq t \leq b} Y(t) = \sup_{r \in Q} Y(r), \quad \text{almost surely.} \qquad (4.5.3)$$

For each k, arrange the numbers in the set $\{r_1, r_2, \ldots, r_k\}$ in increasing order $\{r_1^{(k)} < r_2^{(k)} < \cdots < r_k^{(k)}\}$. Then for any $\epsilon > 0$,

$$\left\{ \sup_{r \in Q} Y(r) \geq \epsilon \right\} = \bigcap_{n=1}^{\infty} \bigcup_{k=1}^{\infty} \left\{ \max_{1 \leq \nu \leq k} Y(r_\nu^{(k)}) \geq \epsilon - \frac{1}{n} \right\}. \qquad (4.5.4)$$

It follows from Equations (4.5.3) and (4.5.4) that

$$P\left\{ \sup_{a \leq t \leq b} Y(t) \geq \epsilon \right\} = P\left\{ \sup_{r \in Q} Y(r) \geq \epsilon \right\}$$

$$= \lim_{n \to \infty} \lim_{k \to \infty} P\left\{ \max_{1 \leq \nu \leq k} Y(r_\nu^{(k)}) \geq \epsilon - \frac{1}{n} \right\}. \qquad (4.5.5)$$

Now, note that $Y(r_1^{(k)}), Y(r_2^{(k)}), \ldots, Y(r_k^{(k)})$ is a discrete submartingale. Hence by the discrete Doob submartingale inequality (see page 330 in the book [5]),

$$P\Big\{ \max_{1 \le \nu \le k} Y(r_\nu^{(k)}) \ge \epsilon - \frac{1}{n} \Big\} \le \frac{1}{\epsilon - \frac{1}{n}} E\big[Y(r_k^{(k)})^+\big]. \qquad (4.5.6)$$

On the other hand, observe that $Y(r_1^{(k)})^+$, $Y(r_2^{(k)})^+$, ..., $Y(r_k^{(k)})^+$, $Y(b)^+$ is a submartingale. Hence

$$E\Big[Y(b)^+ \,\Big|\, \mathcal{F}_{r_k^{(k)}}\Big] \ge Y(r_k^{(k)})^+, \quad \text{almost surely,}$$

which, upon taking expectation, yields

$$E\big[Y(r_k^{(k)})^+\big] \le E\big[Y(b)^+\big]. \qquad (4.5.7)$$

After putting Equations (4.5.5), (4.5.6), and (4.5.7) together, we immediately obtain Equation (4.5.1). □

Example 4.5.2. Let (C, μ) be the Wiener space in Section 3.1. The Brownian motion $B(t, \omega) = \omega(t)$, $0 \le t \le 1$, $\omega \in C$, is a martingale. Apply the Doob submartingale inequality to $X_t = B(t)$ to get

$$P\Big\{ \sup_{0 \le t \le 1} |B(t)| \ge \epsilon \Big\} \le \frac{1}{\epsilon} E|B(1)|.$$

Rewrite the left-hand side in terms of the Wiener measure μ as

$$P\Big\{ \sup_{0 \le t \le 1} |B(t)| \ge \epsilon \Big\} = \mu\Big\{ \sup_{0 \le t \le 1} |\omega(t)| \ge \epsilon \Big\} = \mu\big\{ \omega \in C \,;\, \|\omega\|_\infty \ge \epsilon \big\}.$$

On the other hand, since $B(1)$ is a standard normal random variable,

$$E|B(1)| = \int_{\mathbb{R}} |x| \frac{1}{\sqrt{2\pi}} e^{-x^2/2}\, dx = \sqrt{\frac{2}{\pi}}.$$

Therefore, we get the inequality

$$\mu\big\{ \omega \in C \,;\, \|\omega\|_\infty \ge \epsilon \big\} \le \frac{1}{\epsilon} \sqrt{\frac{2}{\pi}},$$

which gives an estimate for the Wiener measure of the set outside the ball with center at 0 and radius ϵ.

4.6 Stochastic Processes Defined by Itô Integrals

Recall that in the beginning of Section 4.3 we fixed a Brownian motion $B(t)$ and a filtration $\{\mathcal{F}_t ; a \le t \le b\}$ satisfying conditions (a) and (b). Take a stochastic process $f \in L^2_{\mathrm{ad}}([a, b] \times \Omega)$. Then for any $t \in [a, b]$,

$$\int_a^t E(|f(s)|^2)\,ds \le \int_a^b E(|f(s)|^2)\,ds < \infty.$$

Hence $f \in L^2_{\mathrm{ad}}([a,t] \times \Omega)$. This implies that for each $t \in [a,b]$, the stochastic integral $\int_a^t f(s)\,dB(s)$ is defined. Consider a stochastic process given by

$$X_t = \int_a^t f(s)\,dB(s), \quad a \le t \le b.$$

Note that by Theorem 4.3.5 we have

$$E(|X_t|^2) = \int_a^t E(|f(s)|^2)\,ds < \infty$$

and so $E|X_t| \le [E(|X_t|^2)]^{1/2} < \infty$. Hence for each t, the random variable X_t is integrable and so we can take the conditional expectation of X_t with respect to a σ-field, in particular, $\mathcal{F}_s$ of the above filtration.

In Section 4.1 we mentioned that the Itô integral $\int_a^b f(t)\,dB(t)$ is defined in such a way that the stochastic process $X_t = \int_a^t f(s)\,dB(s)$, $a \le t \le b$, is a martingale. For example, we have shown directly that the stochastic processes $L(t) = \int_0^t B(s)\,dB(s)$, $t \ge 0$, in Section 4.1 and $X_t = \int_0^t B(s)^2\,dB(s)$, $t \ge 0$, in Section 4.4 are martingales. The next theorem confirms that this property is true in general.

Theorem 4.6.1. (Martingale property) *Suppose $f \in L^2_{\mathrm{ad}}([a,b] \times \Omega)$. Then the stochastic process*

$$X_t = \int_a^t f(s)\,dB(s), \quad a \le t \le b, \tag{4.6.1}$$

is a martingale with respect to the filtration $\{\mathcal{F}_t; a \le t \le b\}$.

Proof. First consider the case that f is a step stochastic process. We need to show that for any $a \le s < t \le b$,

$$E(X_t \,|\, \mathcal{F}_s) = X_s, \quad \text{almost surely.}$$

But $X_t = X_s + \int_s^t f(u)\,dB(u)$. Hence we need to show that

$$E\left(\int_s^t f(u)\,dB(u) \,\Big|\, \mathcal{F}_s\right) = 0, \quad \text{almost surely.} \tag{4.6.2}$$

Suppose f is given by

$$f(u,\omega) = \sum_{i=1}^n \xi_{i-1}(\omega)1_{(t_{i-1},t_i]}(u),$$

where $s = t_0 < t_1 < \cdots < t_n = t$, and ξ_{i-1} is $\mathcal{F}_{t_{i-1}}$-measurable and belongs to $L^2(\Omega)$. Then

$$\int_s^t f(u)\,dB(u) = \sum_{i=1}^n \xi_{i-1}\big(B(t_i) - B(t_{i-1})\big).$$

For any $i = 1, 2, \ldots, n$, we have

$$
\begin{aligned}
E\big[\xi_{i-1}\big(B(t_i) - B(t_{i-1})\big)\,\big|\,\mathcal{F}_s\big] \\
= E\big[E\{\xi_{i-1}\big(B(t_i) - B(t_{i-1})\big)\,\big|\,\mathcal{F}_{t_{i-1}}\}\,\big|\,\mathcal{F}_s\big] \\
= E\big[\xi_{i-1}E\{B(t_i) - B(t_{i-1})\,\big|\,\mathcal{F}_{t_{i-1}}\}\,\big|\,\mathcal{F}_s\big] \\
= 0,
\end{aligned}
$$

since $E\{B(t_i) - B(t_{i-1})\,\big|\,\mathcal{F}_{t_{i-1}}\} = 0$. Hence Equation (4.6.2) holds.

Now we consider the general case. Let $f \in L^2_{\mathrm{ad}}([a,b] \times \Omega)$. Take a sequence $\{f_n\}$ of step stochastic processes in $L^2_{\mathrm{ad}}([a,b] \times \Omega)$ such that

$$\lim_{n \to \infty} \int_a^b E\big(|f(u) - f_n(u)|^2\big)\,du = 0.$$

For each n, define a stochastic process

$$X_t^{(n)} = \int_a^t f_n(u)\,dB(u).$$

By the first case, $X_t^{(n)}$ is a martingale. For $s < t$, write

$$X_t - X_s = \big(X_t - X_t^{(n)}\big) + \big(X_t^{(n)} - X_s^{(n)}\big) + \big(X_s^{(n)} - X_s\big)$$

and then take the conditional expectation to get

$$E\big(X_t - X_s\,\big|\,\mathcal{F}_s\big) = E\big(X_t - X_t^{(n)}\,\big|\,\mathcal{F}_s\big) + E\big(X_s^{(n)} - X_s\,\big|\,\mathcal{F}_s\big). \qquad (4.6.3)$$

Note that

$$
\begin{aligned}
E\{\big|E\big(X_t - X_t^{(n)}\,\big|\,\mathcal{F}_s\big)\big|^2\} &\le E\{E\big(|X_t - X_t^{(n)}|^2\,\big|\,\mathcal{F}_s\big)\} \\
&= E\{|X_t - X_t^{(n)}|^2\}.
\end{aligned}
$$

Then apply Theorem 4.3.5 to get

$$
\begin{aligned}
E\{\big|E\big(X_t - X_t^{(n)}\,\big|\,\mathcal{F}_s\big)\big|^2\} &\le \int_a^t E\big(|f(u) - f_n(u)|^2\big)\,du \\
&\le \int_a^b E\big(|f(u) - f_n(u)|^2\big)\,du \\
&\to 0, \qquad \text{as } n \to \infty.
\end{aligned}
$$

Thus by taking a subsequence if necessary, we see that $E(X_t - X_t^{(n)}|\mathcal{F}_s)$ converges almost surely to 0. Similarly, $E(X_s - X_s^{(n)}|\mathcal{F}_s) \to 0$ a.s. Hence by Equation (4.6.3) we have $E(X_t - X_s|\mathcal{F}_s) = 0$ a.s. and so $E(X_t|\mathcal{F}_s) = X_s$ a.s. Thus X_t is a martingale. $\qquad\square$

Next we study the continuity property of the stochastic process X_t defined by Equation (4.6.1). Note that the stochastic integral is not defined for each fixed ω as a Riemann, Riemann–Stieltjes, or even Lebesgue integral. Even for the Wiener integral case, it is not defined this way. Therefore, the continuity of the stochastic process in Equation (4.6.1) is not a trivial fact as in elementary real analysis.

Theorem 4.6.2. (Continuity property) *Suppose $f \in L^2_{\mathrm{ad}}([a,b] \times \Omega)$. Then the stochastic process*

$$X_t = \int_a^t f(s)\, dB(s), \quad a \le t \le b,$$

is continuous, namely, almost all of its sample paths are continuous functions on the interval $[a,b]$.

Proof. First consider the case that f is a step stochastic process, say

$$f(s,\omega) = \sum_{i=1}^n \xi_{i-1}(\omega)\, 1_{(t_{i-1},t_i]}(s),$$

where ξ_{i-1} is $\mathcal{F}_{t_{i-1}}$-measurable. In this case, for each fixed $\omega \in \Omega$, the sample path of X_t is given by

$$X_t(\omega) = \sum_{i=1}^{k-1} \xi_{i-1}(\omega)\big(B(t_i,\omega) - B(t_{i-1},\omega)\big)$$

$$+ \xi_{k-1}(\omega)\big(B(t,\omega) - B(t_{k-1},\omega)\big)$$

for $t_{k-1} \le t < t_k$. Recall that for almost all ω, the Brownian motion path $B(\cdot,\omega)$ is a continuous function. Hence for almost all ω, the sample path $X_{(\cdot)}(\omega)$ is a continuous function on $[a,b]$.

Next consider the general case. Let $\{f_n\}$ be a sequence of step stochastic processes in $L^2_{\mathrm{ad}}([a,b] \times \Omega)$ such that

$$\lim_{n\to\infty} \int_a^b E\big(|f(s) - f_n(s)|^2\big)\, ds = 0.$$

By choosing a subsequence if necessary, we may assume that

$$\int_a^b E\big(|f(s) - f_n(s)|^2\big)\, ds \le \frac{1}{n^6}, \quad \forall n \ge 1. \tag{4.6.4}$$

For each n, define a stochastic process

$$X_t^{(n)} = \int_a^t f_n(s) \, dB(s), \quad a \leq t \leq b.$$

Then, as we have proved above, almost all sample paths of $X_t^{(n)}$ are continuous functions. Note that X_t and $X_t^{(n)}$ are martingales by Theorem 4.6.1. Hence $X_t - X_t^{(n)}$ is a martingale, and then by the Doob submartingale inequality in Equation (4.5.2),

$$P\Big\{ \sup_{a \leq t \leq b} \big|X_t - X_t^{(n)}\big| \geq \frac{1}{n} \Big\} \leq nE\big|X_b - X_b^{(n)}\big|.$$

But by the Schwarz inequality, Theorem 4.3.5, and Equation (4.6.4),

$$E\big|X_b - X_b^{(n)}\big| \leq \Big(E\big(|X_b - X_b^{(n)}|^2\big) \Big)^{1/2}$$

$$= \Big(\int_a^b E\big(|f(s) - f_n(s)|^2\big) \, ds \Big)^{1/2}$$

$$\leq \frac{1}{n^3}.$$

Therefore, for all $n \geq 1$,

$$P\Big\{ \sup_{a \leq t \leq b} \big|X_t - X_t^{(n)}\big| \geq \frac{1}{n} \Big\} \leq \frac{1}{n^2}.$$

Since $\sum_n 1/n^2 < \infty$, by the Borel–Cantelli lemma in Theorem 3.2.1 we have

$$P\Big\{ \sup_{a \leq t \leq b} \big|X_t - X_t^{(n)}\big| \geq \frac{1}{n} \text{ i.o.} \Big\} = 0.$$

Take the complement of the event $\{A_n \text{ i.o.}\}$ to get

$$P\Big\{ \sup_{a \leq t \leq b} \big|X_t - X_t^{(n)}\big| \geq \frac{1}{n} \text{ f.o.} \Big\} = 1.$$

Hence there exists an event Ω_0 such that $P(\Omega_0) = 1$ and for each $\omega \in \Omega_0$, there exists a positive integer $N(\omega)$ such that

$$\sup_{a \leq t \leq b} \big|X_t(\omega) - X_t^{(n)}(\omega)\big| < \frac{1}{n}, \quad \forall n \geq N(\omega).$$

Thus for each $\omega \in \Omega_0$, the sequence of functions $X_{(\cdot)}^{(n)}(\omega), n \geq 1$, converges uniformly to $X_{(\cdot)}(\omega)$ on $[a, b]$. But for each n, the stochastic process $X_t^{(n)}$ is continuous and so there exists an event Ω_n with $P(\Omega_n) = 1$ and for any $\omega \in \Omega_n$, the function $X_{(\cdot)}^{(n)}(\omega)$ is continuous.

Finally let $\widetilde{\Omega} = \cap_{n=0}^{\infty} \Omega_n$. Then we have $P(\widetilde{\Omega}_n) = 1$ and for each $\omega \in \widetilde{\Omega}$, the sequence

$$X_{(\cdot)}^{(n)}(\omega), \quad n = 1, 2, 3, \ldots,$$

is a sequence of continuous functions that converges uniformly to $X_{(\cdot)}(\omega)$ on $[a, b]$. It follows that $X_{(\cdot)}(\omega)$ is a continuous function for each $\omega \in \widetilde{\Omega}$. Hence almost all sample paths of the stochastic process X_t are continuous functions on $[a, b]$. Thus we have shown that X_t is a continuous stochastic process.   □

4.7 Riemann Sums and Stochastic Integrals

Let $f \in L^2_{\text{ad}}([a, b] \times \Omega)$. For a partition $\Delta_n = \{t_0, t_1, \ldots, t_{n-1}, t_n\}$ of the interval $[a, b]$, define the Riemann sum of f with respect to $B(t)$ by

$$\sum_{i=1}^{n} f(t_{i-1})\big(B(t_i) - B(t_{i-1})\big). \tag{4.7.1}$$

It is natural to ask whether this sequence of Riemann sums converges to the Itô integral $\int_a^b f(t)\, dB(t)$. Assume that $E\big(f(t)f(s)\big)$ is a continuous function of t and s. Define a stochastic process f_n as in Equation (4.3.7), namely,

$$f_n(t, \omega) = f(t_{i-1}, \omega), \quad t_{i-1} < t \le t_i.$$

As shown in Case 1 in the proof of Lemma 4.3.3, we have

$$\lim_{n \to \infty} \int_a^b E\big\{|f(t) - f_n(t)|^2\big\}\, dt = 0.$$

Hence by Equation (4.3.15),

$$\int_a^b f(t)\, dB(t) = \lim_{n \to \infty} I(f_n), \quad \text{in } L^2(\Omega).$$

But by Equation (4.3.2), $I(f_n)$ is given by

$$I(f_n) = \sum_{i=1}^{n} f_n(t_{i-1})\big(B(t_i) - B(t_{i-1})\big)$$

$$= \sum_{i=1}^{n} f(t_{i-1})\big(B(t_i) - B(t_{i-1})\big),$$

which is exactly the Riemann sum in Equation (4.7.1). Thus we have proved the following theorem.

Theorem 4.7.1. *Suppose $f \in L^2_{\text{ad}}([a, b] \times \Omega)$ and assume that $E\big(f(t)f(s)\big)$ is a continuous function of t and s. Then*

$$\int_a^b f(t)\, dB(t) = \lim_{\|\Delta_n\| \to 0} \sum_{i=1}^{n} f(t_{i-1})\big(B(t_i) - B(t_{i-1})\big), \quad \text{in } L^2(\Omega),$$

where $\Delta_n = \{a = t_0 < t_1 < \cdots < t_n = b\}$ and $\|\Delta_n\| = \max_{1 \le i \le n}(t_i - t_{i-1})$.

Exercises

1. For a partition $\Delta_n = \{a = t_0 < t_1 < \cdots < t_n = b\}$, define

$$M_{\Delta_n} = \sum_{i=1}^{n} B\left(\frac{t_{i-1} + t_i}{2}\right)(B(t_i) - B(t_{i-1})).$$

   Find $\lim_{\|\Delta_n\| \to 0} M_{\Delta_n}$ in $L^2(\Omega)$.

2. Let $X = \int_0^1 B(t) \, dB(t)$. Find the distribution function of the random variable X.

3. Let $f(t) = B\left(\frac{1}{2}\right)1_{[1/2,1]}(t)$. Show that $\int_0^1 f(t) \, dB(t)$ is not a Gaussian random variable. Compare this example with Theorem 2.3.4.

4. Let $X = \int_a^b |B(t)| \, dB(t)$. Find the variance of the random variable X.

5. The *signum function* is defined by $\operatorname{sgn}(0) = 0$ and $\operatorname{sgn}(x) = x/|x|$ if $x \neq 0$. Let $X_t = \int_0^t \operatorname{sgn}(B(s)) \, dB(s)$. Show that for $s < t$, the random variable $X_t - X_s$ has mean 0 and variance $t - s$.

   Remark: We will prove in Example 8.4.4 that X_t is a Brownian motion. Hence $X_b - X_a = \int_a^b \operatorname{sgn}(B(t)) \, dB(t)$ is a Gaussian random variable. This shows that $\int_a^b f(t) \, dB(t)$ can be Gaussian even if $f(t)$ is not a deterministic function; cf. Theorem 2.3.4.

6. Let $X = \int_a^b f(t)\left[\sin(B(t)) + \cos(B(t))\right] dB(t)$, $f \in L^2[a, b]$. Show that $\operatorname{var}(X) = \int_a^b f(t)^2 \, dt$.

7. Find the variance of the random variable $X = \int_a^b \sqrt{t}\, e^{B(t)} \, dB(t)$.

8. Find the variance of the random variable $X = \int_0^1 e^{B(t)^2/8} \, dB(t)$.

9. Find the variance of the random variable $X = \int_a^b \sqrt{t}\, \sin(B(t)) \, dB(t)$.

10. Let $X = \int_a^b \sinh(B(t)) \, dB(t)$ and $Y = \int_a^b \cosh(B(t)) \, dB(t)$. Here $\sinh(x)$ and $\cosh(x)$ are the hyperbolic sine and cosine functions, respectively, i.e., $\sinh(x) = (e^x - e^{-x})/2$ and $\cosh(x) = (e^x + e^{-x})/2$. Find $\operatorname{var}(X)$, $\operatorname{var}(Y)$, and $\operatorname{cov}(X, Y)$.

11. Let f be a deterministic function in $L^2[a, b]$ and $X_t = X_a + \int_a^t f(s) \, dB(s)$. Show that $\int_a^b f(t) X_t \, dB(t) = \frac{1}{2}(X_b^2 - X_a^2 - \int_a^b f(t)^2 \, dt)$.

12. (a) Show that for any $s < t$,

$$3B(s)^2(B(t) - B(s))$$
$$= B(t)^3 - B(s)^3 - (B(t) - B(s))^3 - 3B(s)(B(t) - B(s))^2.$$

   (b) Use this equality to prove Equation (4.4.2).

13. Let $X_t = B(1)B(t)$, $0 \leq t \leq 1$.

(a) Check whether the random variable $B(1)$ is measurable with respect to the σ-field $\mathcal{F}_t = \sigma\{B(s); s \leq t\}$ for any $0 \leq t < 1$.

(b) Show that for each $0 \leq t < 1$, the random variable X_t is not measurable with respect to $\mathcal{F}_t$ given in (a).

(c) For $0 \leq s \leq t \leq 1$, find $E[X_t \,|\, \mathcal{F}_s]$.

14. Prove the equality in Equation (4.4.6).

15. Show that $X_t = e^{B(t)} - 1 - \frac{1}{2} \int_0^t e^{B(s)} \, ds$ is a martingale.

16. Show that $X_t = e^{B(t) - \frac{1}{2}t}$ is a martingale.

17. The Hermite polynomial $H_n(x; c)$ of degree n with parameter $c > 0$ is defined by

$$H_n(x; c) = (-c)^n e^{x^2/2c} D_x^n e^{-x^2/2c}. \tag{4.7.2}$$

For various identities on the Hermite polynomials, see the book [56]. In particular, we have

$$H_n(x; c) = \sum_{k=0}^{[n/2]} \binom{n}{2k} (2k-1)!! (-c)^k x^{n-2k}, \tag{4.7.3}$$

$$e^{\tau x - c\tau^2/2} = \sum_{n=0}^{\infty} \frac{\tau^n}{n!} H_n(x; c), \quad \text{(generating function)}, \tag{4.7.4}$$

where $[n/2]$ is the largest integer less than or equal $n/2$ and $(2k-1)!! = (2k-1)(2k-3)\cdots 3 \cdot 1$ with the convention $(-1)!! = 1$.

(a) Use Equation (4.7.3) to show that for any $n \geq 0$,

$$\int_a^b H_n(B(t); t) \, dB(t) = \frac{1}{n+1} \Big\{ H_{n+1}(B(b); b) - H_{n+1}(B(a); a) \Big\}.$$

(b) Derive $\int_a^b B(t)^2 \, dB(t)$ from this equality with $n = 2$ and show that it gives the same value as in Equation (4.4.2).

(c) Use Equation (4.7.4) to evaluate the stochastic integral $\int_a^b e^{B(s)} \, dB(s)$ and then show that $X_t = \int_a^t e^{B(s)} \, dB(s)$ is a martingale.

18. Let $f, f_n \in L^2_{\mathrm{ad}}([a, b] \times \Omega)$ and assume that

$$\int_a^b |f(t) - f_n(t)| \, dt \to 0 \quad \text{a.s.}$$

Prove the convergence of $\int_a^b f_n(t) \, dB(t)$ to $\int_a^b f(t) \, dB(t)$ in probability.

19. Suppose $f(t)$ is a deterministic function in $L^2[a, b]$.

(a) Check directly from the definition that $X_t = \left(\int_a^t f(s) \, dB(s) \right)^2$ is a submartingale.

(b) Check directly from the definition that $X_t - \int_a^t f(s)^2\, ds$ is a martingale.

20. Let $f \in L^2_{\text{ad}}([a, b] \times \Omega)$.

(a) Use the conditional Jensen's inequality in Section 2.4 to show that $Y_t = \left(\int_a^t f(s)\, dB(s) \right)^2$ is a submartingale.

(b) Show that $Y_t - \int_a^t f(s)^2\, ds$ is a martingale.

21. Let $\theta(t) \in L^1[a, b]$ and $f(t) \in L^2[a, b]$ be deterministic functions. Prove the integration by parts formula

$$\int_a^b \left(\int_a^t \theta(s)\, ds \right) f(t)\, dB(t)$$

$$= \int_a^b \theta(s)\, ds \int_a^b f(s)\, dB(s) - \int_a^b \theta(s) \left(\int_a^t f(s)\, dB(s) \right) dt.$$

22. Let $f \in L^2_{\text{ad}}([a, b] \times \Omega)$. Suppose Ω_0 is an event such that for any $\omega \in \Omega_0$, $f(t, \omega) = 0$ for all $t \in [a, b]$. Show that $\int_a^b f(t)\, dB(t) = 0$ on Ω_0.

5

An Extension of Stochastic Integrals

In Chapter 4, the Itô integral $I(f) = \int_a^b f(t)\, dB(t)$ is defined for $\{\mathcal{F}_t\}$-adapted stochastic processes $f(t)$ satisfying the condition $\int_a^b E\big(|f(t)|^2\big)\, dt < \infty$. The random variable $I(f)$ belongs to $L^2(\Omega)$. Hence it has integrability. Moreover, the stochastic process $X_t = \int_a^t f(s)\, dB(s)$, $a \le t \le b$, is a martingale. In this chapter, we will extend the Itô integral $I(f)$ to $\{\mathcal{F}_t\}$-adapted stochastic processes $f(t)$ satisfying the condition $\int_a^b |f(t)|^2\, dt < \infty$ almost surely. In this case, the Itô integral $I(f)$ is a random variable and in general does not possess integrability. The lack of integrability of a stochastic process leads to the concept of local martingale. We will show that $X_t = \int_a^t f(s)\, dB(s)$, $a \le t \le b$, is a local martingale.

5.1 A Larger Class of Integrands

As in the beginning of Section 4.3, we fix a Brownian motion $B(t)$ and a filtration $\{\mathcal{F}_t; a \le t \le b\}$ such that

(a) For each t, $B(t)$ is $\mathcal{F}_t$-measurable;
(b) For any $s \le t$, the random variable $B(t) - B(s)$ is independent of the σ-field $\mathcal{F}_s$.

In this chapter, we will define the stochastic integral $\int_a^b f(t)\, dB(t)$ for stochastic processes $f(t, \omega)$ satisfying the following conditions:

(1) $f(t)$ is adapted to the filtration $\{\mathcal{F}_t\}$;
(2) $\int_a^b |f(t)|^2\, dt < \infty$ almost surely.

Condition (2) means that almost all sample paths are functions in the Hilbert space $L^2[a, b]$. Hence the map $\omega \mapsto f(\cdot, \omega)$ is a measurable function from Ω into $L^2[a, b]$.

Notation 5.1.1. We will use $\mathcal{L}_{\mathrm{ad}}(\Omega, L^2[a, b])$ to denote the space of stochastic processes $f(t, \omega)$ satisfying the above conditions (1) and (2).

Recall that in Section 4.3 we use $L^2_{\text{ad}}([a,b] \times \Omega)$ to denote the space of all $\{\mathcal{F}_t\}$-adapted stochastic processes $f(t,\omega)$ such that $\int_a^b E(|f(t)|^2)\,dt < \infty$. By the Fubini theorem, $E\int_a^b |f(t)|^2\,dt = \int_a^b E(|f(t)|^2)\,dt < \infty$. It follows that $\int_a^b |f(t)|^2\,dt < \infty$ almost surely. This shows the inclusion relationship

$$L^2_{\text{ad}}([a,b] \times \Omega) \subset \mathcal{L}_{\text{ad}}(\Omega, L^2[a,b]).$$

Thus we have a larger class of integrands $f(t,\omega)$ for the stochastic integral $\int_a^b f(t)\,dB(t)$. The crucial difference is the possible lack of integrability of the integrand $f(t,\omega)$ with respect to the ω-variable.

Example 5.1.2. Consider the stochastic process $f(t) = e^{B(t)^2}$. It can be easily derived that

$$E(|f(t)|^2) = Ee^{2B(t)^2} = \begin{cases} \frac{1}{\sqrt{1-4t}}, & \text{if } 0 \le t < 1/4; \\ \infty, & \text{if } t \ge 1/4. \end{cases} \qquad (5.1.1)$$

Obviously, we have $\int_0^1 E(|f(t)|^2)\,dt = \infty$ and so $f \notin L^2_{\text{ad}}([0,1] \times \Omega)$. Hence $\int_0^1 e^{B(t)^2}\,dB(t)$ is not a stochastic integral as defined in Chapter 4. On the other hand, since $f(t)$ is a continuous function of t, we obviously have $\int_0^1 |f(t)|^2\,dt < \infty$ almost surely. Thus $f \in \mathcal{L}_{\text{ad}}(\Omega, L^2[0,1])$ and the stochastic integral $\int_0^1 e^{B(t)^2}\,dB(t)$ will be defined in this chapter.

Example 5.1.3. Consider the same stochastic process $f(t) = e^{B(t)^2}$ as in the previous example, but on a different time interval. Using Equation (5.1.1) we can easily check that $\int_0^c E(|f(t)|^2)\,dt < \infty$ for any $0 \le c < \frac{1}{4}$. Hence $f \in L^2_{\text{ad}}([0,c] \times \Omega)$ for any $0 \le c < \frac{1}{4}$. Therefore, $\int_0^c e^{B(t)^2}\,dB(t)$ is defined as in Chapter 4 and belongs to $L^2(\Omega)$ for any $0 \le c < \frac{1}{4}$.

Example 5.1.4. Consider the stochastic process $f(t) = e^{B(t)^k}$. Note that for any integer $k \ge 3$, we have

$$E(|f(t)|^2) = Ee^{2B(t)^k} = \int_{-\infty}^{\infty} e^{2x^k} \frac{1}{\sqrt{2\pi t}} e^{-x^2/2t}\,dx = \infty.$$

Hence $f \notin L^2_{\text{ad}}([a,b] \times \Omega)$. On the other hand, observe that almost all sample paths of $f(t)$ are continuous functions and so belong to $L^2[a,b]$. Thus we have $f \in \mathcal{L}_{\text{ad}}(\Omega, L^2[a,b])$ and the stochastic integral $\int_a^b e^{B(t)^k}\,dB(t)$ will be defined in this chapter.

In general it requires hard computation to check whether an $\{\mathcal{F}_t\}$-adapted stochastic process $f(t)$ belongs to $L^2_{\text{ad}}([a,b] \times \Omega)$. On the other hand, it is easier to see whether a stochastic process f belongs to $\mathcal{L}_{\text{ad}}(\Omega, L^2[a,b])$. For example, when $f(t)$ is $\{\mathcal{F}_t\}$-adapted and has continuous sample paths almost surely, then it belongs to $\mathcal{L}_{\text{ad}}(\Omega, L^2[a,b])$.

We prove a lemma on approximation that will be needed in Section 5.3 for the extension of the stochastic integral $\int_a^b f(t)\,dB(t)$ to the larger class $\mathcal{L}_{\mathrm{ad}}(\Omega, L^2[a,b])$ of integrands.

Lemma 5.1.5. *Let $f \in \mathcal{L}_{\mathrm{ad}}(\Omega, L^2[a,b])$. Then there exists a sequence $\{f_n\}$ in $L^2_{\mathrm{ad}}([a,b] \times \Omega)$ such that*

$$\lim_{n\to\infty} \int_a^b |f_n(t) - f(t)|^2\,dt = 0$$

almost surely, and hence also in probability.

Proof. For each n, define

$$f_n(t,\omega) = \begin{cases} f(t,\omega), & \text{if } \int_a^t |f(s,\omega)|^2\,ds \le n; \\ 0, & \text{otherwise.} \end{cases} \tag{5.1.2}$$

Then f_n is adapted to $\{\mathcal{F}_t\}$. Moreover,

$$\int_a^b |f_n(t,\omega)|^2\,dt = \int_a^{\tau_n(\omega)} |f(t,\omega)|^2\,dt, \quad \text{almost surely,}$$

where $\tau_n(\omega)$ is defined by

$$\tau_n(\omega) = \sup\left\{ t; \int_a^t |f(s,\omega)|^2\,ds \le n \right\}. \tag{5.1.3}$$

Therefore, we have

$$\int_a^b |f_n(t)|^2\,dt \le n, \quad \text{almost surely,}$$

which implies that $\int_a^b E(|f_n(t)|^2)\,dt \le n$ and so $f_n \in L^2_{\mathrm{ad}}([a,b] \times \Omega)$.

Let ω be fixed. As soon as n is so large that $n > \int_a^b |f(t,\omega)|^2\,dt$, then by Equation (5.1.2) we have

$$f_n(t,\omega) = f(t,\omega), \quad \forall t \in [a,b],$$

which implies obviously that

$$\lim_{n\to\infty} \int_a^b |f_n(t,\omega) - f(t,\omega)|^2\,dt = 0.$$

Since $\int_a^b |f(t,\omega)|^2\,dt < \infty$ for almost all $\omega \in \Omega$, the convergence holds almost surely. The other conclusion in the lemma holds because convergence almost surely implies convergence in probability. □

5.2 A Key Lemma

In this section we prove a key lemma for the extension of stochastic integrals to be defined in Section 5.3.

Lemma 5.2.1. *Let $f(t)$ be a step stochastic process in $L^2_{\text{ad}}([a,b] \times \Omega)$. Then the inequality*

$$P\left\{\left|\int_a^b f(t)\, dB(t)\right| > \epsilon\right\} \leq \frac{C}{\epsilon^2} + P\left\{\int_a^b |f(t)|^2\, dt > C\right\}$$

holds for any positive constants ϵ and C.

Proof. For each positive constant C, define a stochastic process $f_C(t, \omega)$ by

$$f_C(t, \omega) = \begin{cases} f(t, \omega), & \text{if } \int_a^t |f(s, \omega)|^2\, ds \leq C; \\ 0, & \text{otherwise.} \end{cases}$$

Observe that

$$\left\{\left|\int_a^b f(t)\, dB(t)\right| > \epsilon\right\}$$
$$\subset \left\{\left|\int_a^b f_C(t)\, dB(t)\right| > \epsilon\right\} \cup \left\{\int_a^b f(t)\, dB(t) \neq \int_a^b f_C(t)\, dB(t)\right\}.$$

Hence

$$P\left\{\left|\int_a^b f(t)\, dB(t)\right| > \epsilon\right\}$$
$$\leq P\left\{\left|\int_a^b f_C(t)\, dB(t)\right| > \epsilon\right\} + P\left\{\int_a^b f(t)\, dB(t) \neq \int_a^b f_C(t)\, dB(t)\right\}.$$

On the other hand, since f is a step stochastic process, we have

$$\left\{\int_a^b f(t)\, dB(t) \neq \int_a^b f_C(t)\, dB(t)\right\} \subset \left\{\int_a^b |f(t, \omega)|^2\, dt > C\right\}.$$

Therefore,

$$P\left\{\left|\int_a^b f(t)\, dB(t)\right| > \epsilon\right\}$$
$$\leq P\left\{\left|\int_a^b f_C(t)\, dB(t)\right| > \epsilon\right\} + P\left\{\int_a^b |f(t)|^2\, dt > C\right\}. \qquad (5.2.1)$$

Note that from the definition of f_C, we have $\int_a^b |f_C(t)|^2\, dt \leq C$ almost surely. Hence $E\int_a^b |f_C(t)|^2\, dt \leq C$ and so we can apply the Chebyshev inequality to the first term of Equation (5.2.1) to get

$$P\left\{ \left| \int_a^b f(t)\, dB(t) \right| > \epsilon \right\}$$

$$\leq \frac{1}{\epsilon^2} E \left| \int_a^b f_C(t)\, dB(t) \right|^2 + P\left\{ \int_a^b |f(t)|^2\, dt > C \right\}$$

$$= \frac{1}{\epsilon^2} \int_a^b E|f_C(t)|^2\, dt + P\left\{ \int_a^b |f(t)|^2\, dt > C \right\}$$

$$\leq \frac{C}{\epsilon^2} + P\left\{ \int_a^b |f(t)|^2\, dt > C \right\}.$$

Hence the lemma is proved. $\qquad\qquad\qquad\qquad\qquad\qquad\qquad\qquad$ □

5.3 General Stochastic Integrals

We need the following approximation lemma in order to define the general stochastic integral.

Lemma 5.3.1. *Let $f \in \mathcal{L}_{\mathrm{ad}}(\Omega, L^2[a, b])$. Then there exists a sequence $\{f_n(t)\}$ of step stochastic processes in $L^2_{\mathrm{ad}}([a, b] \times \Omega)$ such that*

$$\lim_{n \to \infty} \int_a^b |f_n(t) - f(t)|^2\, dt = 0, \quad \text{in probability.} \tag{5.3.1}$$

Proof. First use Lemma 5.1.5 to choose a sequence $\{g_n\}$ in $L^2_{\mathrm{ad}}([a, b] \times \Omega)$ such that

$$\lim_{n \to \infty} \int_a^b |g_n(t) - f(t)|^2\, dt = 0, \quad \text{in probability.} \tag{5.3.2}$$

Next, for each $g_n(t)$, apply Lemma 4.3.3 to find a step stochastic process $f_n(t)$ in $L^2_{\mathrm{ad}}([a, b] \times \Omega)$ such that

$$E \int_a^b |f_n(t) - g_n(t)|^2\, dt < \frac{1}{n}. \tag{5.3.3}$$

Then use the inequality $|u + v|^2 \leq 2(|u|^2 + |v|^2)$ to show that for any $\epsilon > 0$,

$$\left\{ \int_a^b |f_n(t) - f(t)|^2\, dt > \epsilon \right\}$$

$$\subset \left\{ \int_a^b |f_n(t) - g_n(t)|^2\, dt > \frac{\epsilon}{4} \right\} \cup \left\{ \int_a^b |g_n(t) - f(t)|^2\, dt > \frac{\epsilon}{4} \right\},$$

which yields that

$$P\left\{ \int_a^b |f_n(t) - f(t)|^2\, dt > \epsilon \right\}$$

$$\leq P\left\{ \int_a^b |f_n(t) - g_n(t)|^2\, dt > \frac{\epsilon}{4} \right\} + P\left\{ \int_a^b |g_n(t) - f(t)|^2\, dt > \frac{\epsilon}{4} \right\}.$$

Therefore, by the Chebyshev inequality and Equation (5.3.3),

$$P\Big\{\int_a^b |f_n(t) - f(t)|^2\, dt > \epsilon\Big\} \le \frac{4}{\epsilon n} + P\Big\{\int_a^b |g_n(t) - f(t)|^2\, dt > \frac{\epsilon}{4}\Big\}.$$

Hence we can use Equation (5.3.2) to conclude that for any $\epsilon > 0$,

$$\lim_{n\to\infty} P\Big\{\int_a^b |f_n(t) - f(t)|^2\, dt > \epsilon\Big\} = 0,$$

which implies Equation (5.3.1).   □

Now we are ready to define the general stochastic integral

$$\int_a^b f(t)\, dB(t)$$

for $f \in \mathcal{L}_{\mathrm{ad}}(\Omega, L^2[a, b])$. Apply Lemma 5.3.1 to choose a sequence $\{f_n(t)\}$ of step stochastic processes in $L^2_{\mathrm{ad}}([a, b] \times \Omega)$ such that Equation (5.3.1) holds. For each n, the stochastic integral

$$I(f_n) = \int_a^b f_n(t)\, dB(t)$$

is defined as in Section 4.3. Apply Lemma 5.2.1 to $f = f_n - f_m$ with $\epsilon > 0$ and $C = \epsilon^3/2$ to get

$$P\{|I(f_n) - I(f_m)| > \epsilon\}$$

$$\le \frac{\epsilon}{2} + P\Big\{\int_a^b |f_n(t) - f_m(t)|^2\, dt > \frac{\epsilon^3}{2}\Big\}. \tag{5.3.4}$$

Use the inequality $|u + v|^2 \le 2(|u|^2 + |v|^2)$ again to check that

$$\Big\{\int_a^b |f_n(t) - f_m(t)|^2\, dt > \frac{\epsilon^3}{2}\Big\}$$

$$\subset \Big\{\int_a^b |f_n(t) - f(t)|^2\, dt > \frac{\epsilon^3}{8}\Big\} \cup \Big\{\int_a^b |f_m(t) - f(t)|^2\, dt > \frac{\epsilon^3}{8}\Big\},$$

which yields the inequality

$$P\Big\{\int_a^b |f_n(t) - f_m(t)|^2\, dt > \frac{\epsilon^3}{2}\Big\}$$

$$\le P\Big\{\int_a^b |f_n(t) - f(t)|^2\, dt > \frac{\epsilon^3}{8}\Big\} + P\Big\{\int_a^b |f_m(t) - f(t)|^2\, dt > \frac{\epsilon^3}{8}\Big\}.$$

Thus by Equation (5.3.1) we get

$$\lim_{n,m\to\infty} P\left\{ \int_a^b |f_n(t) - f_m(t)|^2 \, dt > \frac{\epsilon^3}{2} \right\} = 0.$$

Hence there exists $N > 1$ such that

$$P\left\{ \int_a^b |f_n(t) - f_m(t)|^2 \, dt > \frac{\epsilon^3}{2} \right\} < \frac{\epsilon}{2}, \quad \forall n, m \geq N. \tag{5.3.5}$$

It follows from Equations (5.3.4) and (5.3.5) that

$$P\left\{ |I(f_n) - I(f_m)| > \epsilon \right\} < \epsilon, \quad \forall n, m \geq N,$$

which shows that the sequence $\{I(f_n)\}_{n=1}^\infty$ of random variables converges in probability. Hence we can define

$$\int_a^b f(t) \, dB(t) = \lim_{n\to\infty} I(f_n), \quad \text{in probability.} \tag{5.3.6}$$

It is easy to check that the limit is independent of the choice of the sequence $\{f_n\}_{n=1}^\infty$. Thus the stochastic integral $\int_a^b f(t) \, dB(t)$ is well-defined.

Therefore, we have finally defined the stochastic integral

$$\int_a^b f(t) \, dB(t), \quad f \in \mathcal{L}_{ad}(\Omega, L^2[a, b]).$$

Observe that if $f \in L^2_{ad}([a, b] \times \Omega)$, then we can take $f_n = f$ for all n in Equation (5.3.1). Then the stochastic integral defined by Equation (5.3.6) is obviously the same as the one defined in Section 4.3. This shows that the new stochastic integral defined by Equation (5.3.6) reduces to the old stochastic integral when $f \in L^2_{ad}([a, b] \times \Omega)$. Thus we have extended the stochastic integral $\int_a^b f(t) \, dB(t)$ for the integrands f from $L^2_{ad}([a, b] \times \Omega)$ to $\mathcal{L}_{ad}(\Omega, L^2[a, b])$.

Example 5.3.2. Let $f(t) = e^{B(t)^2}$. Then obviously $f \in \mathcal{L}_{ad}(\Omega, L^2[0, 1])$ and so the stochastic integral $\int_0^1 e^{B(t)^2} \, dB(t)$ is defined. In fact, we will see later in Chapter 7 that Itô's formula can be used to show that

$$\int_0^1 e^{B(t)^2} \, dB(t) = \int_0^{B(1)} e^{t^2} \, dt - \int_0^1 B(t) e^{B(t)^2} \, dt,$$

where the last integral is a Riemann integral for each sample path $B(\cdot, \omega)$.

For comparison, note that if $f \in L^2_{ad}([a, b] \times \Omega)$, then $\int_a^b f(t) \, dB(t)$ belongs to $L^2(\Omega)$. However, if $f \in \mathcal{L}_{ad}(\Omega, L^2[a, b])$, then $\int_a^b f(t) \, dB(t)$ is just a random variable and in general has no finite expectation. The analogue of Theorem 4.7.1 is given by the next theorem.

Theorem 5.3.3. *Suppose f is a continuous $\{\mathcal{F}_t\}$-adapted stochastic process. Then $f \in \mathcal{L}_{\mathrm{ad}}(\Omega, L^2[a, b])$ and*

$$\int_a^b f(t)\,dB(t) = \lim_{\|\Delta_n\| \to 0} \sum_{i=1}^n f(t_{i-1})\big(B(t_i) - B(t_{i-1})\big), \quad \text{in probability,}$$

where $\Delta_n = \{t_0, t_1, \ldots, t_{n-1}, t_n\}$ is a partition of the finite interval $[a, b]$ and $\|\Delta_n\| = \max_{1 \le i \le n} (t_i - t_{i-1})$.

Proof. First observe that if $\theta \in \mathcal{L}_{\mathrm{ad}}(\Omega, L^2[a, b])$ is given by

$$\theta(t, \omega) = \sum_{i=1}^n \xi_{i-1}(\omega)\, 1_{[t_{i-1}, t_i)}(t),$$

where ξ_{i-1} is $\mathcal{F}_{t_{i-1}}$-measurable, then it can be easily verified that

$$\int_a^b \theta(t)\,dB(t) = \sum_{i=1}^n \xi_{i-1}\big(B(t_i) - B(t_{i-1})\big). \tag{5.3.7}$$

Now, for f as given in the theorem, define $f_n(t) = \sum_{i=1}^n f(t_{i-1}) 1_{[t_{i-1}, t_i)}(t)$ for a partition Δ_n. Then the continuity of f implies that

$$\int_a^b |f_n(t) - f(t)|^2\,dt \longrightarrow 0$$

almost surely, hence in probability as $n \to \infty$. On the other hand, we can look through Equations (5.3.4), (5.3.5), and (5.3.6) to see that

$$\int_a^b f(t)\,dB(t) = \lim_{\|\Delta_n\| \to 0} \int_a^b f_n(t)\,dB(t), \quad \text{in probability.} \tag{5.3.8}$$

Moreover, we can apply Equation (5.3.7) to $\theta = f_n$ to get

$$\int_a^b f_n(t)\,dB(t) = \sum_{i=1}^n f(t_{i-1})\big(B(t_i) - B(t_{i-1})\big). \tag{5.3.9}$$

Obviously, the theorem follows from Equations (5.3.8) and (5.3.9).   □

5.4 Stopping Times

In this section we explain the concept of stopping time to be used in the next section. The reason we need this concept is the lack of integrability of the stochastic integral $\int_a^t f(s)\,dB(s)$ for general $f \in \mathcal{L}_{\mathrm{ad}}(\Omega, L^2[a, b])$.

Definition 5.4.1. *A random variable $\tau \colon \Omega \to [a, b]$ is called a stopping time with respect to a filtration $\{\mathcal{F}_t; a \le t \le b\}$ if $\{\omega; \tau(\omega) \le t\} \in \mathcal{F}_t$ for all $t \in [a, b]$.*

In the definition b can be ∞. Intuitively speaking, we can think of τ as the time to stop playing a game. The condition for τ to be a stopping time means that the decision to stop playing the game before or at time t should be determined by the information provided by $\mathcal{F}_t$.

A filtration $\{\mathcal{F}_t; a \leq t \leq b\}$ is *right continuous* if it satisfies the condition

$$\mathcal{F}_t = \bigcap_{n=1}^{\infty} \mathcal{F}_{t+\frac{1}{n}}, \quad \forall t \in [a, b),$$

where by convention $\mathcal{F}_t = \mathcal{F}_b$ when $t > b$.

Sometimes it is easier to work on the event $\{\omega; \tau(\omega) < t\}$ than the event $\{\omega; \tau(\omega) \leq t\}$. In that case the following fact is useful.

Fact. *Let* $\{\mathcal{F}_t; a \leq t \leq b\}$ *be a right continuous filtration. Then a random variable* $\tau: \Omega \to [a, b]$ *is a stopping time with respect to* $\{\mathcal{F}_t; a \leq t \leq b\}$ *if and only if* $\{\omega; \tau(\omega) < t\} \in \mathcal{F}_t$ *for all* $t \in [a, b]$.

To verify this fact, suppose τ is a stopping time. Then for any $t \in (a, b]$,

$$\{\omega; \tau(\omega) < t\} = \bigcup_{n=1}^{\infty} \{\omega; \tau(\omega) \leq t - n^{-1}\} \in \mathcal{F}_t$$

and for $t = a$, the event $\{\omega; \tau(\omega) < a\} = \emptyset$ is in $\mathcal{F}_a$. Conversely, assume that $\{\omega; \tau(\omega) < t\} \in \mathcal{F}_t$ for all $t \in [a, b]$ and the filtration is right continuous. Then for any $t \in [a, b)$,

$$\{\omega; \tau(\omega) \leq t\} = \bigcap_{n=1}^{\infty} \{\omega; \tau(\omega) < t + n^{-1}\} \in \bigcap_{n=1}^{\infty} \mathcal{F}_{t+\frac{1}{n}} = \mathcal{F}_t.$$

Hence τ is a stopping time.

Obviously, if τ takes a constant value c in $[a, b]$, then τ is a stopping time. Here are two interesting examples of stopping times.

Example 5.4.2. Let $B(t)$ be a Brownian motion. Take the filtration $\{\mathcal{F}_t\}$ in Section 4.3 with $a = 0$ and $b = \infty$. Let

$$\tau(\omega) = \inf\{t > 0; |B(t, \omega)| > 1\}.$$

The random variable τ is the first exit time of the Brownian motion from the interval $[-1, 1]$. Observe that by the continuity of $B(t)$,

$$\{\omega; \tau(\omega) < t\} = \bigcup_{0 < r < t, \, r \in \mathbb{Q}} \{\omega; |B(r)| > 1\},$$

which shows that $\{\omega; \tau(\omega) < t\} \in \mathcal{F}_t$ for all $t > 0$. As for $t = 0$, we have $\{\omega; \tau(\omega) < 0\} = \emptyset \in \mathcal{F}_0$. Hence by the above fact, τ is a stopping time.

Example 5.4.3. Let $f \in \mathcal{L}_{\mathrm{ad}}(\Omega, L^2[a, b])$. For each fixed n, define

$$\tau_n(\omega) = \begin{cases} \inf\left\{t; \int_a^t |f(s, \omega)|^2\, ds > n\right\}, & \text{if } \{t; \cdots\} \neq \emptyset; \\ b, & \text{if } \{t; \cdots\} = \emptyset. \end{cases} \tag{5.4.1}$$

Obviously this random variable is equal to the one defined by Equation (5.1.3). Note that for any $t \in (a, b]$,

$$\{\omega; \tau_n(\omega) < t\} = \bigcup_{a < r < t,\, r \in \mathbb{Q}} \left\{\omega; \int_a^r |f(s, \omega)|^2\, ds > n\right\} \in \mathcal{F}_t.$$

For $t = a$, we have $\{\omega; \tau_n(\omega) < a\} = \emptyset \in \mathcal{F}_a$. Hence by the above fact, τ_n is a stopping time.

5.5 Associated Stochastic Processes

Let $f \in \mathcal{L}_{\mathrm{ad}}(\Omega, L^2[a, b])$. Then $f \in \mathcal{L}_{\mathrm{ad}}(\Omega, L^2[a, t])$ for any $t \in [a, b]$ and we get an associated stochastic process

$$X_t = \int_a^t f(s)\, dB(s), \quad a \leq t \leq b. \tag{5.5.1}$$

In general X_t has no finite expectation and so it makes no sense to say that X_t is a martingale. We will generalize the concept of martingales to cover the situation in which X_t has no finite expectation. This is where we need the concept of stopping times. In order to avoid possible confusion when we take stopping times into account, we will interpret the stochastic integral in Equation (5.5.1) as defined by

$$X_t = \int_a^t f(s)\, dB(s) = \int_a^b 1_{[a,t]}(s) f(s)\, dB(s), \quad a \leq t \leq b, \tag{5.5.2}$$

where the integrand $g(s, \omega) = 1_{[a,t]}(s) f(s, \omega)$ belongs to $\mathcal{L}_{\mathrm{ad}}(\Omega, L^2[a, b])$ for any $t \in [a, b]$.

For each n, define a stochastic process f_n by Equation (5.1.2), namely,

$$f_n(t, \omega) = \begin{cases} f(t, \omega), & \text{if } \int_a^t |f(s, \omega)|^2\, ds \leq n; \\ 0, & \text{otherwise.} \end{cases}$$

Let τ_n be defined by Equation (5.4.1). As shown in Example 5.4.3, τ_n is a stopping time for each n. Replace t in Equation (5.5.2) by $t \wedge \tau_n$ to get a stochastic process

$$X_{t \wedge \tau_n} = \int_a^{t \wedge \tau_n} f(s)\, dB(s) = \int_a^b 1_{[a, t \wedge \tau_n]}(s) f(s)\, dB(s), \quad a \leq t \leq b.$$

Now, make a key observation that

$$1_{[a,t\wedge\tau_n(\omega)]}(s)\, f(s,\omega) = 1_{[a,t]}(s)\, f_n(s,\omega), \quad \text{for almost all } \omega.$$

Therefore,

$$X_{t\wedge\tau_n} = \int_a^{t\wedge\tau_n} f(s)\, dB(s) = \int_a^t f_n(s)\, dB(s), \quad a \le t \le b. \tag{5.5.3}$$

Recall from the proof of Lemma 5.1.5 that $\int_a^b |f_n(t)|^2\, dt \le n$ almost surely and so $\int_a^b E(|f_n(t)|^2)\, dt \le n$. Hence $f_n \in L^2_{\mathrm{ad}}([a,b] \times \Omega)$. We then apply Theorem 4.6.1 to conclude that the stochastic process $X_{t\wedge\tau_n}$ is a martingale for each n.

We observe one more thing; namely, the sequence $\{\tau_n\}$ is monotonically increasing and $\tau_n \to b$ almost surely as $n \to \infty$.

Being motivated by the above discussion, we make the following definition of local martingales.

Definition 5.5.1. An $\{\mathcal{F}_t\}$-adapted stochastic process X_t, $a \le t \le b$, is called a local martingale with respect to $\{\mathcal{F}_t\}$ if there exists a sequence of stopping times τ_n, $n = 1, 2, \ldots$, such that

(1) τ_n increases monotonically to b almost surely as $n \to \infty$;

(2) For each n, $X_{t\wedge\tau_n}$ is a martingale with respect to $\{\mathcal{F}_t; a \le t \le b\}$.

Obviously, a martingale is a local martingale since we can simply choose $\tau_n = b$ for all n. However, a local martingale may not be a martingale due to the lack of integrability. Without the integrability of X_t we cannot talk about conditional expectation of X_t and obviously it makes no sense to say that X_t is a martingale.

From the above discussion we have the next theorem.

Theorem 5.5.2. Let $f \in \mathcal{L}_{\mathrm{ad}}(\Omega, L^2[a,b])$. Then the stochastic process

$$X_t = \int_a^t f(s)\, dB(s), \quad a \le t \le b, \tag{5.5.4}$$

is a local martingale with respect to the filtration $\{\mathcal{F}_t; a \le t \le b\}$ specified in the beginning of Section 5.1.

Example 5.5.3. By Example 5.1.2, $f(t) = e^{B(t)^2}$ belongs to $\mathcal{L}_{\mathrm{ad}}(\Omega, L^2[a,b])$. Hence by Theorem 5.5.2 the stochastic process

$$X_t = \int_a^t e^{B(s)^2}\, dB(s), \quad a \le t \le b,$$

is a local martingale. On the other hand, consider the stochastic process

$$Y_t = \int_0^t e^{B(s)^2}\, dB(s), \quad 0 \le t < \frac{1}{4}.$$

By Example 5.1.3 we have the integrability of Y_t. Moreover, by Theorem 4.6.1 Y_t is a martingale.

Next we study the sample path property of the stochastic process defined in Equation (5.5.4). For this purpose we prepare a lemma.

Lemma 5.5.4. *Suppose* $f, g \in L^2_{\mathrm{ad}}([a, b] \times \Omega)$. *Let A be the event*

$$A = \Big\{ \omega; \ f(t, \omega) = g(t, \omega) \ for \ all \ t \in [a, b] \Big\}.$$

Then we have

$$\int_a^b f(t) \, dB(t) = \int_a^b g(t) \, dB(t), \quad almost \ surely \ on \ A.$$

Proof. Without loss of generality we may assume that $g(t, \omega) = 0$ for all t and ω. Define

$$\tau(\omega) = \begin{cases} \inf \{t; \ f(t, \omega) \neq 0\}, & \text{if } \{t; \cdots\} \neq \emptyset; \\ b, & \text{if } \{t; \cdots\} = \emptyset. \end{cases}$$

Then τ is a stopping time. Consider the random variable

$$Y(\tau) = \int_a^\tau f(s) \, dB(s) = \int_a^b 1_{[a,\tau]}(s) f(s) \, dB(s).$$

Note that the integrand $1_{[a,\tau]}(s) f(s)$ belongs to $L^2_{\mathrm{ad}}([a, b] \times \Omega)$ and so the last stochastic integral is defined. Moreover, observe that for each ω,

$$1_{[a,\tau(\omega)]}(s) \, |f(s, \omega)|^2 = 1_{\{\tau(\omega)\}}(s) \, |f(\tau(\omega), \omega)|^2.$$

Therefore,

$$\int_a^b 1_{[a,\tau]}(t) |f(t)|^2 \, dt = 0, \quad \text{almost surely.}$$

Then by Theorem 4.3.5, we have

$$E\big(|Y(\tau)|^2\big) = E \int_a^b 1_{[a,\tau]}(t) |f(t)|^2 \, dt = 0.$$

Hence $Y(\tau) = 0$ almost surely. Note that if $\omega \in A$, then $\tau(\omega) = b$ and so

$$Y(\tau(\omega)) = \int_a^b f(s) \, dB(s).$$

This shows that $\int_a^b f(s) \, dB(s) = 0$ almost surely on the event A. $\quad\square$

Theorem 5.5.5. *Let $f \in \mathcal{L}_{\mathrm{ad}}(\Omega, L^2[a, b])$. Then the stochastic process*

$$X_t = \int_a^t f(s) \, dB(s), \quad a \leq t \leq b, \tag{5.5.5}$$

has a continuous realization.

Proof. For each n, let f_n be the stochastic process defined by Equation (5.1.2) and let

$$X_t^{(n)} = \int_a^t f_n(s)\, dB(s), \quad a \le t \le b.$$

Then by Theorem 4.6.2, $X_t^{(n)}$ is a continuous stochastic process. Let

$$A_n = \left\{ \omega;\ \int_a^b |f(t,\omega)|^2\, dt \le n \right\}.$$

The sequence $\{A_n\}$ is increasing. Let $A = \cup_{n=1}^\infty A_n$. Then $P(A) = 1$ since $\int_a^b |f(t)|^2\, dt < \infty$ almost surely. Note that if $\omega \in A_n$, then

$$f_n(t,\omega) = f_m(t,\omega), \quad \forall\, m \ge n \text{ and } \forall\, t \in [a,b].$$

Therefore, by Lemma 5.5.4, for almost all $\omega \in A_n$,

$$X_t^{(m)}(\omega) = X_t^{(n)}(\omega), \quad \forall\, m \ge n \text{ and } \forall\, t \in [a,b].$$

Since $A = \cup_{n=1}^\infty A_n$, the above equality implies that for almost all $\omega \in A$, the following limit exists for all $t \in [a,b]$:

$$\lim_{m \to \infty} X_t^{(m)}(\omega).$$

Now define a stochastic process $Y_t(\omega)$ by

$$Y_t(\omega) = \begin{cases} \lim_{m \to \infty} X_t^{(m)}(\omega), & \text{if } \omega \in A; \\ 0, & \text{if } \omega \notin A. \end{cases}$$

Then $Y_t(\omega)$ is a continuous stochastic process. On the other hand, from the definition of the stochastic integral in Equation (5.3.6),

$$X_t = \lim_{m \to \infty} X_t^{(m)}, \quad \text{in probability.}$$

Therefore, for each $t \in [a,b]$, $X_t = Y_t$ almost surely. Hence Y_t is a continuous realization of X_t. $\qquad \square$

Exercises

1. Let $f(t,\omega) = \sum_{i=1}^n \xi_{i-1}(\omega) 1_{[t_{i-1}, t_i)}(t)$ with ξ_{i-1} being $\mathcal{F}_{t_{i-1}}$-measurable for each i and $a = t_0 < t_1 < \cdots < t_n = b$. Show that the stochastic integral as defined in Section 5.1 is given by

$$\int_a^b f(t)\, dB(t) = \sum_{i=1}^n \xi_{i-1}\big(B(t_i) - B(t_{i-1})\big).$$

2. Let τ be a stopping time. Prove that $B(t+\tau)-B(\tau)$ is a Brownian motion.

3. Let $\{\tau_n\}$ be a sequence of stopping times. Assume that either the sequence increases to τ or decreases to τ. Show that τ is a stopping time.

4. Let $\{\tau_n\}$ be a sequence of stopping times. Show that $\sup_n \tau_n$, $\inf_n \tau_n$, $\limsup_n \tau_n$, $\liminf_n \tau_n$ are stopping times.

5. Let τ_1 and τ_2 be two stopping times. Show that $\tau_1 \vee \tau_2$, $\tau_1 \wedge \tau_2$, and $\tau_1 + \tau_2$ are stopping times. How about $\tau_1 \tau_2$?

6. Show that $X_t = \int_0^t e^{B(s)^2}\, dB(s)$, $0 \le t \le 1$, is not a martingale.

7. Let $X_\varepsilon = \int_0^1 \varepsilon^{-\lambda} e^{-B(t)^2/2\varepsilon}\, dB(t)$. Show that $X_\varepsilon \to 0$ in $L^2(\Omega)$ as $\varepsilon \downarrow 0$ if and only if $0 < \lambda < \frac{1}{4}$.

8. Let $Y_\varepsilon = \int_0^\varepsilon \varepsilon^{-\lambda} e^{-B(t)^2/2\varepsilon}\, dB(t)$. Show that $Y_\varepsilon \to 0$ in $L^2(\Omega)$ as $\varepsilon \downarrow 0$ if and only if $0 < \lambda < \frac{1}{2}$.

9. Let $Z_\varepsilon = \int_0^{\varepsilon/2} \varepsilon^{-\lambda} e^{-B(t)^2/2\varepsilon}\, dB(t)$. Show that $Z_\varepsilon \to 0$ in $L^2(\Omega)$ as $\varepsilon \downarrow 0$ if and only if $0 < \lambda < \frac{1}{2}$.

10. Let $B_1(t)$ and $B_2(t)$ be independent Brownian motions with starting point $(B_1(0), B_2(0)) \ne (0,0)$. Let $X_t = \log\left(B_1(t)^2 + B_2(t)^2\right)$.

    (a) Show that X_t is a local martingale.

    (b) Show that $E|X_t| < \infty$ for all $t > 0$.

    (c) Find EX_t.

    (d) Show that X_t is not a martingale.

    Remark: This example shows that a local martingale having integrability does not necessarily become a martingale. For more information, see the book by Durrett [11].

6

Stochastic Integrals for Martingales

In Chapters 4 and 5 we defined the Itô integral $\int_a^b f(t)\,dB(t)$ with respect to a Brownian motion $B(t)$. In Chapter 4 the integrand $f(t)$ is adapted and satisfies the condition $E\int_a^b |f(t)|^2\,dt < \infty$, while in Chapter 5 the integrand $f(t)$ is adapted and satisfies the condition $\int_a^b |f(t)|^2\,dt < \infty$ almost surely. In this chapter we will use the same ideas with modified techniques to define the stochastic integral $\int_a^b f(t)\,dM(t)$ with respect to a right continuous, square integrable martingale $M(t)$ with left-hand limits. An essential requirement for the integrand $f(t)$ is its predictability in order to compensate the jumps of the martingale $M(t)$.

6.1 Introduction

Let $g \in L^2_{\mathrm{ad}}([a,b] \times \Omega)$, i.e., $g(t)$ is an adapted stochastic process satisfying the condition that $\int_a^b E(|g(t)|^2)\,dt < \infty$. Then by Theorems 4.6.1 and 4.6.2 the stochastic process

$$M(t) = \int_a^t g(s)\,dB(s), \quad a \le t \le b, \tag{6.1.1}$$

is a continuous martingale. Intuitively, the "stochastic differential" dM_t of M_t can be defined by

$$dM(t) = g(t)\,dB(t).$$

Hence it is reasonable to define a stochastic integral with respect to $M(t)$ by

$$\int_a^b f(t)\,dM(t) = \int_a^b f(t)g(t)\,dB(t). \tag{6.1.2}$$

In general, we can define the stochastic integral $\int_a^b f(t)\,dM(t)$ with respect to a right continuous, square integrable martingale $M(t)$ with left-hand limits.

As it will turn out, when $M(t)$ is given by Equation (6.1.1), then the equality in Equation (6.1.2) does hold. An important noncontinuous martingale is the compensated Poisson process to be discussed in Section 6.2. The key tool for defining martingale stochastic integrals is the Doob–Meyer decomposition theorem, which we will explain in Section 6.4.

We need to point out that right continuous, square integrable martingales with left-hand limits are not the most general integrators. For a more general treatment of stochastic integration, see the books [8] [15] [68].

6.2 Poisson Processes

In this section we will discuss Poisson processes. The compensated Poisson processes are important examples of discontinuous martingales. They will be used as integrators in Section 6.5.

Definition 6.2.1. *A Poisson process with parameter $\lambda > 0$ is a stochastic process $N(t, \omega)$ satisfying the following properties:*

(1) $P\{\omega\,;\, N(0, \omega) = 0\} = 1$.

(2) *For any $0 \leq s < t$, the random variable $N(t) - N(s)$ is a Poisson random variable with parameter $\lambda(t - s)$, i.e.,*

$$P\{N(t) - N(s) = k\} = e^{-\lambda(t-s)} \frac{\left(\lambda(t - s)\right)^k}{k!}, \quad k = 0, 1, 2, \ldots, n, \ldots.$$

(3) *$N(t, \omega)$ has independent increments, i.e., for any $0 \leq t_1 < t_2 < \cdots < t_n$, the random variables*

$$N(t_1),\; N(t_2) - N(t_1),\; \ldots,\, N(t_n) - N(t_{n-1}),$$

are independent.

(4) *Almost all sample paths of $N(t, \omega)$ are right continuous functions with left-hand limits, i.e.,*

$$P\{\omega\,;\, N(\cdot, \omega) \text{ is right continuous with left-hand limits}\} = 1.$$

By comparing this definition with Definition 2.1.1 of a Brownian motion, we see that conditions (1) and (3) are the same, while conditions (2) and (4) are different. Moreover, it can be shown (see, e.g., [72]) that condition (2) in the above definition can be replaced by the following condition: For any $0 \leq s < t$,

$$P\{N(t) - N(s) = 1\} = \lambda(t - s) + o(t - s),$$
$$P\{N(t) - N(s) \geq 2\} = o(t - s),$$

where $o(\delta)$ denotes a quantity such that $\lim_{\delta \to 0} o(\delta)/\delta = 0$.

As in the case of Brownian motion, we may call $N(t)$ a *Poisson process with respect to a filtration* $\{\mathcal{F}_t; t \geq 0\}$ if it satisfies conditions (1), (2), and (4) in Definition 6.2.1 and the assumption in Proposition 4.2.1 with respect to a filtration $\{\mathcal{F}_t; t \geq 0\}$.

Here are some examples of Poisson processes: (i) the number of telephone calls to a certain office during the time period $[0, t]$, (ii) the number of people entering a post office up to and including time t on a business day, (iii) the number of earthquakes in a certain region during the time period $[0, t]$.

The sample paths of a Poisson process are step functions with jumps (or occurrence of events). Let T_1 be the time of the first jump and, for $n \geq 2$, let T_n be the elapsed time between the $(n-1)$st and the nth jumps. The random variables T_n are called *interarrival times*.

The distribution of T_n can be easily derived. First, note that the event $\{T_1 > t\}$ coincides with the event $\{N(t) = 0\}$. Hence by condition (2),

$$P\{T_1 > t\} = P\{N(t) = 0\} = e^{-\lambda t}, \quad t \geq 0.$$

To find the distribution of T_2 we use the conditional expectation to get

$$P\{T_2 > t\} = E\big(P\{T_2 > t \mid T_1\}\big).$$

Now use conditions (2) and (3) to show that

$$P(T_2 > t \mid T_1 = s) = P\{\text{no jump in } (s, s+t] \mid T_1 = s\}$$
$$= P\{\text{no jump in } (s, s+t]\}$$
$$= e^{-\lambda t}.$$

Therefore, T_2 is independent of T_1 and

$$P\{T_2 > t\} = e^{-\lambda t}, \quad t \geq 0.$$

In general, we can use the same arguments as those above to show that T_n is independent of $T_1, T_2, \ldots, T_{n-1}$ and

$$P\{T_n > t\} = e^{-\lambda t}, \quad t \geq 0.$$

Thus $\{T_n; n \geq 1\}$ is a sequence of independent random variables having the same exponential distribution with parameter λ, namely,

$$P\{T_n \leq t\} = \begin{cases} 1 - e^{-\lambda t}, & \text{if } t \geq 0; \\ 0, & \text{if } t < 0. \end{cases}$$

The sum $S_n = T_1 + T_2 + \cdots + T_n$ is called the *waiting time* until the nth jump. Its probability density is given by the gamma density function

$$f(t) = \frac{\lambda}{(n-1)!} (\lambda t)^{n-1} e^{-\lambda t}, \quad t \geq 0.$$

It is an interesting fact that the Poisson process $N(t)$ can be expressed in terms of the waiting times or the interarrival times by

$$N(t) = \max\{k; S_k = T_1 + T_2 + \cdots + T_k \leq t\}.$$

Now let $\{\mathcal{F}_t; t \geq 0\}$ be a filtration satisfying the conditions

(a) For each t, $N(t)$ is $\mathcal{F}_t$-measurable;
(b) For any $s \leq t$, the random variable $N(t) - N(s)$ is independent of the σ-field $\mathcal{F}_s$.

Then $N(t)$ is a Poisson process with respect to the filtration $\{\mathcal{F}_t; t \geq 0\}$. For any $0 \leq s < t$, we have

$$\begin{aligned} E(N(t)\,|\,\mathcal{F}_s) &= E((N(t) - N(s)) + N(s)\,|\,\mathcal{F}_s) \\ &= E(N(t) - N(s)) + N(s) \\ &= \lambda(t - s) + N(s). \end{aligned} \tag{6.2.1}$$

Thus the Poisson process $N(t)$ is not a martingale.

Definition 6.2.2. *The* compensated Poisson process $\widetilde{N}(t)$ *is defined by*

$$\widetilde{N}(t) = N(t) - \lambda t.$$

It follows from Equation (6.2.1) that the compensated Poisson process $\widetilde{N}(t)$ is a martingale with respect to $\{\mathcal{F}_t; t \geq 0\}$. For any $s < t$, the moment generating function of $\widetilde{N}(t) - \widetilde{N}(s)$ is easily checked to be given by

$$Ee^{x(\widetilde{N}(t) - \widetilde{N}(s))} = \exp\left[\lambda(t - s)(e^x - 1 - x)\right].$$

Expand both sides as power series in x and then compare the coefficients of the powers of x to get

$$E\{\widetilde{N}(t) - \widetilde{N}(s)\} = 0,$$
$$E\{[\widetilde{N}(t) - \widetilde{N}(s)]^2\} = \lambda(t - s),$$
$$E\{[\widetilde{N}(t) - \widetilde{N}(s)]^3\} = \lambda(t - s),$$
$$E\{[\widetilde{N}(t) - \widetilde{N}(s)]^4\} = \lambda(t - s) + 3\lambda^2(t - s)^2.$$

Recall that the quadratic variation of a Brownian motion $B(t)$ is given by Theorem 4.1.2. The next theorem gives the quadratic variation of the compensated Poisson process $\widetilde{N}(t)$.

Theorem 6.2.3. *Suppose* $\Delta_n = \{t_0, t_1, \ldots, t_{n-1}, t_n\}$ *is a partition of a finite interval* $[a, b]$. *Then*

$$\sum_{i=1}^{n} \left(\widetilde{N}(t_i) - \widetilde{N}(t_{i-1})\right)^2 \longrightarrow \lambda(b - a) + \widetilde{N}(b) - \widetilde{N}(a) \tag{6.2.2}$$

in $L^2(\Omega)$ *as* $\|\Delta_n\| = \max_{1 \leq i \leq n}(t_i - t_{i-1})$ *tends to* 0.

Proof. For simplicity, let

$$X_i = \left(\widetilde{N}(t_i) - \widetilde{N}(t_{i-1})\right)^2 - \left(\widetilde{N}(t_i) - \widetilde{N}(t_{i-1})\right) - \lambda(t_i - t_{i-1}).$$

Then

$$\sum_{i=1}^{n} \left(\widetilde{N}(t_i) - \widetilde{N}(t_{i-1})\right)^2 - \left(\lambda(b-a) + \widetilde{N}(b) - \widetilde{N}(a)\right) = \sum_{i=1}^{n} X_i.$$

Thus we need to show that $E\left(\left|\sum_{i=1}^{n} X_i\right|^2\right) = \sum_{i,j=1}^{n} E(X_i X_j)$ tends to zero as $\|\Delta_n\| \to 0$. For $i \neq j$, say $i < j$, we can first take the conditional expectation with respect to the σ-field $\mathcal{F}_{t_{j-1}}$ to show that $E(X_i X_j) = 0$. On the other hand, we can use the above moments of $\widetilde{N}(t) - \widetilde{N}(s)$ to show that for any i,

$$E(X_i^2) = 2\lambda^2(t_i - t_{i-1})^2,$$

which implies that $\sum_{i=1}^{n} E(X_i^2) \leq 2\lambda^2(b-a)\|\Delta_n\| \to 0$ as $\|\Delta_n\| \to 0$ and so the theorem is proved. $\square$

6.3 Predictable Stochastic Processes

Let $M(t)$ be a right continuous, square integrable martingale with left-hand limits. In order to define the stochastic integral $\int_a^b f(t)\,dM(t)$, we make a crucial observation about the integrand $f(t)$. Note that such a martingale $M(t)$, e.g., the compensated Poisson process $\widetilde{N}(t)$, may have jumps.

Consider a right continuous step function $g = -1_{[a,c)} + 1_{[c,b]}$ with $a < c < b$. It is well known that a bounded function $f(t)$ is Riemann–Stieltjes integrable with respect to g if and only if $f(t)$ is left continuous at c. This condition leads to the concept of predictability for stochastic processes.

From now on, we fix a filtration $\{\mathcal{F}_t; a \leq t \leq b\}$ and assume that it is right continuous, i.e., $\mathcal{F}_t = \mathcal{F}_{t+}$ for each t, where $\mathcal{F}_{t+}$ is defined by

$$\mathcal{F}_{t+} = \bigcap_{n=1}^{\infty} \mathcal{F}_{t+\frac{1}{n}}.$$

Moreover, all σ-fields $\mathcal{F}_t$ are assumed to be complete by Remark 2.5.2.

Let $\mathbb{L}$ denote the collection of all jointly measurable (which we have always assumed) stochastic processes $X(t, \omega)$ satisfying the following conditions:

(1) $X(t)$ is adapted with respect to $\{\mathcal{F}_t\}$.

(2) Almost all sample paths of $X(t)$ are left continuous.

Note that condition (2) is motivated by the above observation that $f(t)$ must be left continuous at c in order to be Riemann–Stieltjes integrable with respect to $g(t) = -1_{[a,c)}(t) + 1_{[c,b]}(t)$.

Let $\mathcal{P}$ denote the smallest σ-field of subsets of $[a,b] \times \Omega$ with respect to which all stochastic processes in $\mathbb{L}$ are measurable. The σ-field $\mathcal{P}$ is generated by subsets of $[a,b] \times \Omega$ of the form $(s,t] \times A$ with $A \in \mathcal{F}_s$ for $a \leq s < t \leq b$ and $\{a\} \times B$ with $B \in \mathcal{F}_a$.

Definition 6.3.1. *A stochastic process $X(t)$ is said to be* predictable *if the function $(t, \omega) \mapsto X(t, \omega)$ is $\mathcal{P}$-measurable on $[a, b] \times \Omega$.*

Obviously, all stochastic processes in $\mathbb{L}$ are predictable. Here are some more examples of predictable stochastic processes.

Example 6.3.2. Let $X(t)$ be an $\mathcal{F}_t$-adapted, right continuous stochastic process with left-hand limits. Let $X(t-)$ denote its left-hand limit. Then $X(t-)$ is predictable.

Example 6.3.3. Let $f(t, \omega)$ be a step stochastic process given by

$$f(t, \omega) = \sum_{i=1}^{n} \xi_{i-1}(\omega) 1_{(t_{i-1}, t_i]}(t),$$

where ξ_{i-1} is $\mathcal{F}_{t_{i-1}}$-measurable. Then $f(t, \omega)$ is predictable.

Example 6.3.4. Let $g(t, \omega)$ be a step stochastic process given by

$$g(t, \omega) = \sum_{i=1}^{n} \eta_{i-1}(\omega) 1_{[t_{i-1}, t_i)}(t),$$

where η_{i-1} is $\mathcal{F}_{t_{i-1}}$-measurable. Then $g(t, \omega)$ is not predictable. However, the stochastic process $g(t-, \omega) = \sum_{i=1}^{n} \eta_{i-1}(\omega) 1_{(t_{i-1}, t_i]}(t)$ is predictable.

Example 6.3.5. Let $N(t)$ be a Poisson process. Then $N(t-)$ is predictable. However, $N(t)$ is not predictable. The reason is related to the fact that if a sample path $N(t, \omega)$ has a jump in (a, b), then the Riemann–Stieltjes integral $\int_a^b N(t, \omega) \, dN(t, \omega)$ does not exist.

6.4 Doob–Meyer Decomposition Theorem

In this section we briefly explain the submartingale decomposition, which will be used in Section 6.5 for the extension of the Itô integral when the integrator $B(t)$ is replaced by a martingale.

Let $\{X_n\}_{n=0}^{\infty}$ be a submartingale with respect to a filtration $\{\mathcal{F}_n\}_{n=0}^{\infty}$. Define a sequence $\{A_n\}_{n=0}^{\infty}$ of random variables by $A_0 = 0$ and

$$A_n = \sum_{i=1}^{n} \left(E[X_i \,|\, \mathcal{F}_{i-1}] - X_{i-1} \right), \quad n \geq 1.$$

Note that A_n is $\mathcal{F}_{n-1}$-measurable. Moreover, since $\{X_n\}$ is a submartingale, we have $E[X_i \,|\, \mathcal{F}_{i-1}] - X_{i-1} \geq 0$ almost surely. Hence $\{A_n\}$ is an increasing sequence almost surely. Let $M_n = X_n - A_n$. It is easily checked that $\{M_n\}_{n=0}^{\infty}$ is a martingale. Thus we have obtained the *Doob decomposition*

$$X_n = M_n + A_n, \quad n \geq 0. \tag{6.4.1}$$

This decomposition of a submartingale as the sum of a martingale and an adapted increasing sequence is unique if we require that $A_0 = 0$ and that A_n be $\mathcal{F}_{n-1}$-measurable.

For the continuous-time case, the situation is much more complicated. The analogue of Equation (6.4.1) is called the *Doob–Meyer decomposition.* We briefly describe this decomposition and avoid the technical details.

All stochastic processes $X(t)$ in this section will be assumed to be right continuous with left-hand limits $X(t-)$.

Let $X(t)$, $a \leq t \leq b$, be a submartingale with respect to a right continuous filtration $\{\mathcal{F}_t; a \leq t \leq b\}$. If $X(t)$ satisfies certain conditions, then it can be decomposed uniquely as

$$X(t) = M(t) + C(t), \quad a \leq t \leq b, \tag{6.4.2}$$

where $M(t)$, $a \leq t \leq b$, is a martingale with respect to $\{\mathcal{F}_t; a \leq t \leq b\}$, $C(t)$ is predictable, right continuous, and increasing almost surely with $EC(t) < \infty$ for all $t \in [a,b]$, and $C(a) = 0$. Moreover, the decomposition is unique if $C(t)$ is required to satisfy certain additional conditions. For the precise statement of the additional conditions and the proof of the Doob–Meyer decomposition theorem, see the books [7] [8] [15] [30] [45] [61] [68].

Definition 6.4.1. *The stochastic process $C(t)$ in Equation (6.4.2) is called the* compensator *of the submartingale $X(t)$.*

Note that if $M(t)$ is a square integrable martingale, then $E(M(t)^2) < \infty$ and we can apply the conditional Jensen's inequality in Section 2.4 to see that $M(t)^2$ is a submartingale.

What we will need in Section 6.5 is a special case of the Doob–Meyer decomposition for a submartingale $M(t)^2$ stated in the next theorem.

Theorem 6.4.2. *Let $M(t)$, $a \leq t \leq b$, be a right continuous, square integrable martingale with left-hand limits. Then there is a unique decomposition*

$$M(t)^2 = L(t) + A(t), \quad a \leq t \leq b, \tag{6.4.3}$$

where $L(t)$ is a right continuous martingale with left-hand limits and $A(t)$ is a predictable, right continuous, and increasing process such that $A(a) = 0$ and $EA(t) < \infty$ for all $t \in [a,b]$.

Notation 6.4.3. For convenience we will use $\langle M \rangle_t$ to denote the compensator $A(t)$ of $M(t)^2$ in Equation (6.4.3).

On the other hand, for a martingale $M(t)$ as given in the theorem, the following limit can be shown to exist:

$$[M]_t \equiv \lim_{\|\Delta_n\| \to 0} \sum_{i=1}^{n} \left(M(t_i) - M(t_{i-1}) \right)^2, \quad \text{in probability}, \tag{6.4.4}$$

where $\Delta_n = \{t_0, t_1, \ldots, t_{n-1}, t_n\}$ is a partition of the interval $[a, t]$ and $\|\Delta_n\| = \max_{1 \leq i \leq n} (t_i - t_{i-1})$. The limit $[M]_t$ is called the *quadratic variation process* of $M(t)$. It can be proved that $M(t)^2 - [M]_t$ is a martingale.

In general, the quadratic variation process $[M]_t$ of $M(t)$ is not equal to the compensator $\langle M \rangle_t$ of $M(t)^2$. By the above theorem, $M(t)^2 - \langle M \rangle_t$ is a martingale. Hence $[M]_t - \langle M \rangle_t$ is also a martingale. Therefore, $\langle M \rangle_t$ is the unique predictable compensator of the stochastic process $[M]_t$.

Example 6.4.4. Consider a Brownian motion $B(t)$. By Theorem 4.1.2 with $a = 0$ and $b = t$, the quadratic variation process of $B(t)$ is given by $[B]_t = t$. On the other hand, recall that $B(t)^2 - t$ is a martingale. Hence we get the Doob–Meyer decomposition of $B(t)^2$:

$$B(t)^2 = \big(B(t)^2 - t\big) + t.$$

Thus the compensator of $B(t)^2$ is given by $\langle B \rangle_t = t$. Note that $[B]_t = \langle B \rangle_t$.

Example 6.4.5. Consider the martingale $M(t) = B(t)^2 - t$. Let $\{\mathcal{F}_t\}$ be the filtration specified in the beginning of Section 4.3. It can be easily checked that for any $s < t$,

$$E\big(B(t)^4 \,|\, \mathcal{F}_s\big) = B(s)^4 + 6(t - s)B(s)^2 + 3(t - s)^2,$$
$$E\big(B(t)^2 \,|\, \mathcal{F}_s\big) = B(s)^2 + (t - s). \tag{6.4.5}$$

Since $M(t)^2 = B(t)^4 - 2tB(t)^2 + t^2$, we obtain

$$E\big(M(t)^2 \,|\, \mathcal{F}_s\big) = B(s)^4 + (4t - 6s)B(s)^2 + 2(t - s)^2 + s^2. \tag{6.4.6}$$

On the other hand, let $M(t)^2 = L(t) + A(t)$ be the Doob–Meyer decomposition of $M(t)^2$. Then we have

$$
\begin{aligned}
E\big(M(t)^2 \,|\, \mathcal{F}_s\big) &= L(s) + E\big(A(t) \,|\, \mathcal{F}_s\big) \\
&= M(s)^2 - A(s) + E\big(A(t) \,|\, \mathcal{F}_s\big) \\
&= B(s)^4 - 2sB(s)^2 + s^2 + E\big(A(t) - A(s) \,|\, \mathcal{F}_s\big). \tag{6.4.7}
\end{aligned}
$$

It follows from Equations (6.4.6) and (6.4.7) that

$$E\big(A(t) - A(s) \,|\, \mathcal{F}_s\big) = 4(t - s)B(s)^2 + 2(t - s)^2. \tag{6.4.8}$$

Now we see from Equation (6.4.5) that

$$
\begin{aligned}
E\left(\int_s^t B(u)^2 \, du \,\middle|\, \mathcal{F}_s \right) &= \int_s^t E\big(B(u)^2 \,|\, \mathcal{F}_s\big) \, du \\
&= \int_s^t \big(B(s)^2 + u - s\big) \, du \\
&= (t - s)B(s)^2 + \frac{1}{2}(t - s)^2. \tag{6.4.9}
\end{aligned}
$$

By comparing Equations (6.4.8) and (6.4.9), we clearly see that

$$A(t) = 4 \int_0^t B(u)^2 \, du. \tag{6.4.10}$$

Therefore, the compensator of $M(t)^2 = \left(B(t)^2 - t\right)^2$ is given by

$$\langle M \rangle_t = 4 \int_0^t B(u)^2 \, du. \tag{6.4.11}$$

This implies that the stochastic process

$$\left(B(t)^2 - t\right)^2 - 4 \int_0^t B(u)^2 \, du$$

is a martingale. On the other hand, we can easily check by direct computation that the quadratic variation process of $M(t) = B(t)^2 - t$ is also given by the right-hand side of Equation (6.4.10). Hence $[M]_t = \langle M \rangle_t$.

Example 6.4.6. Consider a Poisson process $N(t)$. The compensated Poisson process $\widetilde{N}(t) = N(t) - \lambda t$ is a martingale. By Theorem 6.2.3 with $a = 0$ and $b = t$, the quadratic variation process of $\widetilde{N}(t)$ is given by $[\widetilde{N}]_t = \lambda t + \widetilde{N}(t)$. On the other hand, we can find the compensator $\langle \widetilde{N} \rangle_t$ of $\widetilde{N}(t)^2$ as follows. Use the same arguments as those in the case of Brownian motion to derive the following equality for any $0 \le s < t$:

$$E(N(t)^2 \mid \mathcal{F}_s) = \lambda^2(t-s)^2 + \lambda(t-s) + 2\lambda(t-s)N(s) + N(s)^2. \tag{6.4.12}$$

It follows from Equations (6.2.1) and (6.4.12) that

$$E(\widetilde{N}(t)^2 - \lambda t \mid \mathcal{F}_s) = \widetilde{N}(s)^2 - \lambda s.$$

Thus $\widetilde{N}(t)^2 - \lambda t$ is a martingale and we have the Doob–Meyer decomposition for $\widetilde{N}(t)^2$:

$$\widetilde{N}(t)^2 = \left(\widetilde{N}(t)^2 - \lambda t\right) + \lambda t.$$

Hence the compensator of $\widetilde{N}(t)^2$ is given by $\langle \widetilde{N} \rangle_t = \lambda t$. Clearly, $[\widetilde{N}]_t \ne \langle \widetilde{N} \rangle_t$. Note that $[\widetilde{N}]_t = \lambda t + \widetilde{N}(t)$ contains jumps and observe that when $\lambda = 1$, the compensator of $\widetilde{N}(t)^2$ is the same as that of $B(t)^2$.

Example 6.4.7. Consider the martingale $M(t) = B(t) + \widetilde{N}(t)$. Assume that $B(t)$ and $\widetilde{N}(t)$ are independent. Then for any $s \le t$,

$$E[B(t)\widetilde{N}(t) \mid \mathcal{F}_s] = E[\{(B(t) - B(s)) + B(s)\}\{(\widetilde{N}(t) - \widetilde{N}(s)) + \widetilde{N}(s)\} \mid \mathcal{F}_s]$$
$$= B(s)\widetilde{N}(s). \tag{6.4.13}$$

Therefore, $B(t)\widetilde{N}(t)$ is a martingale. Using this fact, we see that

$$(B(t) + \widetilde{N}(t))^2 - (t + \lambda t) = (B(t)^2 - t) + 2B(t)\widetilde{N}(t) + (\widetilde{N}(t)^2 - \lambda t)$$

is a martingale. This implies that $\langle M \rangle_t = (1 + \lambda)t$. Moreover, we can use the independence of $B(t)$ and $\widetilde{N}(t)$ to check that

$$[M]_t = [B]_t + [\widetilde{N}]_t = (1 + \lambda)t + \widetilde{N}(t).$$

6.5 Martingales as Integrators

Let $M(t)$, $a \leq t \leq b$, be a fixed martingale with respect to the filtration $\{\mathcal{F}_t; a \leq t \leq b\}$ specified in Section 6.3. We will assume that $M(t)$ is right continuous with left-hand limits and $E(M(t)^2) < \infty$ for all t. By Theorem 6.4.2, the compensator $A(t)$ of $M(t)^2$, denoted by $\langle M \rangle_t$, is a predictable, right continuous, and increasing process with $EA(t) < \infty$, and $A(a) = 0$.

In this section we will extend the stochastic integral $\int_a^b f(t)\, dB(t)$ defined in Chapter 4 to $\int_a^b f(t)\, dM(t)$ for the above martingale $M(t)$.

First we prove a lemma on the conditional expectation of $M(t) - M(s)^2$ with respect to $\mathcal{F}_s$.

Lemma 6.5.1. *For any $s < t$, we have*

$$E\big[\big(M(t) - M(s)\big)^2 \,\big|\, \mathcal{F}_s\big] = E\big[\langle M \rangle_t - \langle M \rangle_s \,\big|\, \mathcal{F}_s\big].$$

Proof. Let $M(t)^2 = L(t) + A(t)$ be the Doob–Meyer decomposition of $M(t)^2$ in Equation (6.4.3). Then

$$
\begin{aligned}
E\big[\big(M(t) - M(s)\big)^2 \,\big|\, \mathcal{F}_s\big] &= E\big[M(t)^2 - 2M(t)M(s) + M(s)^2 \,\big|\, \mathcal{F}_s\big] \\
&= E\big[M(t)^2 \,\big|\, \mathcal{F}_s\big] - M(s)^2 \\
&= E\big[L(t) + A(t) \,\big|\, \mathcal{F}_s\big] - L(s) - A(s) \\
&= E\big[A(t) - A(s) \,\big|\, \mathcal{F}_s\big].
\end{aligned}
$$

This proves the lemma since $A(t) = \langle M \rangle_t$.   □

Observe that the compensator $\langle M \rangle_t$ of $M(t)^2$, instead of the quadratic variation process $[M]_t$ of $M(t)$, is used in Lemma 6.5.1. This lemma will play a critical role for the stochastic integral with respect to $M(t)$.

Next we need to specify the integrand $f(t)$. As pointed out in Section 6.3, we need to assume that $f(t)$ is predictable in order to take care of the jumps of the martingale $M(t)$.

Notation 6.5.2. For convenience we will use $L^2_{\text{pred}}([a, b]_{\langle M \rangle} \times \Omega)$ to denote the space of all predictable stochastic processes $f(t, \omega)$, $a \leq t \leq b$, $\omega \in \Omega$, satisfying the condition $E \int_a^b |f(t)|^2 \, d\langle M \rangle_t < \infty$.

Note that in case $M(t)$ is a Brownian motion $B(t)$, then we have $\langle M \rangle_t = t$ and the predictability is just the same as being adapted. Hence the space $L^2_{\text{pred}}([a, b]_{\langle M \rangle} \times \Omega)$ is just the space $L^2_{\text{ad}}([a, b] \times \Omega)$ defined in Section 4.3.

Now we will follow the same procedure as in Chapter 4, but with suitable modification, to define the stochastic integral

$$\int_a^b f(t)\, dM(t)$$

for $f \in L^2_{\text{pred}}([a, b]_{\langle M \rangle} \times \Omega)$. First consider a step stochastic process f in $L^2_{\text{pred}}([a, b]_{\langle M \rangle} \times \Omega)$ given by

$$f(t, \omega) = \sum_{i=1}^n \xi_{i-1}(\omega) 1_{(t_{i-1}, t_i]}(t).$$

In this case, define $I(f)$ by

$$I(f) = \sum_{i=1}^n \xi_{i-1}\big(M(t_i) - M(t_{i-1})\big). \tag{6.5.1}$$

Obviously, we have $I(\alpha f + \beta g) = \alpha I(f) + \beta I(g)$ for any $\alpha, \beta \in \mathbb{R}$ and step stochastic processes $f, g \in L^2_{\text{pred}}([a, b]_{\langle M \rangle} \times \Omega)$. Moreover, we have the following lemma.

Lemma 6.5.3. *Let $I(f)$ be defined by Equation* (6.5.1). *Then*

$$E\big(|I(f)|^2\big) = E \int_a^b |f(t)|^2 \, d\langle M \rangle_t.$$

Proof. Note that for $i \neq j$, say $i < j$, we have

$$E\big\{\xi_{i-1}\xi_{j-1}\big(M(t_i) - M(t_{i-1})\big)\big(M(t_j) - M(t_{j-1})\big)\big\}$$
$$= E\big\{E\big[\xi_{i-1}\xi_{j-1}\big(M(t_i) - M(t_{i-1})\big)\big(M(t_j) - M(t_{j-1})\big) \,|\, \mathcal{F}_{t_{j-1}}\big]\big\}$$
$$= E\big\{\xi_{i-1}\xi_{j-1}\big(M(t_i) - M(t_{i-1})\big)E\big[M(t_j) - M(t_{j-1}) \,|\, \mathcal{F}_{t_{j-1}}\big]\big\}$$
$$= 0.$$

Therefore,

$$E\big(|I(f)|^2\big) = \sum_{i,j=1}^n E\big\{\xi_{i-1}\xi_{j-1}\big(M(t_i) - M(t_{i-1})\big)\big(M(t_j) - M(t_{j-1})\big)\big\}$$
$$= \sum_{i=1}^n E\big\{\xi_{i-1}^2\big(M(t_i) - M(t_{i-1})\big)^2\big\}$$
$$= \sum_{i=1}^n E\big\{\xi_{i-1}^2 E\big[\big(M(t_i) - M(t_{i-1})\big)^2 \,|\, \mathcal{F}_{t_{i-1}}\big]\big\}.$$

Then apply Lemma 6.5.1 to get

$$
\begin{aligned}
E\big(|I(f)|^2\big) &= \sum_{i=1}^{n} E\big\{\xi_{i-1}^2 E\big[\langle M\rangle_{t_i} - \langle M\rangle_{t_{i-1}} \mid \mathcal{F}_{t_{i-1}}\big]\big\} \\
&= \sum_{i=1}^{n} E\big\{\xi_{i-1}^2\big(\langle M\rangle_{t_i} - \langle M\rangle_{t_{i-1}}\big)\big\} \\
&= E\sum_{i=1}^{n} \xi_{i-1}^2\big(\langle M\rangle_{t_i} - \langle M\rangle_{t_{i-1}}\big) \\
&= E\int_a^b |f(t)|^2\, d\langle M\rangle_t.
\end{aligned}
$$

This proves the equality in the lemma.   □

Next we extend $I(f)$ to $f \in L^2_{\text{pred}}([a,b]_{\langle M\rangle} \times \Omega)$. Recall that the σ-field $\mathcal{P}$ is generated by subsets of $[a,b] \times \Omega$ of the form $(s,t] \times A$ with $A \in \mathcal{F}_s$ for $a \le s < t \le b$ and $\{a\} \times B$ with $B \in \mathcal{F}_a$. Hence if $f \in L^2_{\text{pred}}([a,b]_{\langle M\rangle} \times \Omega)$, then there exists a sequence $\{f_n\}$ of step stochastic processes in $L^2_{\text{pred}}([a,b]_{\langle M\rangle} \times \Omega)$ such that

$$
\lim_{n\to\infty} E\int_a^b |f(t) - f_n(t)|^2\, d\langle M\rangle_t = 0.
$$

Thus by Lemma 6.5.3 the sequence $\{I(f_n)\}$ is Cauchy in $L^2(\Omega)$. Define

$$
I(f) = \lim_{n\to\infty} I(f_n), \quad \text{in } L^2(\Omega). \tag{6.5.2}
$$

We can check that the limit is independent of the choice of the sequence $\{f_n\}$. Hence $I(f)$, denoted by $\int_a^b f(t)\, dM(t)$, is well-defined. It is called the *stochastic integral* of f with respect to the martingale $M(t)$. For any $\alpha, \beta \in \mathbb{R}$ and $f, g \in L^2_{\text{pred}}([a,b]_{\langle M\rangle} \times \Omega)$, we have

$$
\int_a^b \big(\alpha f(t) + \beta g(t)\big)\, dM(t) = \alpha \int_a^b f(t)\, dM(t) + \beta \int_a^b g(t)\, dM(t).
$$

Example 6.5.4. Let $B(t)$ be a Brownian motion and let $\{\mathcal{F}_t; a \le t \le b\}$ be a filtration specified in Section 4.3. Take $g \in L^2_{\text{ad}}([a,b] \times \Omega)$ and consider the martingale

$$
M(t) = \int_a^t g(s)\, dB(s), \quad a \le t \le b.
$$

The compensator of $M(t)^2$ is given by

$$
\langle M\rangle_t = \int_a^t |g(s)|^2\, ds.
$$

Therefore, $f \in L^2_{\text{pred}}([a,b]_{\langle M\rangle} \times \Omega)$ if and only if it satisfies the condition

$$E \int_a^b |f(t)|^2 |g(t)|^2 \, dt < \infty,$$

which is valid if and only if $fg \in L^2_{ad}([a, b] \times \Omega)$. Moreover, we have

$$\int_a^b f(t) \, dM(t) = \int_a^b f(t) g(t) \, dB(t).$$

Example 6.5.5. Let $\Delta_n = \{t_0, t_1, \ldots, t_{n-1}, t_n\}$ be a partition of a finite interval $[a, b]$. Then

$$\sum_{i=1}^n \left(\widetilde{N}(t_i) - \widetilde{N}(t_{i-1}) \right)^2$$

$$= \sum_{i=1}^n \left(\left[\widetilde{N}(t_i)^2 - \widetilde{N}(t_{i-1})^2 \right] - 2\widetilde{N}(t_{i-1}) \left(\widetilde{N}(t_i) - \widetilde{N}(t_{i-1}) \right) \right)$$

$$= \widetilde{N}(b)^2 - \widetilde{N}(a)^2 - 2 \sum_{i=1}^n \widetilde{N}(t_{i-1}) \left(\widetilde{N}(t_i) - \widetilde{N}(t_{i-1}) \right). \tag{6.5.3}$$

From Equation (6.5.2) we see that

$$\int_a^b \widetilde{N}(t-) \, d\widetilde{N}(t) = \lim_{\|\Delta_n\| \to 0} \sum_{i=1}^n \widetilde{N}(t_{i-1}) \left(\widetilde{N}(t_i) - \widetilde{N}(t_{i-1}) \right), \quad \text{in } L^2(\Omega).$$

Therefore, by Equation (6.5.3),

$$\int_a^b \widetilde{N}(t-) \, d\widetilde{N}(t) = \frac{1}{2} \left(\widetilde{N}(b)^2 - \widetilde{N}(a)^2 - \lim_{\|\Delta_n\| \to 0} \sum_{i=1}^n \left(\widetilde{N}(t_i) - \widetilde{N}(t_{i-1}) \right)^2 \right).$$

Now use Equation (6.2.2) in Theorem 6.2.3 to conclude that

$$\int_a^b \widetilde{N}(t-) \, d\widetilde{N}(t) = \frac{1}{2} \left(\widetilde{N}(b)^2 - \widetilde{N}(a)^2 - \lambda(b - a) - \widetilde{N}(b) + \widetilde{N}(a) \right). \tag{6.5.4}$$

In particular, we have

$$\int_0^t \widetilde{N}(s-) \, d\widetilde{N}(s) = \frac{1}{2} \left(\widetilde{N}(t)^2 - \lambda t - \widetilde{N}(t) \right).$$

Example 6.5.6. In this example we compute the stochastic integral of $N(t-)$ with respect to $\widetilde{N}(t)$. Since $N(t) = \widetilde{N}(t) + \lambda t$, we have

$$\int_a^b N(t-) \, d\widetilde{N}(t) = \int_a^b \widetilde{N}(t-) \, d\widetilde{N}(t) + \lambda \int_a^b t \, d\widetilde{N}(t). \tag{6.5.5}$$

The first term in the right-hand side is given by Equation (6.5.4). To evaluate the second term, let $\{t_0, t_1, \ldots, t_{n-1}, t_n\}$ be a partition of $[a, b]$. Then

$$\sum_{i=1}^{n} t_{i-1}\big(\widetilde{N}(t_i) - \widetilde{N}(t_{i-1})\big) = b\widetilde{N}(b) - a\widetilde{N}(a) - \sum_{i=1}^{n} \widetilde{N}(t_i)(t_i - t_{i-1}).$$

Take the limit to get

$$\int_a^b t \, d\widetilde{N}(t) = b\widetilde{N}(b) - a\widetilde{N}(a) - \int_a^b \widetilde{N}(t) \, dt. \tag{6.5.6}$$

Then put Equations (6.5.4) and (6.5.6) into Equation (6.5.5) to see that

$$\int_a^b N(t-) \, d\widetilde{N}(t) = \frac{1}{2}\Big(\widetilde{N}(b)^2 - \widetilde{N}(a)^2 - \lambda(b-a) - \widetilde{N}(b) + \widetilde{N}(a)\Big)$$
$$+ \lambda b\widetilde{N}(b) - \lambda a\widetilde{N}(a) - \lambda \int_a^b \widetilde{N}(t) \, dt.$$

We can also use the fact that $\widetilde{N}(t) = N(t) - \lambda t$ to rewrite the right-hand side of this equality in terms of $N(t)$,

$$\int_a^b N(t-) \, d\widetilde{N}(t) = \frac{1}{2}\Big(N(b)^2 - N(a)^2 - N(b) + N(a)\Big) - \lambda \int_a^b N(t) \, dt. \tag{6.5.7}$$

In particular, we have

$$\int_0^t N(s-) \, d\widetilde{N}(s) = \frac{1}{2}\Big(\widetilde{N}(t)^2 - \lambda t - \widetilde{N}(t)\Big) + \lambda t\widetilde{N}(t) - \lambda \int_0^t \widetilde{N}(s) \, ds$$
$$= \frac{1}{2}\Big(N(t)^2 - N(t)\Big) - \lambda \int_0^t N(s) \, ds.$$

Example 6.5.7. Let $M(t)$ be a continuous and square integrable martingale. Then we have the equality

$$\int_0^t M(s) \, dM(s) = \frac{1}{2}\Big(M(t)^2 - M(0)^2 - \langle M \rangle_t\Big),$$

which can be derived in the same way as the Brownian motion case by using the fact that $[M]_t = \langle M \rangle_t$ for such a martingale.

We now state two theorems on stochastic processes defined by stochastic integrals with respect to a martingale integrator. For the proofs, just modify the arguments in Chapter 4 for the case that $M(t)$ is a Brownian motion.

Theorem 6.5.8. *Let* $f \in L^2_{\mathrm{pred}}([a, b]_{\langle M \rangle} \times \Omega)$. *Then the stochastic process*

$$X_t = \int_a^t f(s) \, dM(s), \quad a \le t \le b,$$

is a martingale and the following equality holds:

$$E\big(|X(t)|^2\big) = E \int_a^t |f(s)|^2 \, d\langle M \rangle_s.$$

Theorem 6.5.9. *Let $f \in L^2_{\text{pred}}([a, b]_{\langle M \rangle} \times \Omega)$. Then the stochastic process*

$$X_t = \int_a^t f(s) \, dM(s), \quad a \leq t \leq b,$$

is right continuous with left-hand limits, i.e., almost all of its sample paths are right continuous functions with left-hand limits on $[a, b]$. Furthermore, the compensator of X_t^2 is given by

$$\langle X \rangle_t = \int_a^t |f(s)|^2 \, d\langle M \rangle_s.$$

6.6 Extension for Integrands

Recall that in Section 4.3 we defined the Itô integral $\int_a^b f(t) \, dB(t)$ with respect to a Brownian motion $B(t)$ for $f \in L^2_{\text{ad}}([a, b] \times \Omega)$. In Section 5.3 we extended this integral to $f \in \mathcal{L}_{\text{ad}}(\Omega, L^2[a, b])$.

Suppose $M(t)$ is a right continuous, square integrable martingale with left-hand limits. In the previous section we defined the stochastic integral $\int_a^b f(t) \, dM(t)$ for $f \in L^2_{\text{pred}}([a, b]_{\langle M \rangle} \times \Omega)$. In this section we will briefly explain the extension of this integral to a larger class of integrands. For details, see the books by Kallianpur [45] and by Protter [68].

Notation 6.6.1. We will use $\mathcal{L}_{\text{pred}}(\Omega, L^2[a, b]_{\langle M \rangle})$ to denote the space of all predictable stochastic processes $f(t, \omega)$, $a \leq t \leq b$, $\omega \in \Omega$, satisfying the condition $\int_a^b |f(t)|^2 \, d\langle M \rangle_t < \infty$ almost surely.

Note that when $M(t) = B(t)$, the space $\mathcal{L}_{\text{pred}}(\Omega, L^2[a, b]_{\langle M \rangle})$ is just the space $\mathcal{L}_{\text{ad}}(\Omega, L^2[a, b])$ in Section 5.3.

Let $f \in \mathcal{L}_{\text{pred}}(\Omega, L^2[a, b]_{\langle M \rangle})$. Define a sequence $\{f_n(t)\}_{n=1}^\infty$ of stochastic processes in a similar way as in Section 5.3 by

$$f_n(t, \omega) = \begin{cases} f(t, \omega), & \text{if } \int_a^t |f(s, \omega)|^2 \, d\langle M \rangle_s(\omega) \leq n; \\ 0, & \text{otherwise.} \end{cases}$$

Lemma 6.6.2. *Let $f \in \mathcal{L}_{\text{pred}}(\Omega, L^2[a, b]_{\langle M \rangle})$. Then $f_n \in L^2_{\text{pred}}([a, b]_{\langle M \rangle} \times \Omega)$ for each $n \geq 1$ and the sequence $\int_a^b f_n(t) \, dM(t)$ converges in probability as $n \to \infty$.*

In view of the above lemma, we can define the stochastic integral of f with respect to $M(t)$ by

$$\int_a^b f(t) \, dM(t) = \lim_{n \to \infty} \int_a^b f_n(t) \, dM(t), \quad \text{in probability.} \tag{6.6.1}$$

The stochastic integral in Equation (6.6.1) is indeed well-defined, namely, it is independent of the choice of the sequence $\{f_n\}$.

We state several theorems concerning the stochastic integral $\int_a^b f(t)\, dM(t)$ for $f \in \mathcal{L}_{\mathrm{pred}}(\Omega, L^2[a, b]_{\langle M \rangle})$, the associated stochastic process $\int_a^t f(s)\, dM(s)$, and a stochastic integral with respect to the compensator $\langle M \rangle_t$. For the proofs, see the books [45] and [68].

Theorem 6.6.3. *Let $f \in \mathcal{L}_{\mathrm{pred}}(\Omega, L^2[a, b]_{\langle M \rangle})$ and $g_n \in L^2_{\mathrm{pred}}([a, b]_{\langle M \rangle} \times \Omega)$ for $n \geq 1$. Suppose*

$$\lim_{n \to \infty} \int_a^b |f(t) - g_n(t)|^2 \, d\langle M \rangle_t = 0, \quad \text{almost surely.}$$

Then we have

$$\int_a^b f(t)\, dM(t) = \lim_{n \to \infty} \int_a^b g_n(t)\, dM(t), \quad \text{in probability.}$$

Theorem 6.6.4. *Let $f \in \mathcal{L}_{\mathrm{pred}}(\Omega, L^2[a, b]_{\langle M \rangle})$. Then the stochastic process*

$$X(t) = \int_a^t f(s)\, dM(s), \quad a \leq t \leq b,$$

is a local martingale with respect to the filtration specified in Section 6.3 and has a right continuous realization with left-hand limits. Moreover, if $M(t)$ is continuous, then $X(t)$ has a continuous realization.

Theorem 6.6.5. *Let $M(t)$ be a continuous and square integrable martingale. If $f \in \mathcal{L}_{\mathrm{pred}}(\Omega, L^2[a, b]_{\langle M \rangle})$ is a continuous stochastic process, then*

$$\int_a^b f(t)\, dM(t) = \lim_{\|\Delta_n\| \to 0} \sum_{i=1}^n f(t_{i-1})\big(M(t_i) - M(t_{i-1})\big), \quad \text{in probability,}$$

where $\Delta_n = \{t_0, t_1, \ldots, t_{n-1}, t_n\}$ is a partition of the interval $[a, b]$ and $\|\Delta_n\| = \max_{1 \leq i \leq n}(t_i - t_{i-1})$.

Theorem 6.6.6. *Let $M(t)$ be a continuous and square integrable martingale. Then for any bounded, continuous, and adapted stochastic process $Y(t)$,*

$$\int_a^b Y(t)\, d\langle M \rangle_t = \lim_{\|\Delta_n\| \to 0} \sum_{i=1}^n Y(t_{i-1})\big(M(t_i) - M(t_{i-1})\big)^2, \quad \text{in probability,}$$

where $\Delta_n = \{t_0, t_1, \ldots, t_{n-1}, t_n\}$ is a partition of the interval $[a, b]$ and $\|\Delta_n\| = \max_{1 \leq i \leq n}(t_i - t_{i-1})$. Here the left-hand side is the Riemann–Stieltjes integral $\int_a^b Y(t, \omega)\, d\langle M \rangle_t(\omega)$ defined for each ω in some Ω_0 with $P(\Omega_0) = 1$.

Exercises

1. Let $F(t)$ and $G(t)$ be right continuous nondecreasing functions on $[a, b]$. Prove that

$$
\int_a^b F(t-)\, dG(t) + \int_a^b G(t-)\, dF(t)
$$

$$
= F(b)G(b) - F(a)G(a) - \sum_{a < t \leq b} \Delta F(t)\Delta G(t), \qquad (6.6.2)
$$

where the integrals are Riemann–Stieltjes integrals and $\Delta\varphi(t)$ is the jump of φ at t, i.e., $\Delta\varphi(t) = \varphi(t) - \varphi(t-)$.

2. Put $F(t) = G(t) = \tilde{N}(t, \omega)$, the compensated Poisson process, in Equation (6.6.2) and show that

$$
\int_a^b \tilde{N}(t-)\, d\tilde{N}(t) = \frac{1}{2}\left(\tilde{N}(b)^2 - \tilde{N}(a)^2 - \lambda(b - a) - \tilde{N}(b) + \tilde{N}(a) \right),
$$

where the left-hand side is a Riemann–Stieltjes integral for each ω. Note that this integral coincides with the stochastic integral in Equation (6.5.4).

3. Put $F(t) = G(t) = N(t, \omega)$, the Poisson process, in Equation (6.6.2) and show that

$$
\int_a^b N(t-)\, dN(t) = \frac{1}{2}\left(N(b)^2 - N(a)^2 - N(b) + N(a) \right), \qquad (6.6.3)
$$

where the left-hand side is a Riemann–Stieltjes integral for each ω. Note that since $N(t)$ is not a martingale, the integral $\int_a^b N(t-)\, dN(t)$ is not a stochastic integral as defined in Section 6.5.

4. Use Equation (6.6.3) to show that

$$
\int_a^b N(t-)\, d\tilde{N}(t) = \frac{1}{2}\left(N(b)^2 - N(a)^2 - N(b) + N(a) \right) - \lambda \int_a^b N(t)\, dt,
$$

where the left-hand side is a Riemann–Stieltjes integral for each ω. This integral coincides with the stochastic integral in Equation (6.5.7).

5. Find the quadratic variation process $[N]_t$ of a Poisson process $N(t)$ with parameter $\lambda > 0$.

6. Suppose $\lambda \in \mathbb{R}$. Prove that $M(t) = e^{\lambda B(t) - \lambda^2 t/2}$ is a martingale and the compensator of $M(t)^2$ is given by

$$
\langle M \rangle_t = \lambda^2 \int_0^t e^{2\lambda B(u) - \lambda^2 u}\, du.
$$

7. Let $f \in L^2[a, b]$ and $M(t) = \int_a^t f(s)\, dB(s)$. Find the quadratic variation process $[M]_t$ of $M(t)$ and the compensator $\langle M \rangle_t$ of $M(t)^2$.

8. Let $f(t)$ be adapted and $\int_a^b E|f(t)|^2 \, dt < \infty$. Show that

$$M(t) = \left(\int_a^t f(s) \, dB(s) \right)^2 - \int_a^t f(s)^2 \, ds$$

is a martingale and find the compensator of $M(t)^2$.

9. Let $s \le t$. Show that

$$E\{B(t)^3 \mid \mathcal{F}_s\} = 3(t-s)B(s) + B(s)^3. \tag{6.6.4}$$

10. Use Equation (6.6.4) to derive a martingale $X_t = B(t)^3 - 3tB(t)$. Find the quadratic variation process $[X]_t$ and the compensator $\langle X \rangle_t$.

11. Let $s \le t$. Show that

$$E\{B(t)^4 \mid \mathcal{F}_s\} = 3(t-s)^2 + 6(t-s)B(s)^2 + B(s)^4. \tag{6.6.5}$$

12. Let $s \le t$. Show that

$$E\left\{ \int_0^t B(u)^2 \, du \,\middle|\, \mathcal{F}_s \right\} = \frac{1}{2}(t-s)^2 + (t-s)B(s)^2 + \int_0^s B(u)^2 \, du. \tag{6.6.6}$$

13. Use Equations (6.6.5) and (6.6.6) to show that

$$X_t = B(t)^4 - 6 \int_0^t B(u)^2 \, du$$

is a martingale and then derive the quadratic variation process $[X]_t$ and the compensator $\langle X \rangle_t$.

14. Let $s \le t$. Show that

$$E\{B(t)e^{B(t)} \mid \mathcal{F}_s\} = \left(t - s + B(s) \right) e^{B(s) + \frac{1}{2}(t-s)}. \tag{6.6.7}$$

15. Use Equation (6.6.7) to derive a martingale $X_t = (B(t) - t)\, e^{B(t) - \frac{1}{2}t}$. Find the quadratic variation process $[X]_t$ and the compensator $\langle X \rangle_t$.

16. Suppose a Brownian motion $B(t)$ and a compensated Poisson process $\widetilde{N}(t)$ are independent. By Example 6.4.7, $M(t) = B(t)\widetilde{N}(t)$ is a martingale. Find the quadratic variation process $[M]_t$ and the compensator $\langle M \rangle_t$.

17. Find the variances of $\int_a^b N(t-)\, d\widetilde{N}(t)$ and $\int_a^b \widetilde{N}(t-)^2 \, d\widetilde{N}(t)$.

18. Let $B(t)$ and $\widetilde{N}(t)$ be independent. Find the variances of $\int_a^b B(t)\, d\widetilde{N}(t)$ and $\int_a^b e^{B(t)} \, d\widetilde{N}(t)$.

7

The Itô Formula

The chain rule in the Leibniz–Newton calculus is the formula $\frac{d}{dt}f(g(t)) = f'(g(t))g'(t)$ for differentiable functions f and g. It can be rewritten in the integral form as $f(g(t)) - f(g(a)) = \int_a^t f'(g(s))g'(s)\,ds$. On the other hand, the chain rule in the Itô calculus, for the simplest case, states that

$$f(B(t)) = f(B(a)) + \int_a^t f'(B(s))\,dB(s) + \frac{1}{2}\int_a^t f''(B(s))\,ds$$

for a Brownian motion $B(t)$ and a twice continuously differentiable function f. This formula is often written in a symbolic differential form as

$$df(B(t)) = f'(B(t))\,dB(t) + \frac{1}{2}f''(B(t))\,dt.$$

The appearance of the term $\frac{1}{2}f''(B(t))\,dt$ is a consequence of the nonzero quadratic variation of the Brownian motion $B(t)$. In this chapter we will study the celebrated Itô formula and its generalizations to multidimensional spaces and martingales.

7.1 Itô's Formula in the Simplest Form

The Leibniz–Newton calculus deals with deterministic functions. A basic rule for differentiation is the chain rule, which gives the derivative of a composite function $f(g(t))$. It states that if f and g are differentiable, then $f(g(t))$ is also differentiable and has derivative

$$\frac{d}{dt}f(g(t)) = f'(g(t))g'(t).$$

In terms of the fundamental theorem of calculus this equality says that

$$f(g(t)) - f(g(a)) = \int_a^t f'(g(s))g'(s)\,ds. \tag{7.1.1}$$

The Itô calculus deals with random functions, i.e., stochastic processes. Let $B(t)$ be a Brownian motion and f a differentiable function. Consider the composite function $f(B(t))$. Since almost all sample paths of $B(t)$ are nowhere differentiable, the equality $\frac{d}{dt}f(B(t)) = f'(B(t))B'(t)$ obviously has no meaning. However, when we rewrite $B'(s)\,ds$ as an integrator $dB(s)$ in the Itô integral, Equation (7.1.1) leads to the following question.

Question 7.1.1. Does the equality $f(B(t)) - f(B(a)) = \int_a^t f'(B(s))\,dB(s)$ hold for any differentiable function f?

The integral $\int_a^t f'(B(s))\,dB(s)$ is of course an Itô integral defined as in Chapter 4 or 5 depending on whether $f'(B(t))$ belongs to $L^2_{ad}([a,b] \times \Omega)$ or $\mathcal{L}_{ad}(\Omega, L^2[a,b])$, respectively. For the simple function $f(x) = x^2$, the equality in Question 7.1.1 would give

$$B(t)^2 - B(a)^2 = 2\int_a^t B(s)\,dB(s),$$

which is contradictory to the equality from Example 4.4.1 with $b = t$,

$$B(t)^2 - B(a)^2 - (t-a) = 2\int_a^t B(s)\,dB(s).$$

Therefore, the answer to Question 7.1.1 is negative. But then is there a formula for the composite function $f(B(t))$ to serve as the chain rule in an integral form for the Itô calculus? To figure out the answer, consider a partition $\Delta_n = \{t_0, t_1, \ldots, t_{n-1}, t_n\}$ of $[a, t]$. Then we have

$$f(B(t)) - f(B(a)) = \sum_{i=1}^n \Big(f(B(t_i)) - f(B(t_{i-1})) \Big). \qquad (7.1.2)$$

Let f be a C^2-function, namely, it is twice differentiable and the second derivative f'' is continuous. Then we have the Taylor expansion

$$f(x) - f(x_0) = f'(x_0)(x - x_0) + \frac{1}{2}f''\big(x_0 + \lambda(x - x_0)\big)(x - x_0)^2,$$

where $0 < \lambda < 1$. Therefore, by Equation (7.1.2)

$$f(B(t)) - f(B(a)) = \sum_{i=1}^n f'(B_{t_{i-1}})(B_{t_i} - B_{t_{i-1}})$$

$$+ \frac{1}{2}\sum_{i=1}^n f''\big(B_{t_{i-1}} + \lambda_i(B_{t_i} - B_{t_{i-1}})\big)(B_{t_i} - B_{t_{i-1}})^2, \qquad (7.1.3)$$

where $0 < \lambda_i < 1$ and B_t is used to denote $B(t)$ here and below in order to make parentheses easy to read.

Now, the first summation in Equation (7.1.3) can be taken care of by Theorem 5.3.3, namely,

$$\lim_{\|\Delta_n\|\to 0} \sum_{i=1}^{n} f'(B_{t_{i-1}})(B_{t_i} - B_{t_{i-1}}) = \int_a^t f'(B(s))\, dB(s), \quad \text{in probability.}$$

As for the second summation in Equation (7.1.3), we can easily guess the limit in view of the quadratic variation of $B(t)$ in Theorem 4.1.2, i.e.,

$$\sum_{i=1}^{n} f''(B_{t_{i-1}} + \lambda_i(B_{t_i} - B_{t_{i-1}}))(B_{t_i} - B_{t_{i-1}})^2 \longrightarrow \int_a^t f''(B(s))\, ds \quad (7.1.4)$$

in some sense as $\|\Delta_n\| \to 0$. We will prove in the next section that there exists a subsequence such that Equation (7.1.4) converges almost surely.

It is worthwhile to point out that if the second derivative f'' is bounded, then the proof of Equation (7.1.4) is quite easy. But in general a C^2-function f may have unbounded f''. In that case, the proof of Equation (7.1.4) is somewhat involved. This is the reason why we postpone the proof until the next section.

Summing up the above discussion, we have the following theorem, which K. Itô proved in 1944 [31].

Theorem 7.1.2. *Let $f(x)$ be a C^2-function. Then*

$$f(B(t)) - f(B(a)) = \int_a^t f'(B(s))\, dB(s) + \frac{1}{2} \int_a^t f''(B(s))\, ds, \quad (7.1.5)$$

where the first integral is an Itô integral (as defined in Chapter 5) and the second integral is a Riemann integral for each sample path of $B(s)$.

Observe that the last term in Equation (7.1.5) is a consequence of the nonzero quadratic variation of the Brownian motion $B(t)$. This extra term distinguishes the Itô calculus from the Leibniz–Newton calculus.

Example 7.1.3. Take the function $f(x) = x^2$ for Equation (7.1.5) to get

$$B(t)^2 - B(a)^2 = 2 \int_a^t B(s)\, dB(s) + (t - a).$$

When $t = b$, this equality gives

$$\int_a^b B(t)\, dB(t) = \frac{1}{2}\left(B(b)^2 - B(a)^2 - (b - a)\right),$$

which is the same as Equation (4.4.1). On the other hand, when $a = 0$,

$$B(t)^2 = 2 \int_0^t B(s)\, dB(s) + t,$$

which is the Doob–Meyer decomposition of the submartingale $B(t)^2$.

Example 7.1.4. Put $f(x) = x^4$ for Equation (7.1.5) to get

$$B(t)^4 = \left(B(a)^4 + 4 \int_a^t B(s)^3 \, dB(s)\right) + 6 \int_a^t B(s)^2 \, ds.$$

This is the Doob–Meyer decomposition of the submartingale $B(t)^4$. On the other hand, this equation gives the value of an Itô integral

$$\int_a^t B(s)^3 \, dB(s) = \frac{1}{4}\left(B(t)^4 - B(a)^4\right) - \frac{3}{2}\int_a^t B(s)^2 \, ds.$$

Example 7.1.5. Take $f(x) = e^x$ for Equation (7.1.5) to get

$$e^{B(t)} = \left(e^{B(a)} + \int_a^t e^{B(s)} \, dB(s)\right) + \frac{1}{2}\int_a^t e^{B(s)} \, ds.$$

This equality is the Doob–Meyer decomposition of the submartingale $e^{B(t)}$. We can also rewrite it as the evaluation of an Itô integral

$$\int_a^t e^{B(s)} \, dB(s) = e^{B(t)} - e^{B(a)} - \frac{1}{2}\int_a^t e^{B(s)} \, ds. \qquad (7.1.6)$$

7.2 Proof of Itô's Formula

In this section we will prove the almost sure convergence of a subsequence in Equation (7.1.4). This will complete the proof of Itô's formula in Theorem 7.1.2. First we prove two lemmas.

Lemma 7.2.1. *Let $g(x)$ be a continuous function on $\mathbb{R}$. For each $n \geq 1$, let $\Delta_n = \{t_0, t_1, \ldots, t_{n-1}, t_n\}$ be a partition of $[a, t]$ and let $0 < \lambda_i < 1$ for $1 \leq i \leq n$. Then there exists a subsequence of*

$$\sum_{i=1}^n \left(g\big(B_{t_{i-1}} + \lambda_i(B_{t_i} - B_{t_{i-1}})\big) - g(B_{t_{i-1}})\right)(B_{t_i} - B_{t_{i-1}})^2 \qquad (7.2.1)$$

converging to 0 almost surely as $\|\Delta_n\| \to 0$.

Remark 7.2.2. In case the partitions satisfy (i) $\Delta_1 \subset \Delta_2 \subset \cdots \subset \Delta_n \subset \cdots$ or (ii) $\sum_{n=1}^\infty \|\Delta_n\|^2 < \infty$, then the original sequence in Equation (7.2.1) does converge to 0 almost surely. See the proof below.

Proof. For simplicity, let X_n denote the summation in Equation (7.2.1). For each $n \geq 1$, define a random variable

$$\xi_n = \max_{1 \leq i \leq n, \, 0 < \lambda < 1} \left|g\big(B_{t_{i-1}} + \lambda(B_{t_i} - B_{t_{i-1}})\big) - g(B_{t_{i-1}})\right|.$$

Then obviously we have

$$|X_n| \le \xi_n \sum_{i=1}^{n} (B_{t_i} - B_{t_{i-1}})^2. \tag{7.2.2}$$

By the continuity of $g(x)$ and the Brownian motion $B(t)$, ξ_n converges to 0 almost surely. On the other hand, by Theorem 4.1.2, the summation in Equation (7.2.2) converges to $t - a$ in $L^2(\Omega)$. Hence the right-hand side of Equation (7.2.2) has a subsequence converging to 0 almost surely. It follows that $\{X_n\}$ has a subsequence converging to 0 almost surely.

In case the partitions satisfy condition (i) or (ii) in Remark 7.2.2, then the summation in Equation (7.2.2) converges to $t - a$ almost surely. Hence X_n converges to 0 almost surely.   □

Lemma 7.2.3. *Let $g(x)$ be a continuous function on $\mathbb{R}$. For each $n \ge 1$, let $\Delta_n = \{t_0, t_1, \ldots, t_{n-1}, t_n\}$ be a partition of $[a, t]$. Then the sequence*

$$\sum_{i=1}^{n} g(B_{t_{i-1}}) \Big((B_{t_i} - B_{t_{i-1}})^2 - (t_i - t_{i-1}) \Big) \tag{7.2.3}$$

converges to 0 in probability as $\|\Delta_n\| \to 0$.

Remark 7.2.4. In case $g(x)$ is a bounded measurable function on $\mathbb{R}$, then the convergence can be easily shown to be in $L^2(\Omega)$. This is an exercise at the end of this chapter.

Proof. For each $L > 0$, define the events

$$A_{i-1}^{(L)} \equiv \{|B(t_j)| \le L \text{ for all } j \le i - 1\}, \ 1 \le i \le n. \tag{7.2.4}$$

Let S_n denote the summation in Equation (7.2.3) and define

$$S_{n,L} \equiv \sum_{i=1}^{n} g(B_{t_{i-1}}) 1_{A_{i-1}^{(L)}} \Big((B_{t_i} - B_{t_{i-1}})^2 - (t_i - t_{i-1}) \Big) \tag{7.2.5}$$

$$= \sum_{i=1}^{n} g(B_{t_{i-1}}) 1_{A_{i-1}^{(L)}} X_i = \sum_{i=1}^{n} Y_i, \tag{7.2.6}$$

where $X_i = (B_{t_i} - B_{t_{i-1}})^2 - (t_i - t_{i-1})$ and $Y_i = g(B_{t_{i-1}}) 1_{A_{i-1}^{(L)}} X_i$.

Let $\mathcal{F}_t = \sigma\{B(s); s \le t\}$. Then for $i \ne j$, say $i < j$, we can use the conditional expectation to get

$$E(Y_i Y_j) = E(E[Y_i Y_j \,|\, \mathcal{F}_{t_{j-1}}])$$

$$= E(Y_i g(B_{t_{j-1}}) 1_{A_{j-1}^{(L)}} E[X_j \,|\, \mathcal{F}_{t_{j-1}}])$$

$$= 0. \tag{7.2.7}$$

On the other hand, it follows from the definition of $A_{i-1}^{(L)}$ that

$$Y_i^2 \leq \max_{|x| \leq L} \left(|g(x)|^2 \right) X_i^2.$$

Hence by Equation (4.1.7),

$$E(Y_i^2) \leq 2(t_i - t_{i-1})^2 \max_{|x| \leq L} \left(|g(x)|^2 \right). \tag{7.2.8}$$

From Equations (7.2.6), (7.2.7), and (7.2.8), we see that

$$
\begin{aligned}
E(S_{n,L}^2) &= \sum_{i=1}^n E(Y_i^2) \\
&\leq 2 \max_{|x| \leq L} (|g(x)|^2) \sum_{i=1}^n (t_i - t_{i-1})^2 \\
&\leq 2\|\Delta_n\|(t-a) \max_{|x| \leq L} (|g(x)|^2) \\
&\to 0, \quad \text{as } \|\Delta_n\| \to 0.
\end{aligned}
$$

Hence for any fixed number $L > 0$, the sequence $S_{n,L}$ converges to 0 in $L^2(\Omega)$ and so in probability as $\|\Delta_n\| \to 0$.

Now observe that the events defined in Equation (7.2.4) satisfy

$$A_0^{(L)} \supset A_1^{(L)} \supset \cdots \supset A_{n-1}^{(L)}.$$

This relationship, in view of the definition of $S_{n,L}$ in Equation (7.2.5), implies that $A_{n-1}^{(L)} \subset \{S_n = S_{n,L}\}$. Therefore,

$$\{S_n \neq S_{n,L}\} \subset \left(A_{n-1}^{(L)} \right)^c \subset \left\{ \max_{a \leq s \leq t} |B(s)| > L \right\},$$

which implies that

$$P\{S_n \neq S_{n,L}\} \leq P\left\{ \max_{a \leq s \leq t} |B(s)| > L \right\}.$$

But by the Doob submartingale inequality in Theorem 4.5.1,

$$P\left\{ \max_{a \leq s \leq t} |B(s)| > L \right\} \leq \frac{1}{L} E|B(t)| = \frac{1}{L} \sqrt{\frac{2t}{\pi}}.$$

Therefore, for any $n \geq 1$, we have

$$P\{S_n \neq S_{n,L}\} \leq \frac{1}{L} \sqrt{\frac{2t}{\pi}}. \tag{7.2.9}$$

Finally, observe that for any $\varepsilon > 0$,

$$\{|S_n| > \varepsilon\} \subset \{|S_{n,L}| > \varepsilon\} \cup \{S_n \neq S_{n,L}\},$$

which implies that

$$P\{|S_n| > \varepsilon\} \leq P\{|S_{n,L}| > \varepsilon\} + P\{S_n \neq S_{n,L}\}. \tag{7.2.10}$$

Put Equation (7.2.9) into Equation (7.2.10) to get

$$P\{|S_n| > \varepsilon\} \leq P\{|S_{n,L}| > \varepsilon\} + \frac{1}{L}\sqrt{\frac{2t}{\pi}}.$$

Choose a large number L so that $\frac{1}{L}\sqrt{\frac{2t}{\pi}} < \varepsilon/2$. Then with this number L, we have already proved that $S_{n,L}$ converges to 0 in probability as $\|\Delta_n\| \to 0$. Hence there exists $n_0 \geq 1$ such that $P\{|S_{n,L}| > \varepsilon\} < \varepsilon/2$ for all $n \geq n_0$. Therefore, $P\{|S_n| > \varepsilon\} < \varepsilon$ for all $n \geq n_0$. This shows that S_n converges to 0 in probability as $\|\Delta_n\| \to 0$.   □

Now we go back to consider Equation (7.1.4). Rewrite the summation in this equation as

$$
\begin{aligned}
T_n &\equiv \sum_{i=1}^{n} f''(B_{t_{i-1}} + \lambda_i(B_{t_i} - B_{t_{i-1}}))(B_{t_i} - B_{t_{i-1}})^2 \\
&= \sum_{i=1}^{n} \Big(f''(B_{t_{i-1}} + \lambda_i(B_{t_i} - B_{t_{i-1}})) - f''(B_{t_{i-1}})\Big)(B_{t_i} - B_{t_{i-1}})^2 \\
&\quad + \sum_{i=1}^{n} f''(B_{t_{i-1}})\Big((B_{t_i} - B_{t_{i-1}})^2 - (t_i - t_{i-1})\Big) \\
&\quad + \sum_{i=1}^{n} f''(B_{t_{i-1}})(t_i - t_{i-1}).
\end{aligned}
$$

Obviously, the last summation converges to $\int_a^t f''(B(s))\,ds$ almost surely. By Lemmas 7.2.1 and 7.2.3, the other two summations in the right-hand side converge to 0 in probability. Hence there exists a subsequence $\{T_{n_k}\}$ of $\{T_n\}$ converging to $\int_a^t f''(B(s))\,ds$ almost surely as $\|\Delta_{n_k}\| \to 0$. This completes the proof of Itô's formula in Theorem 7.1.2.

7.3 Itô's Formula Slightly Generalized

Let $f(t, x)$ be a function of t and x. Put $x = B(t)$ to get a stochastic process $f(t, B(t))$. What can we say about this stochastic process? Note that t appears in two places, one as a variable of f and the other in the Brownian motion $B(t)$ in place of x. For the first t, we can apply the Leibniz–Newton calculus. But for the second t in $B(t)$, we need to use the Itô calculus.

Suppose $f(t, x)$ is continuous and has continuous partial derivatives $\frac{\partial f}{\partial t}, \frac{\partial f}{\partial x}$, and $\frac{\partial^2 f}{\partial x^2}$. Then we can use the Taylor expansion to get

$$
\begin{aligned}
f(t, x) - f(s, x_0) &= \left[f(t, x) - f(s, x) \right] + \left[f(s, x) - f(s, x_0) \right] \\
&= \frac{\partial f}{\partial t}(s + \rho(t - s), x)(t - s) + \frac{\partial f}{\partial x}(s, x_0)(x - x_0) \\
&\quad + \frac{1}{2} \frac{\partial^2 f}{\partial x^2}(s, x_0 + \lambda(x - x_0))(x - x_0)^2,
\end{aligned}
\tag{7.3.1}
$$

where $0 < \rho, \lambda < 1$. Then, similar to Equations (7.1.2) and (7.1.3),

$$
\begin{aligned}
f(t, B_t) - f(a, B_a) &= \sum_{i=1}^{n} \left(f(t_i, B_{t_i}) - f(t_{i-1}, B_{t_{i-1}}) \right) \\
&= \sum_1 + \sum_2 + \sum_3,
\end{aligned}
\tag{7.3.2}
$$

where $\sum_1, \sum_2$, and $\sum_3$ are the summations corresponding to $\frac{\partial f}{\partial t}, \frac{\partial f}{\partial x}$, and $\frac{\partial^2 f}{\partial x^2}$, respectively, in Equation (7.3.1). By the continuity of $\frac{\partial f}{\partial t}(t, x)$ and the Brownian motion $B(t)$, we have

$$
\begin{aligned}
\sum_1 &= \sum_{i=1}^{n} \frac{\partial f}{\partial t}\left(t_{i-1} + \rho(t_i - t_{i-1}), B_{t_{i-1}} \right)(t_i - t_{i-1}) \\
&\longrightarrow \int_a^t \frac{\partial f}{\partial t}(s, B(s))\, ds, \quad \text{almost surely,}
\end{aligned}
$$

as $\|\Delta_n\| \to 0$. On the other hand, we can easily modify the arguments in Sections 7.1 and 7.2 to show that there exists a subsequence $\{\Delta_{n_k}\}$ of the partitions $\{\Delta_n\}$ such that

$$
\sum_2 \longrightarrow \int_a^t \frac{\partial f}{\partial x}(s, B(s))\, dB(s), \quad \text{almost surely,}
$$

$$
\sum_3 \longrightarrow \int_a^t \frac{\partial^2 f}{\partial x^2}(s, B(s))\, ds, \quad \text{almost surely,}
$$

as $\|\Delta_{n_k}\| \to 0$. Therefore, we have shown the following Itô's formula.

Theorem 7.3.1. *Let $f(t, x)$ be a continuous function with continuous partial derivatives $\frac{\partial f}{\partial t}, \frac{\partial f}{\partial x}$, and $\frac{\partial^2 f}{\partial x^2}$. Then*

$$
\begin{aligned}
f(t, B(t)) &= f(a, B(a)) + \int_a^t \frac{\partial f}{\partial x}(s, B(s))\, dB(s) \\
&\quad + \int_a^t \left(\frac{\partial f}{\partial t}(s, B(s)) + \frac{1}{2} \frac{\partial^2 f}{\partial x^2}(s, B(s)) \right) ds.
\end{aligned}
\tag{7.3.3}
$$

We give some simple examples to illustrate how to use Equation (7.3.3).

Example 7.3.2. Consider the function $f(t, x) = tx^2$. Then $\frac{\partial f}{\partial t} = x^2$, $\frac{\partial f}{\partial x} = 2tx$, and $\frac{\partial^2 f}{\partial x^2} = 2t$. Hence by Equation (7.3.3),

$$tB(t)^2 = 2 \int_0^t sB(s)\, dB(s) + \left(\int_0^t B(s)^2\, ds + \frac{1}{2}t^2 \right).$$

We can use conditional expectation to show that $tB(t)^2$ is a submartingale. Hence this equality gives the Doob–Meyer decomposition of $tB(t)^2$. It also gives the evaluation of an Itô integral

$$\int_0^t sB(s)\, dB(s) = \frac{1}{2}tB(t)^2 - \frac{1}{4}t^2 - \frac{1}{2}\int_0^t B(s)^2\, ds.$$

Example 7.3.3. The stochastic process $M(t) = B(t)^2 - t$ is a martingale. We will use Itô's formula to find the compensator of $M(t)^2$. Consider the function $f(t, x) = (x^2 - t)^2$. Then

$$\frac{\partial f}{\partial t} = -2(x^2 - t), \qquad \frac{\partial f}{\partial x} = 4x(x^2 - t), \qquad \frac{\partial^2 f}{\partial x^2} = 4(x^2 - t) + 8x^2.$$

Use Equation (7.3.3) to get

$$M(t)^2 = (B(t)^2 - t)^2 = 4 \int_0^t B(s)(B(s)^2 - s)\, dB(s) + 4 \int_0^t B(s)^2\, ds.$$

Notice that the first integral is a martingale and the second integral is an increasing process. Hence by the Doob–Meyer decomposition theorem, the compensator of $M(t)^2$ is given by

$$\langle M \rangle_t = 4 \int_0^t B(s)^2\, ds.$$

Example 7.3.4. Let $f(t, x) = e^{cx - \frac{1}{2}c^2 t}$ for a fixed constant c. Then

$$\frac{\partial f}{\partial t} = -\frac{1}{2}c^2 f(t, x), \qquad \frac{\partial f}{\partial x} = cf(t, x), \qquad \frac{\partial^2 f}{\partial x^2} = c^2 f(t, x).$$

By Equation (7.3.3), we have

$$e^{cB(t) - \frac{1}{2}c^2 t} = 1 + c \int_0^t e^{cB(s) - \frac{1}{2}c^2 s}\, dB(s).$$

Hence $M(t) = e^{cB(t) - \frac{1}{2}c^2 t}$ is a martingale for any constant c. To find the compensator of $M(t)^2$, we can apply Equation (7.3.3) again to the function $f(t, x) = e^{2cx - c^2 t}$ to get

$$M(t)^2 = 1 + 2c \int_0^t e^{2cB(s)-c^2 s} \, dB(s) + c^2 \int_0^t e^{2cB(s)-c^2 s} \, ds.$$

Therefore, the compensator of $M(t)^2$ is given by

$$\langle M \rangle_t = c^2 \int_0^t e^{2cB(s)-c^2 s} \, ds.$$

Example 7.3.5. Let $f(t, x) = e^{x^2 - t}$. Then we have

$$\frac{\partial f}{\partial t} = -f(t, x), \quad \frac{\partial f}{\partial x} = 2x f(t, x), \quad \frac{\partial^2 f}{\partial x^2} = (1 + 2x^2) 2 f(t, x).$$

Apply Equation (7.3.3) to get

$$e^{B(t)^2 - t} = 1 + 2 \int_0^t B(s) e^{B(s)^2 - s} \, dB(s) + 2 \int_0^t B(s)^2 e^{B(s)^2 - s} \, ds.$$

Note that the first integral of the equation is a local martingale by Theorem 5.5.2, while the second integral is an increasing stochastic process. This is similar to the Doob–Meyer decomposition for submartingales.

7.4 Itô's Formula in the General Form

Observe from Itô's formula in Theorems 7.1.2 and 7.3.1 that $f(B(t))$ and $f(t, B(t))$ contain two integrals, i.e., an Itô integral and a Riemann integral. This leads to a very special class of stochastic processes to be defined below.

Let $\{\mathcal{F}_t; a \leq t \leq b\}$ be a filtration specified as in Section 5.1. This filtration will be fixed throughout the rest of this chapter. Recall that we have a class $\mathcal{L}_{\mathrm{ad}}(\Omega, L^2[a, b])$ defined in Notation 5.1.1. Now we introduce another class.

Notation 7.4.1. For convenience, we will use $\mathcal{L}_{\mathrm{ad}}(\Omega, L^1[a, b])$ to denote the space of all $\{\mathcal{F}_t\}$-adapted stochastic processes $f(t)$ such that $\int_a^b |f(t)| \, dt < \infty$ almost surely.

Definition 7.4.2. *An Itô process is a stochastic process of the form*

$$X_t = X_a + \int_a^t f(s) \, dB(s) + \int_a^t g(s) \, ds, \quad a \leq t \leq b, \tag{7.4.1}$$

where X_a is $\mathcal{F}_a$-measurable, $f \in \mathcal{L}_{\mathrm{ad}}(\Omega, L^2[a, b])$, and $g \in \mathcal{L}_{\mathrm{ad}}(\Omega, L^1[a, b])$.

A very useful and convenient shorthand for writing Equation (7.4.1) is the following "stochastic differential"

$$dX_t = f(t) \, dB(t) + g(t) \, dt. \tag{7.4.2}$$

It should be pointed out that the stochastic differential has no meaning by itself since Brownian paths are nowhere differentiable. It is merely a symbolic expression to mean the equality in Equation (7.4.1). Observe that once we have the stochastic differential in Equation (7.4.2), we can immediately convert it into the integral form in Equation (7.4.1).

The next theorem gives the general form of Itô's formula.

Theorem 7.4.3. *Let X_t be an Itô process given by*

$$X_t = X_a + \int_a^t f(s)\, dB(s) + \int_a^t g(s)\, ds, \quad a \le t \le b.$$

Suppose $\theta(t,x)$ is a continuous function with continuous partial derivatives $\frac{\partial \theta}{\partial t}, \frac{\partial \theta}{\partial x}$, and $\frac{\partial^2 \theta}{\partial x^2}$. Then $\theta(t,X_t)$ is also an Itô process and

$$\theta(t, X_t) = \theta(a, X_a) + \int_a^t \frac{\partial \theta}{\partial x}(s, X_s) f(s)\, dB(s)$$

$$+ \int_a^t \left[\frac{\partial \theta}{\partial t}(s, X_s) + \frac{\partial \theta}{\partial x}(s, X_s) g(s) + \frac{1}{2} \frac{\partial^2 \theta}{\partial x^2}(s, X_s) f(s)^2 \right] ds. \quad (7.4.3)$$

This theorem can be proved by modifying the arguments in Sections 7.1 and 7.2. The technical details and notation to carry out the complete proof are rather complicated, but the essential ideas are exactly the same as those in the proof for Theorem 7.1.2.

A good way to obtain Equation (7.4.3) is through the symbolic derivation in terms of a stochastic differential by using the Taylor expansion to the first order for dt and the second order for dX_t and also the following table:

$\times$	$dB(t)$	dt
$dB(t)$	dt	0
dt	0	0

Itô Table (1)

Here is the symbolic derivation of Equation (7.4.3). First apply the Taylor expansion to get

$$d\theta(t, X_t) = \frac{\partial \theta}{\partial t}(t, X_t)\, dt + \frac{\partial \theta}{\partial x}(t, X_t)\, dX_t + \frac{1}{2} \frac{\partial^2 \theta}{\partial x^2}(t, X_t)\, (dX_t)^2.$$

Then use the Itô Table (1) to get $(dX_t)^2 = f(t)^2\, dt$. Therefore,

$$d\theta(t, X_t) = \frac{\partial \theta}{\partial t}\, dt + \frac{\partial \theta}{\partial x}\Big(f(t)\, dB(t) + g(t)\, dt \Big) + \frac{1}{2} \frac{\partial^2 \theta}{\partial x^2} f(t)^2\, dt$$

$$= \frac{\partial \theta}{\partial x} f(t)\, dB(t) + \Big(\frac{\partial \theta}{\partial t} + \frac{\partial \theta}{\partial x} g(t) + \frac{1}{2} \frac{\partial^2 \theta}{\partial x^2} f(t)^2 \Big)\, dt.$$

Here, for simplicity, we have omitted the variables (t, X_t) from all partial derivatives. Finally we convert this equation into Equation (7.4.3).

For computation we can always use this kind of symbolic derivation on stochastic differentials to get the results. However, we need to emphasize that this derivation is not a proof for Itô's formula. It just produces the correct results and that is all.

Example 7.4.4. Let $f \in \mathcal{L}_{ad}(\Omega, L^2[0, 1])$. Consider the Itô process

$$X_t = \int_0^t f(s)\, dB(s) - \frac{1}{2} \int_0^t f(s)^2\, ds, \quad 0 \le t \le 1,$$

and the function $\theta(x) = e^x$. Then $dX_t = f(t)\, dB(t) - \frac{1}{2}f(t)^2\, dt$. Apply the Taylor expansion and the Itô Table (1) to get

$$d\theta(X_t) = e^{X_t}\, dX_t + \frac{1}{2}e^{X_t}\, (dX_t)^2$$

$$= e^{X_t}\left(f(t)\, dB(t) - \frac{1}{2}f(t)^2\, dt\right) + \frac{1}{2}e^{X_t}f(t)^2\, dt$$

$$= f(t)e^{X_t}\, dB(t).$$

Therefore, we have

$$e^{\int_0^t f(s)\, dB(s) - \frac{1}{2}\int_0^t f(s)^2\, ds} = 1 + \int_0^t f(s)e^{\int_0^s f(u)\, dB(u) - \frac{1}{2}\int_0^s f(u)^2\, du}\, dB(s).$$

Hence by Theorem 5.5.2, $Y_t = \exp\left[\int_0^t f(s)\, dB(s) - \frac{1}{2}\int_0^t f(s)^2\, ds\right]$ is a local martingale. But when $f(t)$ is a deterministic function in $L^2[0, 1]$, then Y_t is a martingale by Theorem 4.6.1.

Example 7.4.5. Consider the Langevin equation

$$dX_t = \alpha\, dB(t) - \beta X_t\, dt, \quad X_0 = x_0, \tag{7.4.4}$$

where $\alpha \in \mathbb{R}$ and $\beta > 0$. The quantity X_t is the velocity at time t of a free particle that performs a Brownian motion that is different from $B(t)$ in the equation. This "stochastic differential equation" is interpreted as the following stochastic integral equation:

$$X_t = x_0 + \alpha B(t) - \beta \int_0^t X_s\, ds.$$

We can use the iteration method and integration by parts to derive the solution X_t for each sample path of $B(t)$. The computation is somewhat complicated. But if we use Itô's formula, then it is very easy to derive the solution.

Let $\theta(t, x) = e^{\beta t}x$. Then $\frac{\partial \theta}{\partial t} = \beta e^{\beta t}x$, $\frac{\partial \theta}{\partial x} = e^{\beta t}$, and $\frac{\partial^2 \theta}{\partial x^2} = 0$. Hence by Itô's formula, we have

$$d(e^{\beta t} X_t) = \beta e^{\beta t} X_t \, dt + e^{\beta t} \, dX(t)$$
$$= \beta e^{\beta t} X_t \, dt + e^{\beta t} \big(\alpha \, dB(t) - \beta X_t \, dt \big)$$
$$= \alpha e^{\beta t} \, dB(t).$$

Then convert this stochastic differential into a stochastic integral,

$$e^{\beta t} X_t = e^{\beta s} X_s + \int_s^t \alpha e^{\beta u} \, dB(u), \quad s \le t.$$

Therefore, X_t is given by

$$X_t = e^{-\beta(t-s)} X_s + \alpha \int_s^t e^{-\beta(t-u)} \, dB(u), \quad s \le t. \tag{7.4.5}$$

In particular, when $s = 0$, we get the solution of the Langevin equation in Equation (7.4.4):

$$X_t = e^{-\beta t} x_0 + \alpha \int_0^t e^{-\beta(t-u)} \, dB(u). \tag{7.4.6}$$

Definition 7.4.6. *The solution X_t, given by Equation (7.4.6), of the Langevin equation (7.4.4) is called an Ornstein–Uhlenbeck process.*

Theorem 7.4.7. *Let X_t be the Ornstein–Uhlenbeck process given by Equation (7.4.6). Then we have the conditional probability*

$$P\big[X_t \le y \,\big|\, X_s = x\big] = \int_\infty^y G_{\frac{\alpha^2}{2\beta}(1 - e^{-2\beta(t-s)})}\big(e^{-\beta(t-s)} x, \, v\big) \, dv, \quad s < t, \tag{7.4.7}$$

where $G_t(m, v) = \frac{1}{\sqrt{2\pi t}} e^{-\frac{1}{2t}(v-m)^2}$, the Gaussian density function in v with mean m and variance t. In particular, when $s = 0$ and $x = x_0$, the random variable X_t is Gaussian with mean $x_0 e^{-\beta t}$ and variance $\frac{\alpha^2}{2\beta}(1 - e^{-2\beta t})$.

Remark 7.4.8. By Equation (7.4.7), $P\big[X_t \le y \,\big|\, X_s = x\big]$ is a function of $t - s$. Hence for any $s < t$ and $h \ge 0$, we have

$$P\big[X_t \le y \,\big|\, X_s = x\big] = P\big[X_{t+h} \le y \,\big|\, X_{s+h} = x\big].$$

This shows that X_t is a stationary process. Moreover, for any $s \ge 0$, we have the following interesting limit:

$$\lim_{t \to \infty} G_{\alpha^2(1 - e^{-2\beta(t-s)})/(2\beta)}\big(e^{-\beta(t-s)} x, \, v\big) = G_{\alpha^2/(2\beta)}(0, \, v),$$

where $G_{\alpha^2/(2\beta)}(0, \cdot)$ is the Gaussian density function with mean 0 and variance $\alpha^2/(2\beta)$. It can be shown that the measure $d\rho(v) = G_{\alpha^2/(2\beta)}(0, v) \, dv$ is an invariant measure for the Ornstein–Uhlenbeck process.

Proof. Let $s \leq t$. Given $X_s = x$, we can use Equation (7.4.5) to get

$$X_t = e^{-\beta(t-s)}x + \alpha \int_s^t e^{-\beta(t-u)}\, dB(u).$$

Note that the integral in this equation is a Wiener integral, which is a Gaussian random variable with mean 0 and variance

$$\alpha^2 \int_s^t e^{-2\beta(t-u)}\, du = \frac{\alpha^2}{2\beta}\left(1 - e^{-2\beta(t-s)}\right).$$

Hence the random variable X_t, given $X_s = x$, is Gaussian with mean $e^{-\beta(t-s)}x$ and variance $\alpha^2(1 - e^{-2\beta(t-s)})/(2\beta)$. This proves Equation (7.4.7).   □

7.5 Multidimensional Itô's Formula

Let $B_1(t), B_2(t), \ldots, B_m(t)$ be m independent Brownian motions. Consider n Itô processes $X_t^{(1)}, X_t^{(2)}, \ldots, X_t^{(n)}$ given by

$$X_t^{(i)} = X_a^{(i)} + \sum_{j=1}^m \int_a^t f_{ij}(s)\, dB_j(s) + \int_a^t g_i(s)\, ds, \quad 1 \leq i \leq n, \qquad (7.5.1)$$

where $f_{ij} \in \mathcal{L}_{\mathrm{ad}}(\Omega, L^2[a,b])$ and $g_i \in \mathcal{L}_{\mathrm{ad}}(\Omega, L^1[a,b])$ for all $1 \leq i \leq n$ and $1 \leq j \leq m$. If we introduce the matrices

$$B(t) = \begin{bmatrix} B_1(t) \\ \vdots \\ B_m(t) \end{bmatrix}, \quad X_t = \begin{bmatrix} X_t^{(1)} \\ \vdots \\ X_t^{(n)} \end{bmatrix},$$

$$f(t) = \begin{bmatrix} f_{11}(t) & \cdots & f_{1m}(t) \\ \vdots & \ddots & \vdots \\ f_{n1}(t) & \cdots & f_{nm}(t) \end{bmatrix}, \quad g(t) = \begin{bmatrix} g_1(t) \\ \vdots \\ g_n(t) \end{bmatrix},$$

then Equation (7.5.1) can be written as a matrix equation:

$$X_t = X_a + \int_a^t f(s)\, dB(s) + \int_a^t g(s)\, ds, \quad a \leq t \leq b.$$

We can extend Itô's formula in Theorem 7.4.3 to the multidimensional case. Suppose $\theta(t, x_1, \ldots, x_n)$ is a continuous function on $[a, b] \times \mathbb{R}^n$ and has continuous partial derivatives $\frac{\partial \theta}{\partial t}, \frac{\partial \theta}{\partial x_i}$, and $\frac{\partial^2 \theta}{\partial x_i \partial x_j}$ for $1 \leq i, j \leq n$. Then the stochastic differential of $\theta(t, X_t^{(1)}, \ldots, X_t^{(n)})$ is given by

$$d\theta(t, X_t^{(1)}, \ldots, X_t^{(n)})$$

$$= \frac{\partial \theta}{\partial t}(t, X_t^{(1)}, \ldots, X_t^{(n)}) \, dt + \sum_{i=1}^{n} \frac{\partial \theta}{\partial x_i}(t, X_t^{(1)}, \ldots, X_t^{(n)}) \, dX_t^{(i)}$$

$$+ \frac{1}{2} \sum_{i,j=1}^{n} \frac{\partial^2 \theta}{\partial x_i \partial x_j}(t, X_t^{(1)}, \ldots, X_t^{(n)}) \, dX_t^{(i)} dX_t^{(j)}, \tag{7.5.2}$$

where the product $dX_t^{(i)} dX_t^{(j)}$ should be computed using the following table:

$\times$	$dB_j(t)$	dt
$dB_i(t)$	$\delta_{ij} \, dt$	0
dt	0	0

Itô Table (2)

The product $dB_i(t)dB_j(t) = 0$ for $i \neq j$ is the symbolic expression of the following fact. Let $B_1(t)$ and $B_2(t)$ be two independent Brownian motions and let $\Delta_n = \{t_0, t_1, \ldots, t_{n-1}, t_n\}$ be a partition of $[a, b]$. Then

$$\sum_{i=1}^{n} \big(B_1(t_i) - B_1(t_{i-1})\big)\big(B_2(t_i) - B_2(t_{i-1})\big) \longrightarrow 0$$

in $L^2(\Omega)$ as $\|\Delta_n\| = \max_{1 \leq i \leq n}(t_i - t_{i-1})$ tends to 0. The proof of this fact is left as an exercise at the end of this chapter.

Example 7.5.1. Consider the function $\theta(x, y) = xy$. We have $\frac{\partial \theta}{\partial x} = y$, $\frac{\partial \theta}{\partial y} = x$, $\frac{\partial^2 \theta}{\partial x \partial y} = \frac{\partial^2 \theta}{\partial y \partial x} = 1$, and $\frac{\partial^2 \theta}{\partial x^2} = \frac{\partial^2 \theta}{\partial y^2} = 0$. Apply Equation (7.5.2) to this function $\theta(x, y)$ for two Itô processes X_t and Y_t to get

$$d(X_t Y_t) = Y_t \, dX_t + X_t \, dY_t + \frac{1}{2} dX_t dY_t + \frac{1}{2} dY_t dX_t$$

$$= Y_t \, dX_t + X_t \, dY_t + dX_t dY_t.$$

Therefore,

$$X_t Y_t = X_a Y_a + \int_a^t Y_s \, dX_s + \int_a^t X_s \, dY_s + \int_a^t dX_s dY_s. \tag{7.5.3}$$

This equality is called the *Itô product formula*. If X_t and Y_t are given by $dX_t = f(t) \, dB(t) + \xi(t) \, dt$ and $dY_t = g(t) \, dB(t) + \eta(t) \, dt$, then the product formula means the following equality:

$$X_t Y_t = X_a Y_a + \int_a^t \Big(f(s)Y_s + g(s)X_s\Big) \, dB(s)$$

$$+ \int_a^t \Big(\xi(s)Y_s + \eta(s)X_s + f(s)g(s)\Big) \, ds.$$

In particular, when $\xi = \eta = 0$, we have

$$X_t Y_t = X_a Y_a + \int_a^t \left(f(s) Y_s + g(s) X_s \right) dB(s) + \int_a^t f(s) g(s) \, ds.$$

This implies that the stochastic process

$$\left(\int_a^t f(s) \, dB(s) \right) \left(\int_a^t g(s) \, dB(s) \right) - \int_a^t f(s) g(s) \, ds$$

is a local martingale for any $f, g \in \mathcal{L}_{ad}(\Omega, L^2[a, b])$.

Example 7.5.2. Let $X_t = \cos B(t)$ and $Y_t = \sin B(t)$. The column vector V_t with components X_t and Y_t represents the position at time t of an object moving in the unit circle with angle governed by a Brownian motion. Apply Itô's formula to get

$$dX_t = -\sin B(t) \, dB(t) - \frac{1}{2} \cos B(t) \, dt = -Y_t \, dB(t) - \frac{1}{2} X_t \, dt,$$

$$dY_t = \cos B(t) \, dB(t) - \frac{1}{2} \sin B(t) \, dt = X_t \, dB(t) - \frac{1}{2} Y_t \, dt.$$

Therefore, the stochastic differential of V_t satisfies the linear equation

$$dV_t = \begin{bmatrix} 0 & -1 \\ 1 & 0 \end{bmatrix} V_t \, dB(t) - \frac{1}{2} V_t \, dt, \quad V_0 = \begin{bmatrix} 1 \\ 0 \end{bmatrix}. \tag{7.5.4}$$

The SDE in Equation (7.5.4) is interpreted to mean the corresponding SIE.

Example 7.5.3. Let $X_t = \cos B(t)$, $Y_t = \sin B(t)$, and $Z_t = B(t)$. The column vector V_t with components X_t, Y_t, and Z_t represents the position at time t of an object tracing a helix Brownian curve. Apply Itô's formula to get

$$dX_t = -\sin B(t) \, dB(t) - \frac{1}{2} \cos B(t) \, dt = -Y_t \, dB(t) - \frac{1}{2} X_t \, dt,$$

$$dY_t = \cos B(t) \, dB(t) - \frac{1}{2} \sin B(t) \, dt = X_t \, dB(t) - \frac{1}{2} Y_t \, dt,$$

$$dZ_t = dB(t).$$

Therefore, the stochastic differential of V_t satisfies the linear equation

$$dV_t = \left(K V_t + q \right) dB(t) + L V_t \, dt, \quad V_0 = v_0, \tag{7.5.5}$$

where K, q, L, and v_0 are given by

$$K = \begin{bmatrix} 0 & -1 & 0 \\ 1 & 0 & 0 \\ 0 & 0 & 0 \end{bmatrix}, \quad q = \begin{bmatrix} 0 \\ 0 \\ 1 \end{bmatrix}, \quad L = \begin{bmatrix} -\frac{1}{2} & 0 & 0 \\ 0 & -\frac{1}{2} & 0 \\ 0 & 0 & 0 \end{bmatrix}, \quad v_0 = \begin{bmatrix} 1 \\ 0 \\ 0 \end{bmatrix}.$$

The SDE in Equation (7.5.5) is interpreted to mean the corresponding SIE.

Example 7.5.4. Take two independent Brownian motions $B_1(t)$ and $B_2(t)$. Let $X_t = e^{B_1(t)} \cos B_2(t)$, $Y_t = e^{B_1(t)} \sin B_2(t)$, $Z_t = e^{B_1(t)}$. Then the column vector V_t with components X_t, Y_t, and Z_t represents the position at time t of an object moving in the cone $z^2 = x^2 + y^2$. Apply Itô's formula to check that

$$dX_t = X_t \, dB_1(t) - Y_t \, dB_2(t),$$

$$dY_t = Y_t \, dB_1(t) + X_t \, dB_2(t),$$

$$dZ_t = Z_t \, dB_1(t) + \frac{1}{2} Z_t \, dt.$$

Hence the stochastic differential of V_t satisfies the linear equation

$$dV_t = KV_t \, dB_1(t) + CV_t \, dB_2(t) + LV_t \, dt, \quad V_0 = v_0, \tag{7.5.6}$$

where K, C, L, and v_0 are given by

$$K = \begin{bmatrix} 1 & 0 & 0 \\ 0 & 1 & 0 \\ 0 & 0 & 1 \end{bmatrix}, \quad C = \begin{bmatrix} 0 & -1 & 0 \\ 1 & 0 & 0 \\ 0 & 0 & 0 \end{bmatrix}, \quad L = \begin{bmatrix} 0 & 0 & 0 \\ 0 & 0 & 0 \\ 0 & 0 & \frac{1}{2} \end{bmatrix}, \quad v_0 = \begin{bmatrix} 1 \\ 0 \\ 1 \end{bmatrix}.$$

Of course, the stochastic differential equation in Equation (7.5.6) should be interpreted to mean the corresponding stochastic integral equation.

7.6 Itô's Formula for Martingales

Let $M(t)$ be a right continuous martingale with left-hand limits such that $E(M(t)^2) < \infty$ for each $t \in [a, b]$. We will explain informally how to derive Itô's formula for $M(t)$ without going into the technical details. Unlike the Brownian motion case, the martingale $M(t)$ may have jumps, which need to be taken care of carefully. For example, the compensated Poisson process $\widetilde{N}(t)$ given in Definition 6.2.2 is such a martingale.

Let F be a C^2-function and let $\Delta_n = \{t_0, t_1, \ldots, t_{n-1}, t_n\}$ be a partition of $[a, t]$. Then we have

$$F(M(t)) - F(M(a)) = \sum_{i=1}^{n} \Big(F(M(t_i)) - F(M(t_{i-1})) \Big). \tag{7.6.1}$$

Below we will also use M_t to denote the martingale $M(t)$. Suppose M_t is continuous on $[t_{i-1}, t_i)$ and discontinuous at t_i. Hence M_t has a jump at t_i. Use the Taylor approximation to get

$$F(M_{t_i}) - F(M_{t_{i-1}}) = \big(F(M_{t_i}) - F(M_{t_i-}) \big) + \big(F(M_{t_i-}) - F(M_{t_{i-1}}) \big)$$

$$\approx \big(F(M_{t_i}) - F(M_{t_i-}) \big) + \Big[F'(M_{t_{i-1}})(M_{t_i-} - M_{t_{i-1}})$$

$$+ \frac{1}{2} F''(M_{t_{i-1}})(M_{t_i-} - M_{t_{i-1}})^2 \Big]. \tag{7.6.2}$$

Rewrite the term containing F' as

$$F'(M_{t_{i-1}})(M_{t_i-} - M_{t_{i-1}}) = F'(M_{t_{i-1}})\left\{(M_{t_i} - M_{t_{i-1}}) - (M_{t_i} - M_{t_i-})\right\}$$

and put it into Equation (7.6.2). Then rearrange the terms for the summation in Equation (7.6.1) to get

$$F(M(t)) - F(M(a))$$
$$\approx \sum_{i=1}^n F'(M_{t_{i-1}})(M_{t_i} - M_{t_{i-1}}) + \frac{1}{2}\sum_{i=1}^n C_i + \sum_{i=1}^n J_i, \qquad (7.6.3)$$

where C_i and J_i are given by

$$C_i = F''(M_{t_{i-1}})(M_{t_i-} - M_{t_{i-1}})^2,$$
$$J_i = F(M_{t_i}) - F(M_{t_i-}) - F'(M_{t_{i-1}})(M_{t_i} - M_{t_i-}).$$

Now, for the first summation in Equation (7.6.3), we have

$$\sum_{i=1}^n F'(M_{t_{i-1}})(M_{t_i} - M_{t_{i-1}}) \longrightarrow \int_a^t F'(M_{s-})\,dM_s \qquad (7.6.4)$$

in probability as $\|\Delta_n\| \to 0$. The last summation in Equation (7.6.3) gives the jumps and

$$\sum_{i=1}^n J_i \longrightarrow \sum_{a < s \leq t} \left(F(M_s) - F(M_{s-}) - F'(M_{s-})\Delta M_s\right) \qquad (7.6.5)$$

in probability as $\|\Delta_n\| \to 0$. Here $\Delta M_s = M(s) - M(s-)$ denotes the jump of M at s.

As for the second summation in Equation (7.6.3), recall that we have defined the quadratic variation process $[M]_t$ of $M(t)$ in Equation (6.4.4). Notice that in C_i we have $(M_{t_i-} - M_{t_{i-1}})^2$ as opposed to $(M_{t_i} - M_{t_{i-1}})^2$ in Equation (6.4.4). Since we use the left-hand limit M_{t_i-} in C_i, it is clear that C_i comes from the *continuous part* of $[M]_t$, which is defined by

$$[M]_t^c = [M]_t - [M]_a - \sum_{a < s \leq t} \Delta[M]_s, \qquad a \leq t \leq b, \qquad (7.6.6)$$

where $\Delta[M]_s = [M]_s - [M]_{s-}$, the jump of $[M]_s$ at s. Therefore,

$$\sum_{i=1}^n C_i \longrightarrow \int_a^t F''(M_s)\,d[M]_s^c \qquad (7.6.7)$$

in probability as $\|\Delta_n\| \to 0$. Note that there is no need to use $F''(M_{s-})$ for the integrand because the integrator $d[M]_s^c$ is continuous.

Summing up the above discussion, we have informally derived Itô's formula for martingales in the next theorem.

Theorem 7.6.1. *Let M_t be a right continuous, square integrable martingale with left-hand limits. Suppose F is a C^2-function. Then*

$$F(M_t) = F(M_a) + \int_a^t F'(M_{s-})\, dM_s + \frac{1}{2} \int_a^t F''(M_s)\, d[M]_s^c$$
$$+ \sum_{a < s \leq t} \Big(F(M_s) - F(M_{s-}) - F'(M_{s-})\Delta M_s \Big), \qquad (7.6.8)$$

where $[M]_s^c$ is the continuous part of $[M]_s$ and $\Delta M_s = M_s - M_{s-}$.

Example 7.6.2. Consider the compensated Poisson process $\widetilde{N}(t)$ defined in Section 6.2. Let $F(x) = x^2$. By Example 6.4.6 we have $[\widetilde{N}]_t = \lambda t + \widetilde{N}(t)$. Since $\widetilde{N}(t) = N(t) - \lambda t$, we have $[\widetilde{N}]_t = N(t)$. Hence the continuous part of $[\widetilde{N}]_t$ is given by

$$[\widetilde{N}]_t^c = [\widetilde{N}]_t - [\widetilde{N}]_a - \sum_{a < s \leq t} \Delta [\widetilde{N}]_s = N(t) - N(a) - \sum_{a < s \leq t} \Delta N(s).$$

Note that $\sum_{a < s \leq t} \Delta N(s) - N(t) \quad N(a)$, the total jumps on $[a,t]$. Hence $[\widetilde{N}]_t^c = 0$. On the other hand, it is easily checked that

$$\widetilde{N}(s)^2 - \widetilde{N}(s-)^2 - 2\widetilde{N}(s-)\Delta\widetilde{N}(s) = (\Delta\widetilde{N}(s))^2 = \Delta\widetilde{N}(s) = \Delta N(s).$$

This implies that

$$\sum_{a < s \leq t} \Big(\widetilde{N}(s)^2 - \widetilde{N}(s-)^2 - 2\widetilde{N}(s-)\Delta\widetilde{N}(s) \Big) = \sum_{a < s \leq t} \Delta N(s) = N(t) - N(a).$$

Therefore, by Equation (7.6.8), we have

$$\widetilde{N}(t)^2 = \widetilde{N}(a)^2 + 2 \int_a^t \widetilde{N}(s-)\, d\widetilde{N}(s) + N(t) - N(a),$$

which gives the value of the stochastic integral

$$\int_a^t \widetilde{N}(s-)\, d\widetilde{N}(s) = \frac{1}{2} \Big(\widetilde{N}(t)^2 - \widetilde{N}(a)^2 - N(t) + N(a) \Big).$$

Since $N(t) = \widetilde{N}(t) + \lambda t$, this equation with $t = b$ is exactly the same as the one in Equation (6.5.4).

Example 7.6.3. Consider the martingale $\widetilde{N}(t)$ and the function $F(x) = x^3$. We can compute the summation in Equation (7.6.8) as follows. First use the identity $b^3 - a^3 = (b-a)^3 + 3ab(b-a)$ and the fact that $(\Delta\widetilde{N}(s))^3 = \Delta\widetilde{N}(s) = \Delta N(s)$ to get the equality

$$\widetilde{N}(s)^3 - \widetilde{N}(s-)^3 = \Delta N(s) + 3\widetilde{N}(s)\widetilde{N}(s-)\Delta N(s). \qquad (7.6.9)$$

Next, we use the equalities $(\Delta \widetilde{N}(s))^2 = \Delta \widetilde{N}(s) = \Delta N(s)$ and $\widetilde{N}(s-) = N(s-) - \lambda s$ to show that

$$\widetilde{N}(s)\widetilde{N}(s-)\Delta N(s) - \widetilde{N}(s-)^2 \Delta \widetilde{N}(s) = N(s-)\Delta N(s) - \lambda s \Delta N(s). \quad (7.6.10)$$

It follows from Equations (7.6.9) and (7.6.10) that

$$\widetilde{N}(s)^3 - \widetilde{N}(s-)^3 - 3\widetilde{N}(s-)^2 \Delta \widetilde{N}(s)$$
$$= \Delta N(s) + 3N(s-)\Delta N(s) - 3\lambda s \Delta N(s). \quad (7.6.11)$$

Hence we have

$$\sum_{0<s\leq t} \left(\widetilde{N}(s)^3 - \widetilde{N}(s-)^3 - 3\widetilde{N}(s-)^2 \Delta \widetilde{N}(s) \right)$$
$$= \sum_{0<s\leq t} \Delta N(s) + 3 \sum_{0<s\leq t} N(s-)\Delta N(s) - 3\lambda \sum_{0<s\leq t} s\Delta N(s). \quad (7.6.12)$$

Now, the first summation is obviously given by

$$\sum_{0<s\leq t} \Delta N(s) = N(t). \quad (7.6.13)$$

For the second summation, we have

$$\sum_{0<s\leq t} N(s-)\Delta N(s) = 1 + 2 + 3 + \cdots + (N(t) - 1)$$
$$= \frac{N(t)\left(N(t) - 1\right)}{2}. \quad (7.6.14)$$

The third summation is in fact a Riemann–Stieltjes integral and we can apply the integration by parts formula to see that

$$\sum_{0<s\leq t} s\Delta N(s) = \int_0^t s\, dN(s) = tN(t) - \int_0^t N(s)\, ds. \quad (7.6.15)$$

Put Equations (7.6.13), (7.6.14), and (7.6.15) into Equation (7.6.12) to get

$$\sum_{0<s\leq t} \left(\widetilde{N}(s)^3 - \widetilde{N}(s-)^3 - 3\widetilde{N}(s-)^2 \Delta \widetilde{N}(s) \right)$$
$$= \frac{3}{2}N(t)^2 - \frac{1}{2}N(t) - 3\lambda t N(t) + 3\lambda \int_0^t N(s)\, ds,$$

which is the summation in Equation (7.6.8) for $F(x) = x^3$ and $M_s = \widetilde{N}(s)$. In the previous example we have already shown that $[\widetilde{N}]_t^c = 0$. Therefore, by Itô's formula in Equation (7.6.8),

$$\int_0^t \widetilde{N}(s-)^2 \, d\widetilde{N}(s)$$

$$= \frac{1}{3}\widetilde{N}(t)^3 - \frac{1}{2}N(t)^2 + \frac{1}{6}N(t) + \lambda t N(t) - \lambda \int_0^t N(s) \, ds$$

$$= \frac{1}{3}\widetilde{N}(t)^3 - \frac{1}{2}\widetilde{N}(t)^2 + \frac{1}{6}\widetilde{N}(t) + \frac{1}{6}\lambda t - \lambda \int_0^t \widetilde{N}(s) \, ds.$$

Thus we have derived the value of the stochastic integral $\int_0^t \widetilde{N}(s-)^2 \, d\widetilde{N}(s)$ in terms of $\widetilde{N}(t)$ and the Riemann integrals of its sample paths.

Finally, we consider a continuous and square integrable martingale M_t. In this case, the summation term in Equation (7.6.8) drops out. Moreover, it is a fact that $[M]_t^c = \langle M \rangle_t$, the compensator of M_t^2. Hence Theorem 7.6.1 yields the next theorem.

Theorem 7.6.4. *Let M_t be a continuous, square integrable martingale and let $F(t,x)$ be a continuous function with continuous partial derivatives $\frac{\partial F}{\partial t}$, $\frac{\partial F}{\partial x}$, and $\frac{\partial^2 F}{\partial x^2}$. Then*

$$F(t, M_t) = F(a, M_a) + \int_a^t \frac{\partial F}{\partial t}(s, M_s) \, ds$$

$$+ \int_a^t \frac{\partial F}{\partial x}(s, M_s) \, dM_s + \frac{1}{2}\int_a^t \frac{\partial^2 F}{\partial x^2}(s, M_s) \, d\langle M \rangle_s.$$

The next theorem is the multidimensional Itô's formula for continuous and square integrable martingales.

Theorem 7.6.5. *Suppose $M_1(t), \ldots, M_n(t)$ are continuous, square integrable martingales. Let $F(t, x_1, \ldots, x_n)$ be a continuous function with continuous partial derivatives $\frac{\partial F}{\partial t}$, $\frac{\partial F}{\partial x_i}$, and $\frac{\partial^2 F}{\partial x_i \partial x_j}$ for $1 \leq i, j \leq n$. Then*

$$dF = \frac{\partial F}{\partial t} \, dt + \sum_{i=1}^n \frac{\partial F}{\partial x_i} \, dM_i(t) + \frac{1}{2}\sum_{i,j=1}^n \frac{\partial^2 F}{\partial x_i \partial x_j} \, d\langle M_i, M_j \rangle_t,$$

where $\langle \cdot, \cdot \rangle_t$ is the cross variation process defined by

$$\langle X, Y \rangle_t = \frac{1}{4}\{\langle X + Y \rangle_t - \langle X - Y \rangle_t\}.$$

Exercises

1. Let $B(t)$ be a Brownian motion. Find all deterministic functions $\rho(t)$ such that $e^{B(t)+\rho(t)}$ is a martingale.

2. Let g be a bounded measurable function and let $\Delta_n = \{t_0, t_1, \ldots, t_{n-1}, t_n\}$ be a partition of $[a, b]$. Show that

$$\sum_{i=1}^{n} g(B(t_{i-1})) \left((B(t_i) - B(t_{i-1}))^2 - (t_i - t_{i-1}) \right) \longrightarrow 0$$

in $L^2(\Omega)$ as $\|\Delta_n\| = \max_{1 \le i \le n}(t_i - t_{i-1})$ tends to 0.

3. Let $f(t, x) = H_n(x; t)$, the Hermite polynomial of degree n.

   (a) Use Equation (4.7.4) to show that $\frac{\partial f}{\partial t} = -\frac{1}{2} \frac{\partial^2 f}{\partial x^2}$.

   (b) Apply Itô's formula to the function $f(t, x)$ to show that $M_n(t) = H_n(B(t); t)$ is a martingale.

   (c) Apply Itô's formula again to the function $f(t, x)^2$ to show that the compensator of $M_n(t)^2$ is given by $\langle M_n \rangle_t = n^2 \int_0^t H_{n-1}(B(s); s)^2 \, ds$.

4. Let $B_1(t)$ and $B_2(t)$ be two independent Brownian motions and let $\Delta_n = \{t_0, t_1, \ldots, t_{n-1}, t_n\}$ be a partition of $[a, b]$. Show that

$$\sum_{i=1}^{n} \left(B_1(t_i) - B_1(t_{i-1}) \right) \left(B_2(t_i) - B_2(t_{i-1}) \right) \longrightarrow 0$$

in $L^2(\Omega)$ as $\|\Delta_n\| = \max_{1 \le i \le n}(t_i - t_{i-1})$ tends to 0.

5. Let $X_t = \cos B(t) - \sin B(t)$ and $Y_t = \sin B(t) - \cos B(t)$. Show that the stochastic differential of the column vector V_t with components X_t and Y_t satisfies the linear equation

$$dV_t = \begin{bmatrix} 0 & -1 \\ 1 & 0 \end{bmatrix} V_t \, dB(t) - \frac{1}{2} V_t \, dt, \quad V_0 = \begin{bmatrix} 1 \\ -1 \end{bmatrix}.$$

6. Let $X_t = e^t \cos B(t)$, $Y_t = e^t \sin B(t)$, $Z_t = e^{2t}$. The column vector V_t with components X_t, Y_t, and Z_t represents the position at time t of an object moving in the circular paraboloid $z = x^2 + y^2$. Show that the stochastic differential of V_t satisfies the linear equation

$$dV_t = \begin{bmatrix} 0 & -1 & 0 \\ 1 & 0 & 0 \\ 0 & 0 & 0 \end{bmatrix} V_t \, dB(t) + \begin{bmatrix} \frac{1}{2} & 0 & 0 \\ 0 & \frac{1}{2} & 0 \\ 0 & 0 & 2 \end{bmatrix} V_t \, dt, \quad V_0 = \begin{bmatrix} 1 \\ 0 \\ 1 \end{bmatrix}.$$

7. Use Itô's formula to evaluate the stochastic integral $\int_0^t \widetilde{N}(s-)^3 \, d\widetilde{N}(s)$.

8. Let $B_1(t)$ and $B_2(t)$ be two independent Brownian motions. Evaluate $\langle B_1 + B_2 \rangle_t$, $\langle B_1 - B_2 \rangle_t$, and $\langle B_1, B_2 \rangle_t$.

9. Let $X_t = \int_a^t f(s) \, dB(s)$, $Y_t = \int_a^t g(s) \, dB(s)$, $f, g \in L^2_{ad}([a, b] \times \Omega)$. Show that $\langle X, Y \rangle_t = \int_a^t f(s) g(s) \, ds$.

Applications of the Itô Formula

The Itô formula, like the chain rule in the Leibniz–Newton calculus, plays a fundamental role in the Itô calculus. In this chapter we will give several interesting applications of Itô's formula to some topics in stochastic analysis such as evaluation of stochastic integrals, Doob–Meyer decomposition, the Stratonovich integral, Lévy's characterization theorem of Brownian motion, Tanaka's formula, and the Girsanov theorem.

8.1 Evaluation of Stochastic Integrals

The Fundamental Theorem of Leibniz–Newton Calculus states that if F is an antiderivative of a continuous function f on $[a, b]$, then

$$\int_a^b f(x)\, dx = F(x)\Big]_a^b \equiv F(b) - F(a).$$

This theorem is used to compute the definite integral of f once we have an antiderivative of f.

Do we have such a fundamental theorem in the Itô calculus? More precisely, suppose $g \in \mathcal{L}_{\mathrm{ad}}(\Omega, L^2[a, b])$ (see Notation 5.1.1). Is there a formula that we can use to compute the stochastic integral $\int_a^b g(t)\, dB(t)$? In general, there is no such a formula. However, when $g(t)$ is of the form $g(t) = f(B(t))$ for a continuous function f with continuous derivative, then Itô's formula is such a formula. We state this fact as the next theorem, which is just another way to write Itô's formula.

Theorem 8.1.1. *Suppose $F(t, x)$ is an antiderivative in x of a continuous function $f(t, x)$. Assume that $\frac{\partial F}{\partial t}$ and $\frac{\partial f}{\partial x}$ are continuous. Then*

$$\int_a^b f(t, B(t))\, dB(t) = F(t, B(t))\Big]_a^b - \int_a^b \left(\frac{\partial F}{\partial t}(t, B(t)) + \frac{1}{2}\frac{\partial f}{\partial x}(t, B(t)) \right) dt.$$

Observe that the integral in the right-hand side of the above equality is a Riemann integral for each sample path of the Brownian motion $B(t)$. On the other hand, note that in order to use this theorem, we need to find an antiderivative $F(t, x)$ of $f(t, x)$ in the x variable. Thus in this spirit, the above form of Itô's formula is the fundamental theorem of the Itô calculus. When the integrand f does not depend on t, the above equality becomes

$$\int_a^b f(B(t))\, dB(t) = F(B(t))\Big]_a^b - \frac{1}{2} \int_a^b f'(B(t))\, dt. \qquad (8.1.1)$$

We give several examples below to illustrate this technique of evaluating stochastic integrals.

Example 8.1.2. To evaluate the stochastic integral $\int_0^t B(s)e^{B(s)}\, dB(s)$, note that the integrand is given by $f(B(s))$ with $f(x) = xe^x$. Hence

$$F(x) = \int xe^x\, dx = xe^x - \int e^x\, dx = xe^x - e^x + C,$$
$$f'(x) = xe^x + e^x.$$

Therefore, by Equation (8.1.1),

$$\int_0^t B(s)e^{B(s)}\, dB(s) = \left(B(t) - 1\right)e^{B(t)} + 1 - \frac{1}{2} \int_0^t \left(B(s) + 1\right)e^{B(s)}\, ds.$$

Example 8.1.3. The integrand of the stochastic integral $\int_0^t \frac{1}{1+B(s)^2}\, dB(s)$ is given by $f(B(s))$ with $f(x) = \frac{1}{1+x^2}$. Hence $F(x)$ and $f'(x)$ are given by

$$F(x) = \int \frac{1}{1+x^2}\, dx = \arctan x + C, \quad f'(x) = -\frac{2x}{(1+x^2)^2}.$$

By Equation (8.1.1), we have

$$\int_0^t \frac{1}{1+B(s)^2}\, dB(s) = \arctan B(t) + \int_0^t \frac{B(s)}{(1+B(s)^2)^2}\, ds.$$

Example 8.1.4. Consider the stochastic integral $\int_0^t \frac{B(s)}{1+B(s)^2}\, dB(s)$. Note that the integrand is given by $f(B(s))$ with $f(x) = \frac{x}{1+x^2}$ and we have

$$F(x) = \int \frac{x}{1+x^2}\, dx = \frac{1}{2} \log(1+x^2) + C, \quad f'(x) = \frac{1-x^2}{(1+x^2)^2}.$$

Then apply Equation (8.1.1) to find the value of the stochastic integral

$$\int_0^t \frac{B(s)}{1+B(s)^2}\, dB(s) = \frac{1}{2} \log\left(1 + B(t)^2\right) - \frac{1}{2} \int_0^t \frac{1 - B(s)^2}{(1+B(s)^2)^2}\, ds.$$

Example 8.1.5. To evaluate the stochastic integral $\int_0^t e^{B(s)-\frac{1}{2}s}\,dB(s)$, note that the integrand is given by $f(s, B(s))$ with $f(t, x) = e^{x-\frac{1}{2}t}$. Hence

$$F(t, x) = e^{x-\frac{1}{2}t}, \quad \frac{\partial F}{\partial t} = -\frac{1}{2}e^{x-\frac{1}{2}t}, \quad \frac{\partial f}{\partial x} = e^{x-\frac{1}{2}t}.$$

Then by Theorem 8.1.1 we have

$$\int_0^t e^{B(s)-\frac{1}{2}s}\,dB(s) = e^{B(t)-\frac{1}{2}t} - 1.$$

8.2 Decomposition and Compensators

Itô's formula can be applied to find the Doob–Meyer decomposition for certain submartingales that are functions of a Brownian motion $B(t)$. In particular, we can find the compensator $\langle M \rangle_t$ of M_t^2 for $M_t = \int_a^t f(s)\,dB(s)$.

Let $f \in L^2_{\mathrm{ad}}([a, b] \times \Omega)$ (see Notation 4.3.1) and consider the stochastic process M_t defined by

$$M_t = \int_a^t f(s)\,dB(s), \quad a \leq t \leq b. \tag{8.2.1}$$

By Theorem 4.6.1, the stochastic process M_t is a martingale. Let φ be a C^2-function. Then by Itô's formula,

$$\varphi(M_t) = \varphi(0) + \int_a^t \varphi'(M_s)f(s)\,dB(s) + \frac{1}{2}\int_a^t \varphi''(M_s)f(s)^2\,ds. \tag{8.2.2}$$

Suppose further that φ is a convex function and $E\int_a^b |\varphi'(M_t)f(t)|^2\,dt < \infty$. Then $\varphi(M_t)$ is a submartingale by the conditional Jensen's inequality (see Section 2.4) and the sum of the first two terms in the right-hand side of the above equation is a martingale. Moreover, the last integral is an increasing process since $\varphi''(x) \geq 0$ by the convexity of φ. Hence Equation (8.2.2) gives the Doob–Meyer decomposition of the submartingale $\varphi(M_t)$. In particular, when $\varphi(x) = x^2$, we have

$$M_t^2 = 2\int_a^t M_s f(s)\,dB(s) + \int_a^t f(s)^2\,ds.$$

Hence the compensator of M_t^2 for $M(t)$ in Equation (8.2.1) is given by

$$\langle M \rangle_t = \int_a^t f(s)^2\,ds,$$

or equivalently,

$$d\langle M \rangle_t = f(t)^2\,dt = (dM_t)^2, \quad \langle M \rangle_a = 0. \tag{8.2.3}$$

In general, suppose we start with a reasonably good function $\theta(t, x)$. We can apply Itô's formula to $\theta(t, B(t))$ to get a martingale

$$M_t = \int_a^t \frac{\partial \theta}{\partial x}(s, B(s)) \, dB(s).$$

Then Equation (8.2.3) can be used to find the compensator $\langle M \rangle_t$ of M_t^2. We give several examples below to show this method.

Example 8.2.1. Consider the martingale $M_t = B(t)^2 - t$. There are three ways to find the compensator of M_t^2. The first way was done in Example 6.4.5. The second way is to apply Itô's formula to the function $f(t, x) = (x^2 - t)^2$ in order to find the Doob–Meyer decomposition of M_t^2, which yields the compensator of M_t^2. This is somewhat complicated. The third way is to use the following representation of M_t as an Itô integral:

$$M_t = B(t)^2 - t = 2 \int_0^t B(s) \, dB(s), \tag{8.2.4}$$

so that $dM_t = 2B(t) \, dB(t)$. Then by Equation (8.2.3),

$$d\langle M \rangle_t = (dM_t)^2 = 4B(t)^2 \, dt,$$

or equivalently in the integral form

$$\langle M \rangle_t = 4 \int_0^t B(s)^2 \, ds,$$

which is the same as Equation (6.4.11). But the derivation here is easier due to the use of Equation (8.2.3) and the representation in Equation (8.2.4).

Example 8.2.2. By the last example, the stochastic process

$$K_t = (B(t)^2 - t)^2 - 4 \int_0^t B(s)^2 \, ds$$

is a martingale. In order to find the compensator of K_t^2, let us first find the representation of K_t as an Itô integral. Apply Itô's formula to get

$$dK_t = -2(B(t)^2 - t) \, dt + 2(B(t)^2 - t)2B(t) \, dB(t)$$

$$+ \frac{1}{2} \left[4(B(t)^2 - t) + 8B(t)^2 \right] dt - 4B(t)^2 \, dt$$

$$= 4B(t)(B(t)^2 - t) \, dB(t).$$

Hence we have

$$d\langle K \rangle_t = (dK_t)^2 = 16B(t)^2(B(t)^2 - t)^2 \, dt,$$

which can be written in an integral form to get the compensator of K_t^2:

$$\langle K \rangle_t = 16 \int_0^t B(s)^2(B(s)^2 - s)^2 \, ds.$$

Example 8.2.3. Let M_t be a continuous and square integrable martingale with $M_0 = 0$. Then $G_t \equiv M_t^2 - \langle M \rangle_t$ is a martingale. In fact, we have

$$G_t = 2 \int_0^t M_s \, dM_s.$$

Therefore, the compensator $\langle G \rangle_t$ is given by

$$\langle G \rangle_t = 4 \int_0^t M_s^2 \, d\langle M \rangle_s.$$

Hence the stochastic process H_t defined by

$$H_t = (M_t^2 - \langle M \rangle_t)^2 - 4 \int_0^t M_s^2 \, d\langle M \rangle_s$$

is a martingale. In fact, we can use Theorem 7.6.4 to carry out the same computation as in the previous example to get

$$H_t = 4 \int_0^t M_s (M_s^2 - \langle M \rangle_s) \, dM_s.$$

Thus the compensator $\langle H \rangle_t$ is given by

$$\langle H \rangle_t = 16 \int_0^t M_s^2 (M_s^2 - \langle M \rangle_s)^2 \, d\langle M \rangle_s.$$

8.3 Stratonovich Integral

Consider a simple integral in the Leibniz–Newton calculus:

$$\int_a^b e^x \, dx = e^b - e^a. \tag{8.3.1}$$

The corresponding equality in the Itô calculus is given by Equation (7.1.6) with $t = b$, namely,

$$\int_a^b e^{B(t)} \, dB(t) = e^{B(b)} - e^{B(a)} - \frac{1}{2} \int_a^b e^{B(t)} \, dt. \tag{8.3.2}$$

Obviously, the difference between these two equalities is the extra term in Equation (8.3.2). This extra term is a consequence of the nonzero quadratic variation of the Brownian motion $B(t)$. But it is also a consequence of how an Itô integral is defined, namely, the evaluation points for the integrand are the left endpoints of the intervals in a partition.

Recall the two limits in Equations (4.1.9) and (4.1.10) resulting from the evaluation of the integrand at the right and left endpoints, respectively. If we take the average of these two limits, then the extra term $b-a$ disappears. Thus the resulting integral will behave in the same way as in the Leibniz–Newton calculus. This new integral was introduced by Stratonovich in [76].

Below we use Itô's idea to define Stratonovich integrals within the class of Itô processes.

Definition 8.3.1. *Let X_t and Y_t be Itô processes. The* Stratonovich integral *of X_t with respect to Y_t is defined by*

$$\int_a^b X_t \circ dY_t = \int_a^b X_t \, dY_t + \frac{1}{2} \int_a^b (dX_t)(dY_t), \qquad (8.3.3)$$

or equivalently in the stochastic differential form

$$X_t \circ dY_t = X_t \, dY_t + \frac{1}{2}(dX_t)(dY_t). \qquad (8.3.4)$$

Suppose X_t and Y_t are given by

$$X_t = X_a + \int_a^t f(s) \, dB(s) + \int_a^t \xi(s) \, ds, \quad a \le t \le b,$$

$$Y_t = Y_a + \int_a^t g(s) \, dB(s) + \int_a^t \eta(s) \, ds, \quad a \le t \le b,$$

where X_a and Y_a are $\mathcal{F}_a$-measurable, $f, g \in \mathcal{L}_{\mathrm{ad}}(\Omega, L^2[a,b])$, and $\xi, \eta \in \mathcal{L}_{\mathrm{ad}}(\Omega, L^1[a,b])$. Then we have $(dX_t)(dY_t) = f(t)g(t) \, dt$ and so Equation (8.3.3) means that

$$\int_a^b X_t \circ dY_t = \int_a^b X_t g(t) \, dB(t) + \int_a^b \left(X_t \eta(t) + \frac{1}{2} f(t)g(t) \right) dt. \qquad (8.3.5)$$

Recall that X_t is a continuous stochastic process by Theorem 5.5.5. Hence almost all of its sample paths are continuous functions. Therefore,

$$\int_a^b |X_t g(t)|^2 \, dt \le \sup_{a \le s \le b} |X_s|^2 \int_a^b |g(t)|^2 \, dt < \infty, \quad \text{almost surely.}$$

In addition, $X_t g(t)$ is adapted. Thus $Xg \in \mathcal{L}_{\mathrm{ad}}(\Omega, L^2[a,b])$. Similarly,

$$\int_a^b |X_t \eta(t)| \, dt \le \sup_{a \le s \le b} |X_s| \int_a^b |\eta(t)| \, dt < \infty, \quad \text{almost surely,}$$

and $X_t \eta(t)$ is adapted. Hence $X\eta \in \mathcal{L}_{\mathrm{ad}}(\Omega, L^1[a,b])$. Moreover, $f(t)g(t)$ is adapted and by the Schwarz inequality

$$\int_a^b |f(t)g(t)| \, dt \le \left[\int_a^b |f(t)|^2 \, dt \right]^{1/2} \left[\int_a^b |g(t)|^2 \, dt \right]^{1/2} < \infty, \quad \text{almost surely.}$$

Hence $fg \in \mathcal{L}_{\mathrm{ad}}(\Omega, L^1[a,b])$. It follows from Equation (8.3.5) with $b = t$ that the stochastic process $L_t = \int_a^t X_s \circ dY_s$ is also an Itô process.

We summarize the above discussion as the next theorem.

Theorem 8.3.2. *If X_t and Y_t are Itô processes, then the stochastic process*

$$L_t = \int_a^t X_s \circ dY_s, \quad a \leq t \leq b,$$

is also an Itô process.

This theorem implies that the collection of Itô processes is closed under the operation of taking the Stratonovich integral of two Itô processes.

Example 8.3.3. To evaluate the integral $\int_a^b e^{B(t)} \circ dB(t)$, we use Itô's formula:

$$e^{B(t)} \circ dB(t) = e^{B(t)} dB(t) + \frac{1}{2}(de^{B(t)})(dB(t))$$

$$= e^{B(t)} dB(t) + \frac{1}{2}\left(e^{B(t)} dB(t) + \frac{1}{2}e^{B(t)} dt\right) dB(t)$$

$$= e^{B(t)} dB(t) + \frac{1}{2}e^{B(t)} dt.$$

Therefore, we have

$$\int_a^b e^{B(t)} \circ dB(t) = \int_a^b e^{B(t)} dB(t) + \frac{1}{2}\int_a^b e^{B(t)} dt.$$

Then use Equation (8.3.2) to get

$$\int_a^b e^{B(t)} \circ dB(t) = e^{B(b)} - e^{B(a)},$$

which is the counterpart part of Equation (8.3.1) in the Itô calculus.

In general, consider a function $F(t,x)$. We can use Equation (8.3.4) and Itô's formula to get

$$\frac{\partial F}{\partial x}(t, B(t)) \circ dB(t)$$

$$= \frac{\partial F}{\partial x}(t, B(t)) dB(t) + \frac{1}{2}\left(d\frac{\partial F}{\partial x}(t, B(t))\right)(dB(t))$$

$$= \frac{\partial F}{\partial x} dB(t) + \frac{1}{2}\left(\frac{\partial^2 F}{\partial x \partial t} dt + \frac{\partial^2 F}{\partial x^2} dB(t) + \frac{1}{2}\frac{\partial^3 F}{\partial x^3} dt\right) dB(t)$$

$$= \frac{\partial F}{\partial x}(t, B(t)) dB(t) + \frac{1}{2}\frac{\partial^2 F}{\partial x^2}(t, B(t)) dt. \tag{8.3.6}$$

On the other hand, by Itô's formula, we have

$$dF(t, B(t)) = \frac{\partial F}{\partial t}(t, B(t)) dt + \frac{\partial F}{\partial x}(t, B(t)) dB(t) + \frac{1}{2}\frac{\partial^2 F}{\partial x^2}(t, B(t)) dt.$$

Therefore, we have the equality

$$\frac{\partial F}{\partial x}(t, B(t)) \circ dB(t) = dF(t, B(t)) - \frac{\partial F}{\partial t}(t, B(t)) dt.$$

This equality is converted into an integral form in the next theorem.

Theorem 8.3.4. *Suppose $F(t,x)$ is an antiderivative in x of a continuous function $f(t,x)$. Assume that $\frac{\partial F}{\partial t}, \frac{\partial f}{\partial t}$, and $\frac{\partial f}{\partial x}$ are continuous. Then*

$$\int_a^b f(t, B(t)) \circ dB(t) = F(t, B(t))\Big]_a^b - \int_a^b \frac{\partial F}{\partial t}(t, B(t))\, dt.$$

In particular, when the function f does not depend on t, we have

$$\int_a^b f(B(t)) \circ dB(t) = F(B(t))\Big]_a^b. \tag{8.3.7}$$

This equation shows that the Stratonovich integral behaves like the integral in the Leibniz–Newton calculus. Note that Equations (8.1.1) and (8.3.7) are actually equivalent. Hence if we know the value of an Itô integral, then we can use Equation (8.3.3) to evaluate the corresponding Stratonovich integral. Conversely, we can use Theorem 8.3.4 to evaluate a given Stratonovich integral and then use Equation (8.3.6) with $F'(x) = f(x)$ to find the value of the corresponding Itô integral, namely,

$$\int_a^b f(B(t))\, dB(t) = \int_a^b f(B(t)) \circ dB(t) - \frac{1}{2}\int_a^b f'(B(t))\, dt. \tag{8.3.8}$$

Example 8.3.5. We want to evaluate the Itô integral $\int_a^b \sin B(t)\, dB(t)$ and the Stratonovich integral $\int_a^b \sin B(t) \circ dB(t)$. Use Equation (8.3.7) to get

$$\int_a^b \sin B(t) \circ dB(t) = -\cos B(t)\Big]_a^b = -\cos B(b) + \cos B(a).$$

Then use Equation (8.3.8) to find

$$\int_a^b \sin B(t)\, dB(t) = \int_a^b \sin B(t) \circ dB(t) - \frac{1}{2}\int_a^b \cos B(t)\, dt$$

$$= -\cos B(b) + \cos B(a) - \frac{1}{2}\int_a^b \cos B(t)\, dt.$$

Example 8.3.6. Use Equation (8.3.7) to obtain

$$\int_0^t e^{B(s)} \circ dB(s) = e^{B(s)}\Big]_0^t = e^{B(t)} - 1,$$

which can be written in stochastic differential form as $de^{B(t)} = e^{B(t)} \circ dB(t)$. Hence $X_t = e^{B(t)}$ is a solution of the equation

$$dX_t = X_t \circ dB(t), \quad X_0 = 1.$$

On the other hand, from Example 8.1.5 we see that $Y_t = e^{B(t) - \frac{1}{2}t}$ is a solution of the equation

$$dY_t = Y_t\, dB(t), \quad Y_0 = 1. \tag{8.3.9}$$

Note that $Ee^{B(t)} = e^{\frac{1}{2}t}$. Thus $Y_t = X_t/EX_t$. The stochastic process Y_t, being defined as X_t divided by its expectation, is often referred to as the multiplicative renormalization of X_t.

We have defined a Stratonovich integral in terms of an Itô integral in the collection of Itô processes. It is natural to ask the following questions:

(1) Can a Stratonovich integral $\int_a^b f(t) \circ dB(t)$ be directly defined as the limit of Riemann-like sums?
(2) What is the class of stochastic processes $f(t)$ for which $\int_a^b f(t) \circ dB(t)$ is defined?

For the second question, it is somewhat involved to state the answer, and we refer to the book by Elworthy [14] and the original paper by Stratonovich [76] for more information. The answer to the first question for an integrand of the form $f(t, B(t))$ is given in the next theorem.

Theorem 8.3.7. *Let $f(t, x)$ be a continuous function with continuous partial derivatives $\frac{\partial f}{\partial t}$, $\frac{\partial f}{\partial x}$, and $\frac{\partial^2 f}{\partial x^2}$. Then*

$$\int_a^b f(t, B(t)) \circ dB(t)$$

$$= \lim_{\|\Delta_n\| \to 0} \sum_{i=1}^n f\left(t_i^*, \frac{1}{2}\big(B(t_{i-1}) + B(t_i)\big)\right)\big(B(t_i) - B(t_{i-1})\big) \tag{8.3.10}$$

$$= \lim_{\|\Delta_n\| \to 0} \sum_{i=1}^n f\left(t_i^*, B\Big(\frac{t_{i-1} + t_i}{2}\Big)\right)\big(B(t_i) - B(t_{i-1})\big) \tag{8.3.11}$$

in probability, where $t_{i-1} \leq t_i^* \leq t_i$, $\Delta_n = \{t_0, t_1, \ldots, t_{n-1}, t_n\}$ is a partition of the finite interval $[a, b]$, and $\|\Delta_n\| = \max_{1 \leq i \leq n} (t_i - t_{i-1})$.

Proof. We only sketch the ideas of the proof. The arguments for a rigorous proof are the same as those in Sections 7.1 and 7.2. For simplicity, we assume that the function f does not depend on t. For the first limit, note that for each $i = 1, 2, \ldots, n$,

$$\left[f\left(\frac{1}{2}\big(B(t_{i-1}) + B(t_i)\big)\right) - f(B(t_{i-1}))\right]\big(B(t_i) - B(t_{i-1})\big)$$

$$\approx \frac{1}{2}f'(B(t_{i-1}))\big(B(t_i) - B(t_{i-1})\big)^2 \approx \frac{1}{2}f'(B(t_{i-1}))(t_i - t_{i-1}).$$

Therefore, summing up over i, we have

$$\sum_{i=1}^n f\left(\frac{1}{2}\big(B(t_{i-1}) + B(t_i)\big)\right)\big(B(t_i) - B(t_{i-1})\big)$$

$$\approx \sum_{i=1}^n f(B(t_{i-1}))\big(B(t_i) - B(t_{i-1})\big) + \frac{1}{2}\sum_{i=1}^n f'(B(t_{i-1}))(t_i - t_{i-1}).$$

Note that the limit in probability of the right-hand side of this equation is

$$\int_a^b f(B(t))\, dB(t) + \frac{1}{2}\int_a^b f'(B(t))\, dt,$$

which, by Equation (8.3.8), equals the Stratonovich integral $\int_a^b f(B(t)) \circ dB(t)$. Hence Equation (8.3.10) is proved.

For the limit in Equation (8.3.11), note that for each $i = 1, 2, \ldots, n$,

$$\left[f\left(B\left(\frac{t_{i-1}+t_i}{2}\right)\right) - f(B(t_{i-1}))\right]\left(B(t_i) - B(t_{i-1})\right)$$

$$\approx f'(B(t_{i-1}))\left[B\left(\frac{t_{i-1}+t_i}{2}\right) - B(t_{i-1})\right]\left(B(t_i) - B(t_{i-1})\right)$$

$$= f'(B(t_{i-1}))\left[B\left(\frac{t_{i-1}+t_i}{2}\right) - B(t_{i-1})\right]^2$$

$$+ f'(B(t_{i-1}))\left[B\left(\frac{t_{i-1}+t_i}{2}\right) - B(t_{i-1})\right]\left[B(t_i) - B\left(\frac{t_{i-1}+t_i}{2}\right)\right]$$

$$\approx \frac{1}{2}f'(B(t_{i-1}))(t_i - t_{i-1}).$$

Then we can use the same arguments as those above for proving Equation (8.3.10) to show that the limit in Equation (8.3.11) is the Stratonovich integral $\int_a^b f(B(t)) \circ dB(t)$. □

8.4 Lévy's Characterization Theorem

In Chapter 2 a Brownian motion is defined to be a stochastic process $B(t)$ satisfying conditions (1), (2), (3), and (4) in Definition 2.1.1. Obviously, the probability measure P in the underlying probability space $(\Omega, \mathcal{F}, P)$ plays a crucial role in these conditions. To emphasize this fact, if necessary, we will say that $B(t)$ is a Brownian motion with respect to P.

Question 8.4.1. Given a stochastic process, how can we tell whether it is a Brownian motion with respect to some probability measure?

Consider a simple example. Let $B(t), 0 \le t \le 1$, be a given Brownian motion with respect to P. Is the stochastic process $M_t = B(t) - t, 0 \le t \le 1$, a Brownian motion with respect to some probability measure Q?

Define a function $Q : \mathcal{F} \longrightarrow [0, \infty)$ by

$$Q(A) = \int_A e^{B(1) - \frac{1}{2}}\, dP, \quad A \in \mathcal{F}. \tag{8.4.1}$$

Let E_P denote the expectation with respect to P. We have $E_P e^{B(1)} = e^{\frac{1}{2}}$, which implies that $Q(\Omega) = 1$. Thus Q is a probability measure. We will show that the stochastic process $M_t = B(t) - t$, $0 \leq t \leq 1$, is indeed a Brownian motion with respect to Q. (In Section 8.9 we will show how to find Q.)

It is easily seen that $Q(A) = 0$ if and only if $P(A) = 0$. Hence Q and P are equivalent. It follows that $Q\{\omega\,;\, M_0(\omega) = 0\} = P\{\omega\,;\, B(0,\omega) = 0\} = 1$ and $Q\{\omega\,;\, M_t(\omega) \text{ is continuous in t}\} = P\{\omega\,;\, B(t,\omega) \text{ is continuous in t}\} = 1$. Thus conditions (1) and (4) in Definition 2.1.1 are satisfied for M_t with respect to the probability measure Q.

Next we show that condition (2) in Definition 2.1.1 is satisfied for M_t with respect to Q. Let E_Q denote the expectation with respect to Q and let $0 \leq s < t \leq 1$. Then for any $\lambda \in \mathbb{R}$,

$$E_Q\, e^{i\lambda(M_t - M_s)} = e^{-i\lambda(t-s)} E_Q\, e^{i\lambda(B(t) - B(s))}. \tag{8.4.2}$$

By the definition of Q in Equation (8.4.1) we have

$$\begin{aligned}
E_Q\, e^{i\lambda(B(t) - B(s))} &= E_P\big(e^{i\lambda(B(t) - B(s))} e^{B(1) - \frac{1}{2}}\big) \\
&= E_P\big(E_P\big[e^{i\lambda(B(t) - B(s))} e^{B(1) - \frac{1}{2}} \,\big|\, \mathcal{F}_t\big]\big) \\
&= E_P\big(e^{i\lambda(B(t) - B(s))} E_P\big[e^{B(1) - \frac{1}{2}} \,\big|\, \mathcal{F}_t\big]\big), \tag{8.4.3}
\end{aligned}$$

where $\mathcal{F}_t = \sigma\{B(s)\,;\, s \leq t\}$ and $E_P[\,\cdot\,|\,\mathcal{F}_t]$ denotes the conditional expectation given $\mathcal{F}_t$ with respect to P. By Example 7.3.4, the stochastic process $e^{B(t) - \frac{1}{2}t}$ is a martingale. Hence

$$E_P\big[e^{B(1) - \frac{1}{2}} \,\big|\, \mathcal{F}_t\big] = e^{B(t) - \frac{1}{2}t}. \tag{8.4.4}$$

Put this equality into Equation (8.4.3) to get

$$E_Q\, e^{i\lambda(B(t) - B(s))} = e^{-\frac{1}{2}t} E_P\big(e^{(i\lambda+1)(B(t) - B(s))} \cdot e^{B(s)}\big).$$

Note that the factors inside the parentheses are independent. Moreover, we have $E_P e^{c(B(t) - B(s))} = e^{\frac{1}{2}c^2(t-s)}$ for $s \leq t$. Therefore,

$$E_Q\, e^{i\lambda(B(t) - B(s))} = e^{(-\frac{1}{2}\lambda^2 + i\lambda)(t-s)}. \tag{8.4.5}$$

It follows from Equations (8.4.2) and (8.4.5) that

$$E_Q\, e^{i\lambda(M_t - M_s)} = e^{-\frac{1}{2}\lambda^2(t-s)}.$$

Therefore, the random variable $M_t - M_s$ is normally distributed with mean 0 and variance $t - s$ with respect to Q. This shows that condition (2) in Definition 2.1.1 is satisfied for M_t under the probability measure Q.

Now let $0 \leq t_1 < t_2 \leq 1$. Then for any $\lambda_1, \lambda_2 \in \mathbb{R}$,

$$\begin{aligned}
&E_Q\, e^{i\lambda_1 M_{t_1} + i\lambda_2(M_{t_2} - M_{t_1})} \\
&\qquad = e^{-i\lambda_1 t_1 - i\lambda_2(t_2 - t_1)} E_Q\, e^{i\lambda_1 B(t_1) + i\lambda_2(B(t_2) - B(t_1))}. \tag{8.4.6}
\end{aligned}$$

We can use the same arguments as those in the above derivation of Equations (8.4.3), (8.4.4), and (8.4.5) to show that

$$E_Q \, e^{i\lambda_1 B(t_1) + i\lambda_2 (B(t_2) - B(t_1))}$$

$$= e^{-\frac{1}{2} t_2} E_P \big(e^{(i\lambda_2 + 1)(B(t_2) - B(t_1))} \cdot e^{(i\lambda_1 + 1)B(t_1)} \big)$$

$$= e^{-\frac{1}{2}\lambda_1^2 t_1 - \frac{1}{2}\lambda_2^2 (t_2 - t_1) + i\lambda_1 t_1 + i\lambda_2 (t_2 - t_1)}. \tag{8.4.7}$$

Then we put Equation (8.4.7) into Equation (8.4.6) to get

$$E_Q \, e^{i\lambda_1 M_{t_1} + i\lambda_2 (M_{t_2} - M_{t_1})} = e^{-\frac{1}{2}\lambda_1^2 t_1 - \frac{1}{2}\lambda_2^2 (t_2 - t_1)}.$$

In general, we can repeat the above arguments inductively to show that for any $0 \leq t_1 < t_2 < \cdots < t_n \leq 1$,

$$E_Q \, e^{i\lambda_1 M_{t_1} + i\lambda_2 (M_{t_2} - M_{t_1}) + \cdots + i\lambda_n (M_{t_n} - M_{t_{n-1}})}$$

$$= e^{-\frac{1}{2}\lambda_1^2 t_1 - \frac{1}{2}\lambda_2^2 (t_2 - t_1) - \cdots - \frac{1}{2}\lambda_n^2 (t_n - t_{n-1})}, \quad \forall \lambda_1, \lambda_2, \ldots, \lambda_n \in \mathbb{R}.$$

This equality implies that $M_{t_1}, M_{t_2} - M_{t_1}, \ldots, M_{t_n} - M_{t_{n-1}}$ are independent and normally distributed under Q. In particular, M_t satisfies condition (3) in Definition 2.1.1 with respect to Q. This completes the proof that the stochastic process $M_t = B(t) - t$ is a Brownian motion with respect to the probability measure Q defined in Equation (8.4.1).

Obviously, the stochastic process $M_t = B(t) - t$ is not a Brownian motion with respect to P since $E_P(M_t) = -t \neq 0$. However, as shown above, M_t is indeed a Brownian motion with respect to Q. In general, the answer to Question 8.4.1 is given by the next theorem. Recall that $\langle M \rangle_t$ denotes the compensator of M_t^2 for a right continuous, square integrable martingale M_t with left-hand limits.

Theorem 8.4.2. (Lévy's characterization theorem) *A stochastic process M_t is a Brownian motion if and only if there exist a probability measure Q and a filtration $\{\mathcal{F}_t\}$ such that M_t is a continuous martingale with respect to $\{\mathcal{F}_t\}$ under Q, $Q\{M_0 = 0\} = 1$, and $\langle M \rangle_t = t$ almost surely with respect to Q for each t.*

Remark 8.4.3. Recall that the compensated Poisson process $\widetilde{N}(t) = N(t) - \lambda t$ is a martingale and $\langle \widetilde{N} \rangle_t = \lambda t$. Hence the compensated Poisson process with $\lambda = 1$ satisfies all conditions in the above theorem except for the continuity. Therefore, it is the continuity property that distinguishes a Brownian motion from a compensated Poisson process with $\lambda = 1$. We also mention that the theorem remains valid when "martingale" is replaced by "local martingale." For the proof, see the book by Durrett [11].

Proof. The necessity part of this theorem is obvious. We prove the sufficiency part. The underlying probability space is $(\Omega, \mathcal{F}, Q)$ with the filtration $\{\mathcal{F}_t\}$.

By assumption, M_t satisfies conditions (1) and (4) in Definition 2.1.1. Hence we only need to verify conditions (2) and (3). Apply Itô's formula in Theorem 7.6.4 to the martingale M_t and the function $F(t,x) = e^{i\lambda x + \frac{1}{2}\lambda^2 t}$ to get

$$dF(t, M_t) = \frac{1}{2}\lambda^2 F(t, M_t)\, dt + i\lambda F(t, M_t)\, dM_t - \frac{1}{2}\lambda^2 F(t, M_t)\, d\langle M\rangle_t.$$

By assumption, $d\langle M\rangle_t = dt$. Hence we have $dF(t, M_t) = i\lambda F(t, M_t)\, dM_t$, which can be written in the integral form as

$$e^{i\lambda M_t + \frac{1}{2}\lambda^2 t} = 1 + i\lambda \int_0^t e^{i\lambda M_s + \frac{1}{2}\lambda^2 s}\, dM_s.$$

Hence by Theorem 6.5.8, the stochastic process $e^{i\lambda M_t + \frac{1}{2}\lambda^2 t}$ is a martingale with respect to Q and $\{\mathcal{F}_t\}$. Then for any $0 \le s \le t$,

$$E_Q\big[e^{i\lambda M_t + \frac{1}{2}\lambda^2 t} \,\big|\, \mathcal{F}_s\big] = e^{i\lambda M_s + \frac{1}{2}\lambda^2 s}, \tag{8.4.8}$$

or equivalently,

$$E_Q\big[e^{i\lambda(M_t - M_s)} \,\big|\, \mathcal{F}_s\big] = e^{-\frac{1}{2}\lambda^2(t-s)}.$$

Upon taking the expectation in both sides we get

$$E_Q\, e^{i\lambda(M_t - M_s)} = e^{-\frac{1}{2}\lambda^2(t-s)}, \quad \forall \lambda \in \mathbb{R}. \tag{8.4.9}$$

This equality shows that $M_t - M_s$, $s \le t$, is normally distributed with mean 0 and variance $t - s$ under Q. Hence M_t satisfies condition (2) in Definition 2.1.1 with respect to Q.

Next, suppose $0 \le t_1 < t_2$. For any $\lambda_1, \lambda_2 \in \mathbb{R}$,

$$E_Q\, e^{i\lambda_1 M_{t_1} + i\lambda_2(M_{t_2} - M_{t_1})} = E_Q\big(E_Q\big[e^{i\lambda_1 M_{t_1} + i\lambda_2(M_{t_2} - M_{t_1})} \,\big|\, \mathcal{F}_{t_1}\big]\big)$$
$$= E_Q\big(e^{i(\lambda_1 - \lambda_2)M_{t_1}} E_Q\big[e^{i\lambda_2 M_{t_2}} \,\big|\, \mathcal{F}_{t_1}\big]\big). \tag{8.4.10}$$

But by Equation (8.4.8) with $s = t_1$ and $t = t_2$, we have

$$E_Q\big[e^{i\lambda_2 M_{t_2}} \,\big|\, \mathcal{F}_{t_1}\big] = e^{i\lambda_2 M_{t_1} - \frac{1}{2}\lambda_2^2(t_2 - t_1)}. \tag{8.4.11}$$

Put Equation (8.4.11) into Equation (8.4.10) to get

$$E_Q\, e^{i\lambda_1 M_{t_1} + i\lambda_2(M_{t_2} - M_{t_1})} = e^{-\frac{1}{2}\lambda_2^2(t_2 - t_1)} E_Q\, e^{i\lambda_1 M_{t_1}}.$$

By Equation (8.4.9), $E_Q\, e^{i\lambda_1 M_{t_1}} = e^{-\frac{1}{2}\lambda_1^2 t_1}$. Therefore, for all $\lambda_1, \lambda_2 \in \mathbb{R}$,

$$E_Q\, e^{i\lambda_1 M_{t_1} + i\lambda_2(M_{t_2} - M_{t_1})} = e^{-\frac{1}{2}\lambda_1^2 t_1 - \frac{1}{2}\lambda_2^2(t_2 - t_1)}.$$

In general, we can repeat the same arguments inductively to show that for any $0 \le t_1 < t_2 < \cdots < t_n \le 1$,

$$E_Q \, e^{i\lambda_1 M_{t_1} + i\lambda_2 (M_{t_2} - M_{t_1}) + \cdots + i\lambda_n (M_{t_n} - M_{t_{n-1}})}$$
$$= e^{-\frac{1}{2}\lambda_1^2 t_1 - \frac{1}{2}\lambda_2^2 (t_2 - t_1) - \cdots - \frac{1}{2}\lambda_n^2 (t_n - t_{n-1})}, \quad \forall \lambda_1, \lambda_2, \dots, \lambda_n \in \mathbb{R}.$$

It follows that the random variables $M_{t_1}, M_{t_2} - M_{t_1}, \dots, M_{t_n} - M_{t_{n-1}}$ are independent and normally distributed. In particular, M_t satisfies condition (3) in Definition 2.1.1 with respect to Q. This completes the proof that M_t is a Brownian motion with respect to Q. $\qquad\square$

Example 8.4.4. Let $B(t)$ be a Brownian motion with respect to P. Consider the stochastic process

$$X_t = \int_0^t \mathrm{sgn}(B(s)) \, dB(s). \tag{8.4.12}$$

Obviously, $P\{X_0 = 0\} = 1$ and X_t is a continuous martingale with respect to P and the filtration $\mathcal{F}_t = \sigma\{B(s); s \le t\}$. Moreover, the compensator of X_t^2 is given by

$$\langle X \rangle_t = \int_0^t |\mathrm{sgn}(B(s))|^2 \, ds = \int_0^t 1 \, ds = t.$$

Therefore, by Theorem 8.4.2, the stochastic process X_t is a Brownian motion with respect to the same probability measure P.

Example 8.4.5. Let $(\Omega, \mathcal{F}, P)$ be a probability space and let $B(t)$, $0 \le t \le 1$, be a Brownian motion with respect to P. Consider the stochastic process $M_t = B(t) - ct$, where c is a fixed constant. Define $Q : \mathcal{F} \longrightarrow [0, \infty)$ by

$$Q(A) = \int_A e^{cB(1) - \frac{1}{2}c^2} \, dP, \quad A \in \mathcal{F}. \tag{8.4.13}$$

We have $E_P \, e^{cB(1)} = e^{\frac{1}{2}c^2}$. Hence Q is a probability measure on $(\Omega, \mathcal{F})$. Let $\{\mathcal{F}_t; 0 \le t \le 1\}$ be a filtration given by $\mathcal{F}_t = \sigma\{B(s); s \le t\}$.

Note that Q and P are equivalent. Hence $Q\{M_0 = 0\} = P\{B(0) = 0\} = 1$. To show that M_t is a martingale with respect to Q, first note that $e^{cB(t) - \frac{1}{2}c^2 t}$ is a martingale by Example 7.3.4. Thus for any $A \in \mathcal{F}_t$,

$$\int_A M_t \, dQ = E_Q(1_A M_t) = E_P\big(1_A M_t e^{cB(1) - \frac{1}{2}c^2}\big)$$
$$= E_P\big(E_P\big[1_A M_t e^{cB(1) - \frac{1}{2}c^2} \,\big|\, \mathcal{F}_t\big]\big)$$
$$= E_P\big(1_A M_t e^{cB(t) - \frac{1}{2}c^2 t}\big)$$
$$= \int_A M_t e^{cB(t) - \frac{1}{2}c^2 t} \, dP. \tag{8.4.14}$$

Equation (8.4.14) implies that for any $0 \le s \le t$,

$$E_Q[M_t \,|\, \mathcal{F}_s] = M_s \iff E_P\big[M_t e^{cB(t) - \frac{1}{2}c^2 t} \,\big|\, \mathcal{F}_s\big] = M_s e^{cB(s) - \frac{1}{2}c^2 s},$$

or equivalently, M_t is a martingale with respect to Q if and only if $X_t = M_t e^{cB(t)-\frac{1}{2}c^2 t}$ is a martingale with respect to P. Apply Itô's formula to X_t to find that

$$X_t = \int_0^t (1 + cB(s) - c^2 s)e^{cB(s)-\frac{1}{2}c^2 s} \, dB(s),$$

which implies that X_t is a martingale with respect to P. Therefore, M_t is a martingale with respect to Q. Moreover, note that $dM_t = dB(t) - c\,dt$. Hence $d\langle M \rangle_t = (dM_t)^2 = dt$ or $\langle M \rangle_t = t$. Thus by Theorem 8.4.2 the stochastic process $M_t = B(t) - ct$ is a Brownian motion with respect to the probability measure Q defined in Equation (8.4.13).

8.5 Multidimensional Brownian Motions

In this section a Brownian motion on $\mathbb{R}$ will mean a stochastic process $B(t)$ satisfying conditions (2), (3), and (4) in Definition 2.1.1. If $P\{B(0) = x\} = 1$, then we say that $B(t)$ is a Brownian motion starting at x.

By a Brownian motion on $\mathbb{R}^n$, we mean a stochastic process

$$B(t) = (B_1(t), B_2(t), \ldots, B_n(t)),$$

where $B_1(t), B_2(t), \ldots, B_n(t)$ are n independent Brownian motions on $\mathbb{R}$. This Brownian motion starts at a point $x \in \mathbb{R}^n$.

Suppose $f : \mathbb{R}^n \to \mathbb{R}$ is a C^2-function. Then by Itô's formula in Equation (7.5.2), we have

$$\begin{aligned}
df(B(t)) &= \sum_{i=1}^n \frac{\partial f}{\partial x_i}(B(t)) \, dB_i(t) + \frac{1}{2} \sum_{i,j=1}^n \frac{\partial^2 f}{\partial x_i \partial x_j}(B(t)) \, dB_i(t)dB_j(t) \\
&= \sum_{i=1}^n \frac{\partial f}{\partial x_i}(B(t)) \, dB_i(t) + \frac{1}{2} \sum_{i=1}^n \frac{\partial^2 f}{\partial x_i^2}(B(t)) \, dt,
\end{aligned}$$

where we have used Itô Table (2) in Section 7.5 to get $dB_i(t)dB_j(t) = \delta_{ij} \, dt$. The *Laplacian* of f is defined to be the function $\Delta f = \sum_{i=1}^n \frac{\partial^2 f}{\partial x_i^2}$. Hence

$$f(B(t)) = f(x) + \sum_{i=1}^n \int_0^t \frac{\partial f}{\partial x_i}(B(s)) \, dB_i(s) + \frac{1}{2} \int_0^t \Delta f(B(s)) \, ds. \quad (8.5.1)$$

In this section we will need the following fact about the transience of a Brownian motion $B(t)$ on $\mathbb{R}^n$. For the proof and more information, see page 100 of the book by Durrett [11].

Fact. Let $n \geq 2$. Then $P\{B(t) = 0 \text{ for some } t > 0 \,|\, B(0) = x\} = 0$ for any $x \in \mathbb{R}^n$.

This fact is related to the property of the corresponding Newton potential function: $\lim_{x\to 0} p(x) = \infty$ for $n \geq 2$. Here $p(x)$ is defined by

$$p(x) = \begin{cases} -\log|x|, & \text{if } n = 2; \\ |x|^{2-n}, & \text{if } n \geq 3. \end{cases}$$

Suppose $B(t)$ starts at a point $a \neq 0$. Then Equation (8.5.1) holds for any C^2-function f on the domain $\mathbb{R}^n \setminus \{0\}$ for $n \geq 2$.

Now consider $n = 3$ and let $f(x) = |x|^{-1} = (x_1^2 + x_2^2 + x_3^2)^{-1/2}$. The function $f(x)$ is a C^2-function on the domain $\mathbb{R}^3 \setminus \{0\}$. By direct computation, we can easily check that

$$\frac{\partial f}{\partial x_i} = -\frac{x_i}{|x|^3}, \quad \frac{\partial^2 f}{\partial x_i^2} = -\frac{1}{|x|^3} + 3\frac{x_i^2}{|x|^5}, \quad i = 1, 2, 3.$$

Therefore,

$$\Delta f(x) = \sum_{i=1}^{3} \left(-\frac{1}{|x|^3} + 3\frac{x_i^2}{|x|^5} \right) = -3\frac{1}{|x|^3} + 3\frac{|x|^2}{|x|^5} = 0.$$

Then by Equation (8.5.1) we have

$$\frac{1}{|B(t)|} = \frac{1}{|a|} - \sum_{i=1}^{3} \int_0^t \frac{B_i(s)}{|B(s)|^3} \, dB_i(s).$$

By the above fact, the integrand $\frac{B_i(s)}{|B(s)|^3}$ is a continuous function of s almost surely. This implies that the integrand belongs to $\mathcal{L}_{\mathrm{ad}}(\Omega, L^2[0,t])$ for any $t > 0$. Hence by Theorem 5.5.2, the stochastic process $\frac{1}{|B(t)|}$ is a local martingale. We state this conclusion as the next theorem.

Theorem 8.5.1. *Let $B(t)$ be a Brownian motion on $\mathbb{R}^3$ starting at a point $a \neq 0$. Then the stochastic process $\frac{1}{|B(t)|}$, $t \geq 0$, is a local martingale.*

Next, consider $n \geq 2$ and let $B(t)$ be a Brownian motion on $\mathbb{R}^n$ starting at a point $a \neq 0$. The function $f(x) = |x| = (x_1^2 + x_2^2 + \cdots + x_n^2)^{1/2}$ is a C^2-function on the domain $\mathbb{R}^n \setminus \{0\}$ with partial derivatives given by

$$\frac{\partial f}{\partial x_i} = \frac{x_i}{|x|}, \quad \frac{\partial^2 f}{\partial x_i^2} = \frac{1}{|x|} - \frac{x_i^2}{|x|^3}, \quad i = 1, 2, \ldots, n.$$

Therefore, we have

$$\Delta f(x) = \sum_{i=1}^{n} \left(\frac{1}{|x|} - \frac{x_i^2}{|x|^3} \right) = n\frac{1}{|x|} - \frac{|x|^2}{|x|^3} = (n-1)\frac{1}{|x|}.$$

Then we can apply Equation (8.5.1) to the function $f(x)$ to get

$$|B(t)| = |a| + \sum_{i=1}^{n} \int_0^t \frac{B_i(s)}{|B(s)|}\, dB_i(s) + \frac{n-1}{2}\int_0^t \frac{1}{|B(s)|}\, ds. \tag{8.5.2}$$

Now we make a crucial observation that we can take the limit as $a \to 0$. For the summation term, taking the limit is justified since each integrand is dominated by 1. For the last term, the justification is a little bit involved. Suppose $B(t)$ starts at 0. Then

$$E\frac{1}{|B(t)|} = \int_{\mathbb{R}^n} \frac{1}{\sqrt{x_1^2 + \cdots + x_n^2}}\left(\frac{1}{\sqrt{2\pi t}}\right)^n e^{-\frac{1}{2t}(x_1^2 + \cdots + x_n^2)}\, dx.$$

Then we use the polar coordinates $x = ru$ with $r \geq 0$ and $|u| = 1$. The Lebesgue measure becomes $dx = r^{n-1}\, dr\, d\sigma$, where σ is the surface measure on the unit sphere $S^{n-1} = \{u \in \mathbb{R}^n \,;\, |u| = 1\}$. Hence we get

$$E\frac{1}{|B(t)|} = \int_{S^{n-1}} \int_0^\infty \frac{1}{r}\left(\frac{1}{\sqrt{2\pi t}}\right)^n e^{-\frac{r^2}{2t}} r^{n-1}\, dr\, d\sigma$$

$$= \sigma(S^{n-1})\left(\frac{1}{\sqrt{2\pi t}}\right)^n \int_0^\infty r^{n-2} e^{-\frac{r^2}{2t}}\, dr. \tag{8.5.3}$$

Make a change of variables $r = \sqrt{2ty}$ in order to evaluate the last integral,

$$\int_0^\infty r^{n-2} e^{-\frac{r^2}{2t}}\, dr = \frac{1}{2}\left(\sqrt{2t}\right)^{n-1} \int_0^\infty y^{\frac{n-1}{2}-1} e^{-y}\, dy$$

$$= \frac{1}{2}\left(\sqrt{2t}\right)^{n-1} \Gamma((n-1)/2), \tag{8.5.4}$$

where Γ is the gamma function defined by

$$\Gamma(\alpha) = \int_0^\infty x^{\alpha-1} e^{-x}\, dx, \quad \alpha > 0. \tag{8.5.5}$$

On the other hand, the total surface measure $\sigma(S^{n-1})$ is known to be

$$\sigma(S^{n-1}) = \frac{2\pi^{n/2}}{\Gamma(n/2)}, \tag{8.5.6}$$

see, e.g., page 1427 of the the the encyclopedia [42]. Put Equations (8.5.4) and (8.5.6) into Equation (8.5.3) to get

$$E\frac{1}{|B(t)|} = \frac{\Gamma((n-1)/2)}{\Gamma(n/2)}\frac{1}{\sqrt{2t}}, \tag{8.5.7}$$

where $\Gamma(m/2)$ for a positive integer m has the value:

$$\Gamma\left(\frac{m}{2}\right) = \begin{cases} \left(\frac{m}{2} - 1\right)!, & \text{if } m \text{ is even;} \\ \left(\frac{m}{2} - 1\right)\left(\frac{m}{2} - 2\right)\cdots\frac{3}{2}\frac{1}{2}\sqrt{\pi}, & \text{if } m \text{ is odd.} \end{cases}$$

From Equation (8.5.7), we see that the dependence of $E\frac{1}{|B(t)|}$ on t is given by $1/\sqrt{2t}$ for any dimension $n \geq 2$ (Observation: when $n = 1$, the equation breaks down! cf. Section 8.6.)

Note that $\int_0^t 1/\sqrt{2s}\,ds = \sqrt{2t} < \infty$ for any positive real number t. Hence Equation (8.5.7) implies that $\int_0^t E\frac{1}{|B(s)|}\,ds < \infty$. This shows that the last integral in Equation (8.5.2) is defined almost surely for a Brownian motion $B(t)$ starting at 0. Therefore, we can let $a \to 0$ in Equation (8.5.2) to get

$$|B(t)| = \sum_{i=1}^{n} \int_0^t \frac{B_i(s)}{|B(s)|}\,dB_i(s) + \frac{n-1}{2} \int_0^t \frac{1}{|B(s)|}\,ds, \tag{8.5.8}$$

where $B(t)$ is a Brownian motion on $\mathbb{R}^n$, $n \geq 2$, starting at 0. In the next lemma we show that the summation term is indeed a Brownian motion.

Lemma 8.5.2. *Let* $B(t) = (B_1(t), B_2(t), \ldots, B_n(t))$ *be a Brownian motion on* $\mathbb{R}^n$, $n \geq 2$, *starting at 0. Then the stochastic process*

$$W(t) = \sum_{i=1}^{n} \int_0^t \frac{B_i(s)}{|B(s)|}\,dB_i(s) \tag{8.5.9}$$

is a Brownian motion.

Proof. We will use Theorem 8.4.2 to show that $W(t)$ is a Brownian motion. For this purpose, take Q to be the same probability measure P with respect to which the $B_i(t)$'s are Brownian motions and let $\{\mathcal{F}_t\}$ be the filtration given by $\mathcal{F}_t = \sigma\{B_i(s)\,;\, s \leq t, 1 \leq i \leq n\}$. Since each integrand in Equation (8.5.9) has absolute value less than or equal to 1 almost surely with respect to P, the stochastic process $W(t)$ is a continuous martingale with respect to $\{\mathcal{F}_t\}$ by Theorems 4.6.1 and 4.6.2.

Obviously, we have $P\{W(0) = 0\} = 1$. Moreover, the compensator of $W(t)^2$ is given by

$$\langle W \rangle_t = \sum_{i=1}^{n} \int_0^t \frac{B_i(s)^2}{|B(s)|^2}\,ds = \int_0^t \frac{1}{|B(s)|^2} \sum_{i=1}^{n} B_i(s)^2\,ds = \int_0^t 1\,ds = t.$$

Thus by Theorem 8.4.2, $W(t)$ is a Brownian motion. $\square$

Now, by using the Brownian motion $W(t)$ in Equation (8.5.9), we can rewrite Equation (8.5.8) as follows:

$$|B(t)| = W(t) + \frac{n-1}{2} \int_0^t \frac{1}{|B(s)|}\,ds. \tag{8.5.10}$$

Definition 8.5.3. *Let* $B(t) = (B_1(t), B_2(t), \ldots, B_n(t))$ *be a Brownian motion on* $\mathbb{R}^n$, $n \geq 2$, *starting at 0. The stochastic process*

$$L_{n-1}(t) = |B(t)| = \sqrt{B_1(t)^2 + B_2(t)^2 + \cdots + B_n(t)^2}$$

is called a Bessel process *with index* $n - 1$.

From the above discussion, we have proved Equation (8.5.10), which can be stated as the next theorem.

Theorem 8.5.4. *The Bessel process $L_{n-1}(t)$ with index $n-1$ is a solution of the stochastic integral equation*

$$X(t) = W(t) + \frac{n-1}{2} \int_0^t \frac{1}{X(s)} \, ds, \qquad (8.5.11)$$

where $W(t)$ is given by Equation (8.5.9).

Theorem 8.5.5. *For each $t > 0$, the density function of $L_{n-1}(t)$ is given by*

$$f(x) = \begin{cases} \dfrac{2}{(2t)^{n/2}\Gamma(n/2)} \, x^{n-1} e^{-x^2/2t}, & \text{if } x \geq 0; \\ 0, & \text{if } x < 0, \end{cases}$$

where Γ is the gamma function defined in Equation (8.5.5).

Proof. For $x \geq 0$, we have

$$P\{L_{n-1}(t) \leq x\} = P\{B_1(t)^2 + \cdots + B_n(t)^2 \leq x^2\}$$

$$= \int_{x_1^2 + \cdots + x_n^2 \leq x^2} \left(\frac{1}{\sqrt{2\pi t}}\right)^n e^{-(x_1^2 + \cdots + x_n^2)/2t} \, dx.$$

Then we use polar coordinates as in the derivation of Equation (8.5.3) to check that

$$P\{L_{n-1}(t) \leq x\} = \sigma(S^{n-1})\left(\frac{1}{\sqrt{2\pi t}}\right)^n \int_0^x r^{n-1} e^{-r^2/2t} \, dr.$$

This equation together with Equation (8.5.6) gives the density function of the random variable $L_{n-1}(t)$ for $t > 0$. $\qquad\square$

8.6 Tanaka's Formula and Local Time

In applying Itô's formula to the function $f(x) = |x|$ on $\mathbb{R}^n$ in the previous section, we mentioned that Equation (8.5.7) breaks down when $n = 1$. In fact, a one-dimensional Brownian motion $B(t)$ is recurrent. Hence we cannot apply Itô's formula to $f(B(t))$ for the function $f(x) = |x|$, $x \in \mathbb{R}$. However, we can modify the function $f(x) = |x|$ a little bit so that Itô's formula can be applied to derive an interesting property of the one-dimensional Brownian motion $B(t)$.

Let $a \in \mathbb{R}$ be a fixed real number in this section. For each $\varepsilon > 0$, define a function f_ε on $\mathbb{R}$ by

$$f_\varepsilon(x) = \begin{cases} |x - a| - \dfrac{\varepsilon}{2}, & \text{if } |x - a| > \varepsilon; \\ \dfrac{1}{2\varepsilon}(x - a)^2, & \text{if } |x - a| \le \varepsilon. \end{cases}$$

The first two derivatives of f_ε are given by

$$f'_\varepsilon(x) = \begin{cases} 1, & \text{if } x > a + \varepsilon; \\ \dfrac{1}{\varepsilon}(x - a), & \text{if } |x - a| \le \varepsilon; \\ -1, & \text{if } x < a - \varepsilon, \end{cases} \qquad f''_\varepsilon(x) = \begin{cases} 0, & \text{if } |x - a| > \varepsilon; \\ \dfrac{1}{\varepsilon}, & \text{if } |x - a| < \varepsilon. \end{cases}$$

Note that the function f_ε is not a C^2-function since $f''_\varepsilon(x)$ is discontinuous at $x = a \pm \varepsilon$. However, we can still apply Itô's formula to the function f_ε due to the following fact.

Fact. Let $B(t)$ be a one-dimensional Brownian motion. Then for any real number c, the set $\{t \in \mathbb{R}; B(t) = c\}$ has Lebesgue measure 0 almost surely.

Therefore, we get the equality

$$f_\varepsilon(B(t)) = f_\varepsilon(B(0)) + \int_0^t f'_\varepsilon(B(s)) \, dB(s) + \frac{1}{2} \int_0^t f''_\varepsilon(B(s)) \, ds$$

$$= f_\varepsilon(0) + \int_0^t f'_\varepsilon(B(s)) \, dB(s) + \frac{1}{2} \int_0^t f''_\varepsilon(B(s)) \, ds. \qquad (8.6.1)$$

Obviously, $\lim_{\varepsilon \to 0} f_\varepsilon(B(t)) = |B(t) - a|$ almost surely, $\lim_{\varepsilon \to 0} f_\varepsilon(0) = |a|$, and

$$\int_0^t f''_\varepsilon(B(s)) \, ds = \frac{1}{\varepsilon} \int_0^t 1_{(a-\varepsilon, a+\varepsilon)}(B(s)) \, ds.$$

Moreover, for each Brownian motion path $B(\cdot, \omega)$, we can apply the Lebesgue dominated convergence theorem to see that

$$\lim_{\varepsilon \to 0} \int_0^t \left| f'_\varepsilon(B(s, \omega)) - \text{sgn}(B(s, \omega) - a) \right|^2 ds = 0.$$

Therefore, we have

$$\lim_{\varepsilon \to 0} \int_0^t \left| f'_\varepsilon(B(s)) - \text{sgn}(B(s) - a) \right|^2 ds = 0, \qquad \text{almost surely.}$$

Note that convergence almost surely implies convergence in probability and that the stochastic process $f'_\varepsilon(B(\cdot))$ belongs to $L^2_{\text{ad}}([a, t] \times \Omega)$. Hence by the definition of general stochastic integrals in Section 5.3,

$$\lim_{\varepsilon \to 0} \int_0^t f'_\varepsilon(B(s)) \, dB(s) = \int_0^t \text{sgn}(B(s) - a) \, dB(s), \qquad \text{in probability.}$$

Thus from Equation (8.6.1) we get the equality

$$|B(t) - a| = |a| + \int_0^t \text{sgn}(B(s) - a)\, dB(s)$$

$$+ \lim_{\varepsilon \to 0} \frac{1}{2\varepsilon} \int_0^t 1_{(a-\varepsilon, a+\varepsilon)}(B(s))\, ds. \qquad (8.6.2)$$

Note that the equality implies implicitly that the limit in probability exists.

Definition 8.6.1. *The* local time *of a Brownian motion $B(t)$ at a, up to and including t, is defined to be the random variable*

$$L_a(t)(\omega) = \lim_{\varepsilon \to 0} \frac{1}{2\varepsilon} \int_0^t 1_{(a-\varepsilon, a+\varepsilon)}(B(s, \omega))\, ds, \quad \text{in probability.} \qquad (8.6.3)$$

Observe that

$$\frac{1}{2\varepsilon} \int_0^t 1_{(a-\varepsilon, a+\varepsilon)}(B(s))\, ds = \frac{1}{2\varepsilon} \text{Leb}\{s \in [0, t]; \, B(s) \in (a - \varepsilon, a + \varepsilon)\},$$

where $\text{Leb}(\cdot)$ denotes the Lebesgue measure. Hence we can interpret the local time $L_a(t, \omega)$ as the amount of time a Brownian motion path $B(\cdot, \omega)$ spends at a up to and including time t. Thus $L_a(t, \omega)$ is also given by

$$L_a(t)(\omega) = \lim_{\varepsilon \to 0} \frac{1}{2\varepsilon} \text{Leb}\{s \in [0, t]; \, B(s, \omega) \in (a - \varepsilon, a + \varepsilon)\}. \qquad (8.6.4)$$

From Equation (8.6.2) we have the next theorem.

Theorem 8.6.2. *(Tanaka's formula) Let $B(t)$ be a Brownian motion on the real line starting at 0. Then for any $a \in \mathbb{R}$,*

$$|B(t) - a| = |a| + \int_0^t \text{sgn}(B(s) - a)\, dB(s) + L_a(t). \qquad (8.6.5)$$

Note that $B(t) - a$ is a martingale. Hence by the conditional Jensen's inequality, the stochastic process $|B(t) - a|$ is a submartingale. Thus Equation (8.6.5) gives the Doob–Meyer decomposition of $|B(t) - a|$. Moreover, it is easy to see that the stochastic process

$$M_t = \int_0^t \text{sgn}(B(s) - a)\, dB(s)$$

is a Brownian motion by the same arguments as those in Example 8.4.4.

We make some comments on Equation (8.6.3). It is a fact in the theory of generalized functions that

$$\lim_{\varepsilon \to 0} \frac{1}{2\varepsilon} 1_{(a-\varepsilon, a+\varepsilon)} = \delta_a, \qquad (8.6.6)$$

where δ_a is the Dirac delta function at a. This equation means that for any continuous function $f(x)$,

$$\lim_{\varepsilon \to 0} \int_{-\infty}^{\infty} f(x) \frac{1}{2\varepsilon} 1_{(a-\varepsilon, a+\varepsilon)}(x)\, dx = f(a),$$

which is true by the fundamental theorem of calculus. Hence the local time $L_a(t)$ can be expressed as

$$L_a(t) = \int_0^t \delta_a(B(s))\, ds.$$

The integrand $\delta_a(B(s))$ is called the *Donsker delta function*. It is a generalized function on an infinite-dimensional space. Although $\delta_a(B(s))$ is not a random variable, its integral $L_a(t)$ is a random variable. For more information, see the book [56].

8.7 Exponential Processes

In this section we study exponential processes, which will play a crucial role for the Girsanov theorem in Section 8.9.

Let h be a deterministic function in $L^2[0, T]$. Then for any $t \in [0, T]$, the Wiener integral $\int_0^t h(s)\, dB(s)$ has a normal distribution with mean 0 and variance $\sigma^2 = \int_0^t h(s)^2\, ds$. Hence by elementary computation,

$$E e^{\int_0^t h(s)\, dB(s)} = \frac{1}{\sqrt{2\pi}\,\sigma} \int_{-\infty}^{\infty} e^x e^{-\frac{x^2}{2\sigma^2}}\, dx = e^{\frac{1}{2} \int_0^t h(s)^2\, ds}. \tag{8.7.1}$$

The multiplicative renormalization of $e^{\int_0^t h(s)\, dB(s)}$ is defined by

$$Y_t = \frac{e^{\int_0^t h(s)\, dB(s)}}{E e^{\int_0^t h(s)\, dB(s)}} = e^{\int_0^t h(s)\, dB(s) - \frac{1}{2} \int_0^t h(s)^2\, ds}.$$

By Example 7.4.4, Y_t has the representation

$$Y_t = 1 + \int_0^t h(s) e^{\int_0^s h(u)\, dB(u) - \frac{1}{2} \int_0^s h(u)^2\, du}\, dB(s). \tag{8.7.2}$$

Observe that the integrand of the Itô integral in Equation (8.7.2) belongs to the space $L^2_{\mathrm{ad}}([0, T] \times \Omega)$. Hence by Theorem 4.6.1 the stochastic process Y_t, $0 \le t \le T$, is a martingale.

For convenience, we introduce the following definition and notation.

Definition 8.7.1. *The* exponential process *given by* $h \in \mathcal{L}_{\mathrm{ad}}(\Omega, L^2[0, T])$ *is defined to be the stochastic process*

$$\mathcal{E}_h(t) = \exp\left[\int_0^t h(s)\, dB(s) - \frac{1}{2} \int_0^t h(s)^2\, ds\right], \quad 0 \le t \le T. \tag{8.7.3}$$

By applying Itô's formula, we see that $d\mathcal{E}_h(t) = h(t)\mathcal{E}_h(t)\,dB(t)$. Hence

$$\mathcal{E}_h(t) = 1 + \int_0^t h(s)\mathcal{E}_h(s)\,dB(s), \quad 0 \le t \le T,$$

which shows that $\mathcal{E}_h(t)$ is a local martingale by Theorem 5.5.2. This leads to the following question.

Question 8.7.2. Let $h \in \mathcal{L}_{ad}(\Omega, L^2[0,T])$. Is the exponential process given by h a martingale with respect to the filtration $\mathcal{F}_t = \sigma\{B(s); 0 \le s \le t\}$?

The answer to this question is given by the next theorem.

Theorem 8.7.3. *If $h \in \mathcal{L}_{ad}(\Omega, L^2[0,T])$ satisfies the condition that*

$$E(\mathcal{E}_h(t)) = 1, \quad \forall\, t \in [0,T], \tag{8.7.4}$$

then the exponential process $\mathcal{E}_h(t)$, $0 \le t \le T$, given by h is a martingale.

Remark 8.7.4. Obviously, the converse is valid because

$$E(\mathcal{E}_h(t)) = E\big(E[\mathcal{E}_h(t)\,|\,\mathcal{F}_0]\big) = E\mathcal{E}_h(0) = E1 = 1, \quad \forall\, 0 \le t \le T.$$

In general, the condition in Equation (8.7.4) is very hard to verify. A stronger condition, due to Novikov [64], is the following:

$$E \exp\left[\frac{1}{2}\int_0^T h(t)^2\,dt\right] < \infty. \tag{8.7.5}$$

For more information, see the books by Elliott [13], Ikeda and Watanabe [30], Øksendal [66], and Protter [68].

Proof. If $h(t)$ is a deterministic function in $L^2[0,T]$, then Equation (8.7.4) is satisfied in view of Equation (8.7.1). In this case, we have already shown that $\mathcal{E}_h(t)$ (denoted by Y_t above) is a martingale.

By assumption, $dQ = \mathcal{E}_h(T)\,dP$ defines a probability measure. Let Q_t be the restriction of Q to $\mathcal{F}_t$. The condition in Equation (8.7.4) implies that $dQ_t = \mathcal{E}_h(t)\,dP$ (a fact that seems to be obvious, but the verification is rather involved and hence omitted). Let $s \le t$ and let $E_P[\cdot|\mathcal{F}_s]$ denote the conditional expectation with respect to P. Then for any $A \in \mathcal{F}_s$,

$$\int_A E_P[\mathcal{E}_h(t)\,|\,\mathcal{F}_s]\,dP = \int_A \mathcal{E}_h(t)\,dP = \int_A dQ_t = Q(A). \tag{8.7.6}$$

On the other hand, we also have

$$\int_A \mathcal{E}_h(s)\,dP = \int_A dQ_s = Q(A). \tag{8.7.7}$$

It follows from Equations (8.7.6) and (8.7.7) that $E_P[\mathcal{E}_h(t)|\mathcal{F}_s] = \mathcal{E}_h(s)$. Thus $\mathcal{E}_h(t)$, $0 \le t \le T$, is a martingale. $\square$

Example 8.7.5. Let $h(t) = \text{sgn}(B(t))$. The corresponding exponential process is given by

$$\mathcal{E}_h(t) = \exp\left[\int_0^t \text{sgn}(B(s))\,dB(s) - \frac{1}{2}t\right].$$

It is proved in Example 8.4.4 that $X_t = \int_0^t \text{sgn}(B(s))\,dB(s)$ is a Brownian motion. Hence

$$E(\mathcal{E}_h(t)) = e^{-\frac{1}{2}t}Ee^{X_t} = e^{-\frac{1}{2}t}e^{\frac{1}{2}t} = 1, \quad \forall 0 \le t \le T.$$

Thus the condition in Equation (8.7.4) is satisfied. By the above theorem, $\mathcal{E}_h(t)$ is a martingale. On the other hand, since X_t is a Brownian motion, we know from Equation (8.7.2) with $h \equiv 1$ that $\mathcal{E}_h(t) = e^{X_t - \frac{1}{2}t}$ is a martingale.

Example 8.7.6. Suppose $h \in \mathcal{L}_{\text{ad}}(\Omega, L^2[0,T])$ is L^2-bounded, i.e., there exists a finite number $C > 0$ such that $\int_0^T h(t)^2\,dt \le C$ almost surely. In general, it is impossible to check whether h satisfies the condition in Equation (8.7.4) by evaluating the expectation of $\mathcal{E}_h(t)$ directly. However, it is obvious that h satisfies the Novikov condition in Equation (8.7.5). Hence the associated stochastic process $\mathcal{E}_h(t)$, $0 \le t \le T$, is a martingale. For example, when $h(t) = \sin(B(t))$, we get the martingale

$$\mathcal{E}_h(t) = \exp\left[\int_0^t \sin(B(s))\,dB(s) - \frac{1}{2}\int_0^t \sin^2(B(s))\,ds\right].$$

Example 8.7.7. Consider the stochastic process $h(t)$ defined by

$$h(t) = \begin{cases} 0, & \text{if } 0 \le t < \frac{1}{2}; \\ B\left(\frac{1}{2}\right), & \text{if } \frac{1}{2} \le t \le 1. \end{cases} \tag{8.7.8}$$

We have $\int_0^1 h(t)^2\,dt = \frac{1}{2}B\left(\frac{1}{2}\right)^2$. Hence h is not L^2-bounded in the sense of the previous example. However, it satisfies the Novikov condition since

$$Ee^{\int_0^1 h(t)^2\,dt} = Ee^{\frac{1}{2}B\left(\frac{1}{2}\right)^2} = \int_{-\infty}^{\infty} e^{\frac{1}{2}x^2}\frac{1}{\sqrt{\pi}}e^{-x^2}\,dx = \sqrt{2} < \infty.$$

Therefore, we have the following martingale:

$$\mathcal{E}_h(t) = \begin{cases} 1, & \text{if } 0 \le t < \frac{1}{2}; \\ \exp\left[B\left(\frac{1}{2}\right)B(t) - \frac{1}{2}\left(t + \frac{3}{2}\right)B\left(\frac{1}{2}\right)^2\right], & \text{if } \frac{1}{2} \le t \le 1. \end{cases} \tag{8.7.9}$$

8.8 Transformation of Probability Measures

In this section we will give two examples to explain the background for the Girsanov theorem in the next section.

Example 8.8.1. Consider a random variable X having normal distribution with mean μ and variance 1. Suppose we want to find the value of $E(X^4)$. The first method is to compute the integral

$$E(X^4) = \int_{-\infty}^{\infty} x^4 \frac{1}{\sqrt{2\pi}} e^{-(x-\mu)^2/2} \, dx. \tag{8.8.1}$$

The second method is to make a change of random variables. Let $Y = X - \mu$. Then Y is a standard normal random variable. Express X^4 in terms of Y as

$$X^4 = (Y + \mu)^4 = Y^4 + 4\mu Y^3 + 6\mu^2 Y^2 + 4\mu^3 Y + \mu^4. \tag{8.8.2}$$

Hence we get

$$E(X^4) = E(Y^4) + 4\mu E(Y^3) + 6\mu^2 E(Y^2) + 4\mu^3 E(Y) + \mu^4$$
$$= 3 + 6\mu^2 + \mu^4. \tag{8.8.3}$$

Note that taking the expectation in Equation (8.8.2) corresponds to evaluating the integral in Equation (8.8.1) by making a change of variables $y = x - \mu$.

Example 8.8.2. Let $B(t)$ be a Brownian motion and consider the problem to compute the expectation $E\big(B(t)^3 e^{B(1)-\frac{1}{2}}\big)$ for $0 < t \leq 1$. The first method to solve this problem is to express the random variable as the product of two independent random variables,

$$E\big(B(t)^3 e^{B(1)-\frac{1}{2}}\big) = e^{-\frac{1}{2}} E\big[\big(B(t)^3 e^{B(t)}\big)\big(e^{B(1)-B(t)}\big)\big]$$
$$= e^{-\frac{1}{2}} E\big(B(t)^3 e^{B(t)}\big) E\big(e^{B(1)-B(t)}\big), \tag{8.8.4}$$

and then compute the expectations separately. The second method to solve this problem is to use the probability measure Q defined by Equation (8.4.1). Observe the following equality:

$$E\big(B(t)^3 e^{B(1)-\frac{1}{2}}\big) = \int_{\Omega} B(t)^3 e^{B(1)-\frac{1}{2}} \, dP = \int_{\Omega} B(t)^3 \, dQ = E_Q(B(t)^3),$$

where E_Q is the expectation with respect to the probability measure Q. It was shown in Section 8.4 that $M_t = B(t) - t$, $0 \leq t \leq 1$, is a Brownian motion with respect to the probability measure Q. Therefore,

$$E\big(B(t)^3 e^{B(1)-\frac{1}{2}}\big) = E_Q(B(t)^3) = E_Q\big((M_t + t)^3\big)$$
$$= E_Q(M_t^3 + 3t M_t^2 + 3t^2 M_t + t^3)$$
$$= 3t^2 + t^3. \tag{8.8.5}$$

In fact, we can further explore the ideas in Example 8.8.2 to find the intrinsic relationship between the two methods. Let $\mathcal{F}_t = \sigma\{B(s); s \leq t\}$. Recall that the stochastic process $e^{B(t)-\frac{1}{2}t}$, $0 \leq t \leq 1$, is a martingale. Suppose $a(t)$ is a deterministic function. Then

$$\int_\Omega f(B(t)-a(t))e^{B(1)-\frac{1}{2}}\,dP = E\big(f(B(t)-a(t))e^{B(1)-\frac{1}{2}}\big)$$

$$= E\big(E\big[f(B(t)-a(t))e^{B(1)-\frac{1}{2}}\,\big|\,\mathcal{F}_t\big]\big)$$

$$= E\big(f(B(t)-a(t))E\big[e^{B(1)-\frac{1}{2}}\,\big|\,\mathcal{F}_t\big]\big)$$

$$= E\big(f(B(t)-a(t))e^{B(t)-\frac{1}{2}t}\big)$$

$$= \int_\Omega f(B(t)-a(t))e^{B(t)-\frac{1}{2}t}\,dP. \qquad (8.8.6)$$

The last integral can be evaluated by

$$\int_\Omega f(B(t)-a(t))e^{B(t)-\frac{1}{2}t}\,dP = \int_{-\infty}^{\infty} f(x-a(t))e^{x-\frac{1}{2}t}\frac{1}{\sqrt{2\pi t}}e^{-\frac{1}{2t}x^2}\,dx.$$

Make a change of variables $y = x - a(t)$ to obtain

$$\int_{-\infty}^{\infty} f(x-a(t))e^{x-\frac{1}{2}t}\frac{1}{\sqrt{2\pi t}}e^{-\frac{1}{2t}x^2}\,dx$$

$$= \int_{-\infty}^{\infty} f(y)e^{\frac{1}{t}(t-a(t))y-\frac{1}{2t}(t-a(t))^2}\frac{1}{\sqrt{2\pi t}}e^{-\frac{1}{2t}y^2}\,dy$$

$$= \int_\Omega f(B(t))e^{\frac{1}{t}(t-a(t))B(t)-\frac{1}{2t}(t-a(t))^2}\,dP.$$

Therefore, we get

$$\int_\Omega f(B(t)-a(t))e^{B(t)-\frac{1}{2}t}\,dP = \int_\Omega f(B(t))e^{\frac{1}{t}(t-a(t))B(t)-\frac{1}{2t}(t-a(t))^2}\,dP.$$

$$(8.8.7)$$

It follows from Equations (8.8.6) and (8.8.7) that

$$\int_\Omega f(B(t)-a(t))e^{B(1)-\frac{1}{2}}\,dP = \int_\Omega f(B(t))e^{\frac{1}{t}(t-a(t))B(t)-\frac{1}{2t}(t-a(t))^2}\,dP.$$

In particular, take $a(t) = t$ to get

$$\int_\Omega f(B(t)-t)e^{B(1)-\frac{1}{2}}\,dP = \int_\Omega f(B(t))\,dP.$$

Thus we have proved the next theorem.

Theorem 8.8.3. *Let $B(t)$, $0 \le t \le 1$, be a Brownian motion with respect to a probability measure P. Let Q be the probability measure $dQ = e^{B(1)-\frac{1}{2}}\,dP$. Then for any function f such that $E_P|f(B(t))| < \infty$, we have*

$$\int_\Omega f(B(t)-t)\,dQ = \int_\Omega f(B(t))\,dP, \qquad (8.8.8)$$

which can also be expressed as $E_Q f(B(t)-t) = E_P f(B(t))$.

The formula in Equation (8.8.8) involves the transformation of probability measures from P to Q. Take the function $f(x) = e^{i\lambda x}$ with $\lambda \in \mathbb{R}$ to get

$$\int_{\Omega} e^{i\lambda(B(t)-t)} \, dQ = \int_{\Omega} e^{i\lambda B(t)} \, dP = e^{-\frac{1}{2}\lambda^2 t}, \quad \forall \lambda \in \mathbb{R}.$$

This implies that $B(t) - t$ is normally distributed with mean 0 and variance t with respect to the probability measure Q. Actually, one can easily guess that $B(t) - t$ is a Brownian motion under Q. This fact has also been shown in Section 8.4 using Lévy's characterization theorem of a Brownian motion.

8.9 Girsanov Theorem

Let $(\Omega, \mathcal{F}, P)$ be a fixed probability space and let $B(t)$ be a Brownian motion with respect to P. Recall from the previous section that $B(t) - t$, $0 \le t \le 1$, is a Brownian motion with respect to the probability measure Q given by $dQ = e^{B(1)-\frac{1}{2}} \, dP$. This leads to the following question.

Question 8.9.1. Are there other stochastic processes $\varphi(t)$ such that $B(t) - \varphi(t)$ is a Brownian motion with respect to some probability measure?

We will show that if $\varphi(t) = \int_0^t h(s) \, ds$, $0 \le t \le T$, with $h \in \mathcal{L}_{ad}(\Omega, L^2[0,T])$ satisfying the condition in Theorem 8.7.3, then $W_t = B(t) - \varphi(t)$ is indeed a Brownian motion with respect to the probability measure $dQ = \mathcal{E}_h(T) \, dP$. Here $\mathcal{E}_h$ is the exponential process given by h as defined by Equation (8.7.3). The main idea in the proof of this important theorem, due to Girsanov [23], is the transformation of probability measures described in the previous section and Lévy's characterization theorem of Brownian motion in Section 8.4.

In this section we will take the expectation and the conditional expectation of a random variable with respect to several probability measures. Therefore, we will use a subindex Q to denote the relevant quantities with respect to the probability measure Q. For convenience, a martingale under Q (and with respect to a fixed filtration) will be called a *Q-martingale*.

Lemma 8.9.2. *Let $\theta \in L^1(P)$ be nonnegative such that $d\mu = \theta \, dP$ defines a probability measure. Then for any σ-field $\mathcal{G} \subset \mathcal{F}$ and $X \in L^1(\mu)$, we have*

$$E_{\mu}[X \,|\, \mathcal{G}] = \frac{E_P[X\theta \,|\, \mathcal{G}]}{E_P[\theta \,|\, \mathcal{G}]}, \quad \mu\text{-almost surely.}$$

Proof. First note that $E_P|X\theta| = \int_{\Omega} |X| \, \theta \, dP = \int_{\Omega} |X| \, d\mu < \infty$, so that the conditional expectation $E_P[X\theta \,|\, \mathcal{G}]$ is defined. For any $G \in \mathcal{G}$, use condition (2) in Definition 2.4.1 and the definition of μ to get

$$\int_G E_P[X\theta \,|\, \mathcal{G}] \, dP = \int_G X\theta \, dP = \int_G X \, d\mu = \int_G E_{\mu}[X \,|\, \mathcal{G}] \, d\mu. \quad (8.9.1)$$

Moreover, write $d\mu = \theta \, dP$ and then use condition (2) in Definition 2.4.1 and Property 4 of the conditional expectation in Section 2.4 to derive

$$\int_G E_\mu[X\,|\,\mathcal{G}]\,d\mu = \int_G E_\mu[X\,|\,\mathcal{G}]\,\theta\,dP = \int_G E_P\{E_\mu[X\,|\,\mathcal{G}]\,\theta\,|\,\mathcal{G}\}\,dP$$

$$= \int_G E_\mu[X\,|\,\mathcal{G}] \cdot E_P[\theta\,|\,\mathcal{G}]\,dP. \qquad (8.9.2)$$

Equations (8.9.1) and (8.9.2) imply the conclusion of the lemma. $\qquad \square$

Now suppose $h(t)$ is a stochastic process in $\mathcal{L}_{\mathrm{ad}}(\Omega, L^2[0,T])$ satisfying the condition in Equation (8.7.4). Then by Theorem 8.7.3, the exponential process $\mathcal{E}_h(t)$ is a P-martingale. Let Q be the probability measure on $(\Omega, \mathcal{F})$ given by $dQ = \mathcal{E}_h(T)\,dP$, namely,

$$Q(A) = \int_A \mathcal{E}_h(T)\,dP, \quad A \in \mathcal{F}. \qquad (8.9.3)$$

Lemma 8.9.3. *Assume that $h \in \mathcal{L}_{\mathrm{ad}}(\Omega, L^2[0,T])$ satisfies the condition in Equation (8.7.4). Then an $\{\mathcal{F}_t\}$-adapted stochastic process $X(t), 0 \le t \le T$, is a Q-martingale if and only if $X(t)\mathcal{E}_h(t)$ is a P-martingale.*

Proof. First we make a loose remark on the integrability. As we mentioned in the proof of Theorem 8.7.3, the restriction of Q to $\mathcal{F}_t$ is the probability measure $\mathcal{E}_h(t)\,dP$. This implies that $X(t)\mathcal{E}_h(T)$ is P-integrable if and only if $X(t)\mathcal{E}_h(t)$ is P-integrable. Since $dQ = \mathcal{E}_h(T)\,dP$, this means that $X(t)$ is Q-integrable if and only if $X(t)\mathcal{E}_h(t)$ is P-integrable. This provides the integrability below for taking the conditional expectation.

Suppose $X(t)\mathcal{E}_h(t)$ is a P-martingale and let $s \le t$. Apply Lemma 8.9.2 with $\mu = Q$ and $\theta = \mathcal{E}_h(T)$ to get

$$E_Q[X(t)\,|\,\mathcal{F}_s] = \frac{E_P[X(t)\mathcal{E}_h(T)\,|\,\mathcal{F}_s]}{E_P[\mathcal{E}_h(T)\,|\,\mathcal{F}_s]}. \qquad (8.9.4)$$

Since $\mathcal{E}_h(t)$ is a P-martingale by Theorem 8.7.3, we have $E_P[\mathcal{E}_h(T)\,|\,\mathcal{F}_s] = \mathcal{E}_h(s)$. On the other hand, we can use Properties 4 and 5 of the conditional expectation in Section 2.4 to show that

$$E_P[X(t)\mathcal{E}_h(T)\,|\,\mathcal{F}_s] = E_P\{E_P[X(t)\mathcal{E}_h(T)\,|\,\mathcal{F}_t]\,|\,\mathcal{F}_s\}$$

$$= E_P\{X(t)E_P[\mathcal{E}_h(T)\,|\,\mathcal{F}_t]\,|\,\mathcal{F}_s\}$$

$$= E_P\{X(t)\mathcal{E}_h(t)\,|\,\mathcal{F}_s\}$$

$$= X(s)\mathcal{E}_h(s).$$

Therefore, from Equation (8.9.4),

$$E_Q[X(t)\,|\,\mathcal{F}_s] = \frac{X(s)\mathcal{E}_h(s)}{\mathcal{E}_h(s)} = X(s).$$

This shows that $X(t)$ is a Q-martingale.

Conversely, suppose $X(t)$ is a Q-martingale and let $s \leq t$. Then as shown above we have

$$E_P\{X(t)\mathcal{E}_h(t) \,|\, \mathcal{F}_s\} = E_P\big[X(t)\mathcal{E}_h(T) \,|\, \mathcal{F}_s\big]. \tag{8.9.5}$$

Moreover, we use Lemma 8.9.2 to get

$$E_P\big[X(t)\mathcal{E}_h(T) \,|\, \mathcal{F}_s\big] = E_Q\big[X(t) \,|\, \mathcal{F}_s\big] \cdot E_P\big[\mathcal{E}_h(T) \,|\, \mathcal{F}_s\big] = X(s)\mathcal{E}_h(s). \tag{8.9.6}$$

Equations (8.9.5) and (8.9.6) show that $X(t)\mathcal{E}_h(t)$ is a P-martingale.   □

Now, for a stochastic process $h(t)$ in the space $\mathcal{L}_{\mathrm{ad}}(\Omega, L^2[0,T])$, consider the following translation of a Brownian motion:

$$W(t) = B(t) - \int_0^t h(s)\, ds, \quad 0 \leq t \leq T. \tag{8.9.7}$$

Theorem 8.9.4. (Girsanov theorem) *Let $h \in \mathcal{L}_{\mathrm{ad}}(\Omega, L^2[0,T])$ and assume that $E_P(\mathcal{E}_h(t)) = 1$ for all $t \in [0,T]$. Then the stochastic process*

$$W(t) = B(t) - \int_0^t h(s)\, ds, \quad 0 \leq t \leq T,$$

is a Brownian motion with respect to the probability measure $dQ = \mathcal{E}_h(T)\, dP$ defined by Equation (8.9.3).

Proof. First note that the probability measures P and Q are equivalent. Hence $Q\{W(0) = 0\} = 1$ and $W(t)$ is a continuous stochastic process. Let $\{\mathcal{F}_t\}$ be the filtration given by $\mathcal{F}_t = \sigma\{B(s); 0 \leq s \leq t\}$, $0 \leq t \leq T$.

First apply Itô's formula to check that

$$d\mathcal{E}_h(t) = h(t)\mathcal{E}_h(t)\, dB(t). \tag{8.9.8}$$

Then apply the Itô product formula in Equation (7.5.3) to get

$$\begin{aligned}
d\{W(t)\mathcal{E}_h(t)\} &= \{dW(t)\}\mathcal{E}_h(t) + W(t)\{d\mathcal{E}_h(t)\} + \{dW(t)\}\{d\mathcal{E}_h(t)\} \\
&= \{dB(t) - h(t)\, dt\}\mathcal{E}_h(t) + W(t)h(t)\mathcal{E}_h(t)\, dB(t) + h(t)\mathcal{E}_h(t)\, dt \\
&= \{1 + h(t)W(t)\}\mathcal{E}_h(t)\, dB(t),
\end{aligned}$$

or, in the stochastic integral form,

$$W(t)\mathcal{E}_h(t) = \int_0^t \{1 + h(s)W(s)\}\mathcal{E}_h(s)\, dB(s). \tag{8.9.9}$$

Therefore, by Theorem 5.5.2, the stochastic process $W(t)\mathcal{E}_h(t)$, $0 \leq t \leq T$, is a local P-martingale. It turns out that $W(t)\mathcal{E}_h(t)$ is indeed a P-martingale (a fact that can be easily verified when $h(t)$ is a deterministic function in $L^2[0,T]$, but whose proof in the general case is very involved). Hence by Lemma 8.9.3, $W(t)$ is a Q-martingale.

Next, replace $W(t)$ by $W(t)^2 - t$ and repeat the same arguments as those in the derivation of Equation (8.9.9) to show that

$$(W(t)^2 - t)\mathcal{E}_h(t) = \int_0^t \{2W(t) + (W(t)^2 - t)\}\mathcal{E}_h(s)\, dB(s). \qquad (8.9.10)$$

Hence $(W(t)^2 - t)\mathcal{E}_h(t)$ is a P-martingale by the same reason as above for the stochastic process $W(t)\mathcal{E}_h(t)$. Then use Lemma 8.9.3 to see that $W(t)^2 - t$ is a Q-martingale. It follows that the Doob–Meyer decomposition of $W(t)^2$ is given by

$$W(t)^2 = (W(t)^2 - t) + t,$$

which implies that $\langle W \rangle_t = t$ almost surely under the probability measure Q for each t. Hence by Lévy's characterization theorem (Theorem 8.4.2), $W(t)$ is a Brownian motion with respect to Q. ☐

Corollary 8.9.5. *Under the same assumption of Theorem 8.9.4, the following equality holds for any measurable function f such that $E_P|f(B(t))| < \infty$:*

$$\int_\Omega f\big(B(t) - \int_0^t h(s)\, ds\big)\, dQ = \int_\Omega f\big(B(t)\big)\, dP. \qquad (8.9.11)$$

Example 8.9.6. Let $h(t)$ be a deterministic function in $L^2[0,T]$. As pointed out in the beginning of the proof of Theorem 8.7.3, h satisfies the condition in the Girsanov theorem. Hence the stochastic process

$$W(t) = B(t) - \int_0^t h(s)\, ds, \quad 0 \le t \le T,$$

is a Brownian motion with respect to the probability measure Q given by

$$dQ = e^{\int_0^T h(s)\, dB(s) - \frac{1}{2} \int_0^T h(s)^2\, ds}\, dP.$$

The probability measure Q is the translation of P by the function h. In this case, the equality in Equation (8.9.11) is called the Cameron and Martin formula [4] for the translation of Wiener measure.

Example 8.9.7. Consider $h(t) = \mathrm{sgn}(B(t))$. As shown in Example 8.7.6, the function h satisfies the condition in the Girsanov theorem. Therefore, the stochastic process

$$W(t) = B(t) - \int_0^t \mathrm{sgn}(B(s))\, dB(s), \quad 0 \le t \le T,$$

is a Brownian motion with respect to the probability measure Q given by

$$dQ = e^{\int_0^T \mathrm{sgn}(B(s))\, dB(s) - \frac{1}{2}T}\, dP.$$

Example 8.9.8. Let $h(t)$ be the stochastic process given by Equation (8.7.8). As shown in Example 8.7.7, h satisfies the condition in the Girsanov theorem. Hence the stochastic process

$$
W(t) = \begin{cases} B(t), & \text{if } 0 \leq t < 1/2; \\ B(t) - (t - 1/2)B(1/2), & \text{if } 1/2 \leq t \leq 1, \end{cases}
$$

is a Brownian motion with respect to the probability measure Q given by

$$
dQ = e^{B(1/2)\,B(1)-5B(1/2)^2/4}\, dP.
$$

Exercises

1. Evaluate the stochastic integral $\int_0^t \arctan B(s)\, dB(s)$.

2. Let $M_t = \int_0^t \operatorname{sgn}(B(s))\, dB(s)$.

   (a) Show that the compensator of M_t^2 is given by $\langle M \rangle_t = t$.

   (b) Show that $L_t = M_t^2 - t$ is a martingale without using the Doob–Meyer decomposition theorem.

   (c) Find the compensator of L_t^2.

3. Let $X(t)$ and $Y(t)$ be two Itô processes. Prove the following equality for the Stratonovich integral:

   $$
   \int_a^b X(t) \circ dY(t) = \lim_{\|\Delta_n\| \to 0} \sum_{i=1}^n \frac{X(t_{i-1}) + X(t_i)}{2}\big(Y(t_i) - Y(t_{i-1})\big),
   $$

   in probability, where $\Delta_n = \{t_0, t_1, \ldots, t_{n-1}, t_n\}$ is a partition of $[a, b]$.

4. Suppose $X_1, \ldots, X_n$ are random variables such that for all $\lambda_1, \ldots, \lambda_n \in \mathbb{R}$,

   $$
   E e^{i(\lambda_1 X_1 + \cdots + \lambda_n X_n)} = e^{i(\lambda_1 \mu_1 + \cdots + \lambda_n \mu_n) - \frac{1}{2}(\lambda_1^2 \sigma_1^2 + \cdots + \lambda_n^2 \sigma_n^2)},
   $$

   where $\mu_j \in \mathbb{R}$ and $\sigma_j > 0$ for $j = 1, \ldots, n$. Prove that the random variables $X_1, \ldots, X_n$ are independent and normally distributed.

5. Check whether $X(t) = \int_0^t \operatorname{sgn}(B(s) - s)\, dB(s)$ is a Brownian motion.

6. Use Example 8.4.5 to compute $E\big(B(t)^4 e^{B(1)}\big)$ for $0 < t \leq 1$. On the other hand, compute this expectation directly from the definition of a Brownian motion. Moreover, compute $E\big(B(t)^4 e^{B(1)}\big)$ for $t > 1$.

7. Consider the local time $L_a(t)$ defined by Equation (8.6.3).

   (a) Find the expectation $E L_a(t)$.

   (b) Check that $\lim_{a \to \infty} \big(L_a(t) + |a|\big) = \lim_{a \to -\infty} \big(L_a(t) + |a|\big) = 1$.

   (c) Show that $E\big[L_a(t) \,\big|\, \mathcal{F}_s\big] \geq |B(s)| - |a| - \int_0^s \operatorname{sgn}(B(u) - a)\, dB(u)$ for any $s \leq t$. Here $\mathcal{F}_s = \sigma\{B(u); 0 \leq u \leq s\}$.

8. Prove that if h satisfies the Novikov condition, then h belongs to the space $L^2_{ad}([a, b] \times \Omega)$.

9. Verify that the value of the integral in Equation (8.8.1) coincides with that of Equation (8.8.3).

10. Verify without using Theorem 8.7.3 that the exponential process given by Equation (8.7.9) is a martingale.

11. Verify that the value of the right-hand side of Equation (8.8.4) coincides with that of Equation (8.8.5).

12. Let $h(t)$ be a deterministic function in $L^2[0, T]$.

    (a) Show that the stochastic process $W(t)\mathcal{E}_h(t)$ in Equation (8.9.9) is a P-martingale.

    (b) Show that the stochastic process $(W(t)^2 - t)\mathcal{E}_h(t)$ in Equation (8.9.10) is a P-martingale.

13. Let $B(t)$ be a Brownian motion with respect to a probability measure P. Find the density function of $B\left(\frac{2}{3}\right)$ with respect to the probability measure $dQ = e^{B(1) - \frac{1}{2}} dP$.

14. Let $B(t)$ be a Brownian motion with respect to a probability measure P. Find a probability measure with respect to which the stochastic process $W(t) = B(t) + t - t^3$, $0 \le t \le 2$, is a Brownian motion.

15. Let $B(t)$ be a Brownian motion with respect to a probability measure P. Find a probability measure with respect to which the stochastic process $W(t) = B(t) + \int_0^t \min\{1, B(s)\} \, ds$, $0 \le t \le 3$, is a Brownian motion.

Multiple Wiener–Itô Integrals

In 1938, N. Wiener introduced polynomial and homogeneous chaos in his study of statistical mechanics. He defined polynomial chaos as sums of finitely many multiple integrals with respect to a Brownian motion. The polynomial chaoses of different orders are not orthogonal. On the other hand, the homogeneous chaoses (which are defined in terms of polynomial chaos) of different orders are orthogonal. However, Wiener did not directly define homogeneous chaos as integrals. In 1951, K. Itô introduced new multiple integrals that turn out to be exactly homogeneous chaos. The new integrals are nowadays referred to as multiple Wiener–Itô integrals. They are related to the stochastic integral that K. Itô introduced in 1944. We will follow the original idea of K. Itô to define multiple Wiener–Itô integrals. Two important theorems in this chapter are the Wiener–Itô theorem and the martingale representation theorem for square integrable Brownian martingales.

9.1 A Simple Example

Let $B(t)$ be a Brownian motion. The Wiener integral $I(f) = \int_a^b f(t) \, dB(t)$ is defined in Section 2.3 for deterministic functions f in $L^2[a, b]$. The random variable $I(f)$ is normally distributed with mean 0 and variance $\int_a^b f(t)^2 \, dt$. The mapping $I : L^2[a, b] \to L^2(\Omega)$ is an isometry. In view of Riemann integrals and double integrals one studies in calculus, we ask the following question.

Question 9.1.1. How can we define a double integral $\int_a^b \int_a^b f(t, s) \, dB(t) dB(s)$ for a deterministic function $f(t, s)$?

(a) **Wiener's Idea**

The process to define the integral consists of two steps. The first step is to define the integral for step functions. The second step is to approximate an $L^2([a, b]^2)$ function by step functions and take the limit of the corresponding integrals. Suppose f is a step function given by

$$f = \sum_{i=1}^{n} \sum_{j=1}^{m} a_{ij} 1_{[t_{i-1},t_i) \times [s_{j-1},s_j)},$$

where $\{a = t_0, t_1, \ldots, t_{n-1}, t_n = b\}$ and $\{a = s_0, s_1, \ldots, s_{m-1}, s_m = b\}$ are partitions of $[a, b]$. Then define the double Wiener integral of f by

$$(W) \int_a^b \int_a^b f(t,s) \, dB(t) dB(s) = \sum_{i,j} a_{ij} \big(B(t_i) - B(t_{i-1})\big)\big(B(s_j) - B(s_{j-1})\big).$$

Consider a simple example, $f = 1_{[0,1] \times [0,1]}$. Obviously, we have

$$(W) \int_0^1 \int_0^1 1 \, dB(t) dB(s) = B(1)^2. \tag{9.1.1}$$

The double Wiener integral has the separation property, namely,

$$(W) \int_a^b \int_a^b f(t)g(s) \, dB(t)dB(s) = \int_a^b f(t) \, dB(t) \int_a^b g(s) \, dB(s), \tag{9.1.2}$$

where the integrals in the right-hand side of the equation are Wiener integrals defined in Section 2.3. In general, the expectation of a double Wiener integral is nonzero. For example, the one in Equation (9.1.1), being equal to $B(1)^2$, has expectation 1. Thus a double Wiener integral may not be orthogonal to constant functions.

(b) Itô's Idea

There are also two steps in defining a double Wiener–Itô integral. The first step is to define the integral for "off-diagonal step functions." The second step is to approximate an $L^2([a, b]^2)$ function by "off-diagonal step functions" and take the limit of the corresponding integrals.

To motivate the notion of an "off-diagonal step function" and its necessity, consider the example of defining the double integral $\int_0^1 \int_0^1 1 \, dB(t)dB(s)$.

Let $\Delta_n = \{t_0, t_1, \ldots, t_n\}$ be a partition of the interval $[0, 1]$. If we take the partition of the unit square $[0, 1)^2$,

$$[0, 1)^2 = \bigcup_{i,j=1}^{n} [t_{i-1}, t_i) \times [t_{j-1}, t_j), \tag{9.1.3}$$

then we get the following Riemann sum for the integrand $f \equiv 1$:

$$\sum_{i,j=1}^{n} \big(B(t_i) - B(t_{i-1})\big)\big(B(t_j) - B(t_{j-1})\big) = \left[\sum_{i=1}^{n} \big(B(t_i) - B(t_{i-1})\big)\right]^2 = B(1)^2,$$

which is the value of the double Wiener integral in Equation (9.1.1).

Here is the crucial idea of K. Itô in the paper [34]. Remove the diagonal squares from $[0, 1)^2$ in Equation (9.1.3),

$$[0, 1)^2 \setminus \bigcup_{i=1}^{n} [t_{i-1}, t_i)^2 = \bigcup_{i \neq j} [t_{i-1}, t_i) \times [t_{j-1}, t_j). \qquad (9.1.4)$$

Then use the remaining squares in the right-hand side of this equation to form the sum from the increments of the Brownian motion,

$$S_n = \sum_{i \neq j} \big(B(t_i) - B(t_{i-1}) \big) \big(B(t_j) - B(t_{j-1}) \big). \qquad (9.1.5)$$

Obviously, we have

$$S_n = \sum_{i,j=1}^{n} \big(B(t_i) - B(t_{i-1}) \big) \big(B(t_j) - B(t_{j-1}) \big) - \sum_{i=1}^{n} \big(B(t_i) - B(t_{i-1}) \big)^2$$

$$= B(1)^2 - \sum_{i=1}^{n} \big(B(t_i) - B(t_{i-1}) \big)^2.$$

Notice that the limit of the last summation is the quadratic variation of the Brownian motion $B(t)$ on the interval $[0, 1]$ and equals 1. Hence

$$\lim_{\|\Delta_n\| \to 0} S_n = B(1)^2 - 1, \quad \text{in } L^2(\Omega). \qquad (9.1.6)$$

The limit is defined to be the double Wiener–Itô integral of $f \equiv 1$,

$$\int_0^1 \int_0^1 1 \, dB(t) dB(s) = B(1)^2 - 1, \qquad (9.1.7)$$

which is different from the value in Equation (9.1.1). Can this double integral be written as an iterated integral? We can try to write

$$\int_0^1 \int_0^1 1 \, dB(t) dB(s) = \int_0^1 \left[\int_0^1 1 \, dB(s) \right] dB(t) = \int_0^1 B(1) \, dB(t).$$

But the last integral $\int_0^1 B(1) \, dB(t)$ is not defined as an Itô integral since the integrand $B(1)$ is not adapted with respect to the filtration $\{\mathcal{F}_t; 0 \leq t \leq 1\}$, where $\mathcal{F}_t = \sigma\{B(s); 0 \leq s \leq t\}$. We can try another way to write

$$\int_0^1 \int_0^1 1 \, dB(t) dB(s) = 2 \iint_{0 \leq s \leq t \leq 1} 1 \, dB(t) dB(s)$$

$$= 2 \int_0^1 \left[\int_0^t 1 \, dB(s) \right] dB(t)$$

$$= 2 \int_0^1 B(t) \, dB(t)$$

$$= B(1)^2 - 1,$$

which is the double Wiener–Itô integral in Equation (9.1.7).

In order to find out how in general double Wiener–Itô integrals should be defined, let us examine more closely the sum in Equation (9.1.5). It is the double integral of the following "off-diagonal step function"

$$f_n = \sum_{i \neq j} 1_{[t_{i-1},t_i) \times [t_{i-1},t_i)}. \tag{9.1.8}$$

Observe that as $\|\Delta_n\| \to 0$,

$$\int_0^1 \int_0^1 |1 - f_n(t,s)|^2 \, dt \, ds = \sum_{i=1}^n (t_i - t_{i-1})^2 \leq \|\Delta_n\| \to 0. \tag{9.1.9}$$

Hence the function $f \equiv 1$ is approximated by a sequence $\{f_n\}$ of "off-diagonal step functions" that are supported by squares off the diagonal line. This is the crucial idea for multiple Wiener–Itô integrals. We will start with double Wiener–Itô integrals in the next section.

9.2 Double Wiener–Itô Integrals

The object in this section is to define the double Wiener–Itô integral

$$\int_a^b \int_a^b f(t,s) \, dB(t) dB(s), \quad f \in L^2([a,b]^2).$$

This will be done in two steps as in the case of Wiener integrals in Section 2.3. However, there is the new crucial notion of "off-diagonal step functions" mentioned at the end of the previous section.

Let $D = \{(t,s) \in [a,b]^2; t = s\}$ denote the diagonal of the square $[a,b]^2$. By a rectangle in this section we will mean a subset of $[a,b]^2$ of the form $[t_1,t_2) \times [s_1,s_2)$.

Step 1. *Off-diagonal step functions*

Definition 9.2.1. *An* off-diagonal step function *on the square $[a,b]^2$ is defined to be a function of the form*

$$f = \sum_{i \neq j} a_{ij} 1_{[t_{i-1},t_i) \times [t_{j-1},t_j)}, \tag{9.2.1}$$

where $a = t_0 < t_1 < t_2 < \cdots < t_{n-1} < t_n = b$.

Note that an off-diagonal step function vanishes on the diagonal D. Hence the function $f \equiv 1$ on $[a,b]^2$ is not an off-diagonal step function. If $A = [t_1,t_2) \times [s_1,s_2)$ is a rectangle disjoint from the diagonal D, then 1_A can be written in the form of Equation (9.2.1) by taking the set $\{t_1,t_2,s_1,s_2\}$ as the partition points of $[a, b]$. Hence 1_A is an off-diagonal step function. More generally, suppose $A_1, A_2, \ldots, A_n$ are rectangles disjoint from the diagonal D. Then the function $f = \sum_{i=1}^{n} a_i 1_{A_i}$ is an off-diagonal step function for any $a_1, a_2, \ldots, a_n \in \mathbb{R}$. This fact implies that if f and g are off-diagonal step functions, then $af + bg$ is an off-diagonal step function for any $a, b \in \mathbb{R}$. Hence the set of off-diagonal step functions is a vector space.

For an off-diagonal step function f given by Equation (9.2.1), define

$$I_2(f) = \sum_{i \neq j} a_{ij} \big(B(t_i) - B(t_{i-1})\big)\big(B(t_j) - B(t_{j-1})\big). \tag{9.2.2}$$

Note that the representation of an off-diagonal step function f by Equation (9.2.1) is not unique. But it is easily seen that $I_2(f)$ is uniquely defined. Moreover, I_2 is linear, i.e., $I_2(af + bg) = aI_2(f) + bI_2(g)$ for any off-diagonal step functions f and g and $a, b \in \mathbb{R}$.

The *symmetrization* $\widehat{f}(t, s)$ of a function $f(t, s)$ is defined by

$$\widehat{f}(t, s) = \frac{1}{2}\big(f(t, s) + f(s, t)\big).$$

Obviously, $\widehat{f}$ is a symmetric function. If f is a symmetric function, then $\widehat{f} = f$. In general, $\|\widehat{f}\| \leq \|f\|$. The strict inequality can happen. For example, for the function $f(t, s) = t$ on $[0, 1]^2$, we have $\widehat{f}(t, s) = \frac{1}{2}(t + s)$ and

$$\|f\|^2 = \int_0^1 \int_0^1 f(t, s)^2 \, dt \, ds = \frac{1}{3}, \quad \|\widehat{f}\|^2 = \int_0^1 \int_0^1 \widehat{f}(t, s)^2 \, dt \, ds = \frac{7}{24}.$$

Notice that the symmetrization operation is linear, i.e., $(af + bg)\widehat{} = a\widehat{f} + b\widehat{g}$. If f is an off-diagonal step function, then $\widehat{f}$ is also an off-diagonal step function.

Lemma 9.2.2. *Let f be an off-diagonal step function. Then $I_2(f) = I_2(\widehat{f})$.*

Proof. Since I_2 and the symmetrization operation are linear, it suffices to prove the lemma for the case $f = 1_{[t_1, t_2) \times [s_1, s_2)}$ with $[t_1, t_2) \cap [s_1, s_2) = \emptyset$. The symmetrization $\widehat{f}$ of f is given by

$$\widehat{f} = \frac{1}{2}\big(1_{[t_1, t_2) \times [s_1, s_2)} + 1_{[s_1, s_2) \times [t_1, t_2)}\big).$$

Hence by the definition of I_2 in Equation (9.2.2),

$$I_2(f) = \big(B(t_2) - B(t_1)\big)\big(B(s_2) - B(s_1)\big),$$
$$I_2(\widehat{f}) = \frac{1}{2}\big\{\big(B(t_2) - B(t_1)\big)\big(B(s_2) - B(s_1)\big)$$
$$+ \big(B(s_2) - B(s_1)\big)\big(B(t_2) - B(t_1)\big)\big\}$$
$$= \big(B(t_2) - B(t_1)\big)\big(B(s_2) - B(s_1)\big).$$

Therefore, $I_2(f) = I_2(\widehat{f})$ and the lemma is proved. $\qquad\square$

Lemma 9.2.3. *If f is an off-diagonal step function, then $E[I_2(f)] = 0$ and*

$$E[I_2(f)^2] = 2 \int_a^b \int_a^b \widehat{f}(t,s)^2 \, dt \, ds. \tag{9.2.3}$$

Proof. Suppose f is represented by Equation (9.2.1). Then $I_2(f)$ is given by Equation (9.2.2). Since the intervals $[t_{i-1}, t_i)$ and $[t_{j-1}, t_j)$ are disjoint when $i \neq j$, the expectation of each term in the summation of Equation (9.2.2) is zero. Hence $E[I_2(f)] = 0$.

To prove Equation (9.2.3), first assume that f is symmetric. In this case, $a_{ij} = a_{ji}$ for all $i \neq j$. For convenience, let $\xi_i = B(t_i) - B(t_{i-1})$. Then

$$I_2(f) = \sum_{i \neq j} a_{ij} \xi_i \xi_j = 2 \sum_{i<j} a_{ij} \xi_i \xi_j.$$

Therefore, we have

$$E[I_2(f)^2] = 4 \sum_{i<j} \sum_{p<q} a_{ij} a_{pq} E(\xi_i \xi_j \xi_p \xi_q). \tag{9.2.4}$$

Let $i < j$ be fixed. By observing the position of intervals, we can easily see the following implications:

$$p \neq i \implies E(\xi_i \xi_j \xi_p \xi_q) = 0 \quad \forall q > p,$$
$$q \neq j \implies E(\xi_i \xi_j \xi_p \xi_q) = 0 \quad \forall p < q.$$

Hence for fixed $i < j$, the summation over $p < q$ in Equation (9.2.4) reduces to only one term given by $p = i$ and $q = j$. Therefore,

$$\begin{aligned}
E[I_2(f)^2] &= 4 \sum_{i<j} a_{ij}^2 E(\xi_i^2 \xi_j^2) \\
&= 4 \sum_{i<j} a_{ij}^2 (t_i - t_{i-1})(t_j - t_{j-1}) \\
&= 2 \sum_{i \neq j} a_{ij}^2 (t_i - t_{i-1})(t_j - t_{j-1}) \\
&= 2 \int_a^b \int_a^b f(t,s)^2 \, dt \, ds.
\end{aligned}$$

Finally, for any off-diagonal step function f, we have $I_2(f) = I_2(\widehat{f})$ by Lemma 9.2.2. Hence

$$E[I_2(f)^2] = E[I_2(\widehat{f})^2] = 2 \int_a^b \int_a^b \widehat{f}(t,s)^2 \, dt \, ds,$$

which proves Equation (9.2.3). □

Step 2. *Approximation by off-diagonal step functions*

Recall that $\|f\|$ denotes the L^2-norm of a function f defined on the square $[a, b]^2$, i.e., $\|f\|^2 = \int_a^b \int_a^b f(t, s)^2 \, dt \, ds$. By Lemma 9.2.3, $E\left[I_2(f)^2\right] = 2\|\widehat{f}\|^2$ for any off-diagonal step function f. But $\|\widehat{f}\| \leq \|f\|$. Hence $E\left[I_2(f)^2\right] \leq 2\|f\|^2$ for all off-diagonal step functions. This inequality shows that we can extend the mapping I_2 to $L^2([a, b]^2)$ provided that each function in $L^2([a, b]^2)$ can be approximated by a sequence of off-diagonal step functions.

Suppose f is a function in $L^2([a, b]^2)$. Let D_δ denote the set of points in $[a, b]^2$ having a distance $< \delta$ from the diagonal D. For a given $\varepsilon > 0$, we can choose $\delta > 0$ small enough such that

$$\iint_{D_\delta} f(t, s)^2 \, dt \, ds < \frac{\varepsilon}{2}. \qquad (9.2.5)$$

On the other hand, let $D_\delta^c = [a, b]^2 \setminus D_\delta$ and consider the restriction of f to D_δ^c. A fact from measure theory says that there exists a function ϕ of the form $\phi = \sum_{i=1}^n a_i 1_{A_i}$ with rectangles $A_i \subset D_\delta^c$ for all $1 \leq i \leq n$ such that

$$\iint_{D_\delta^c} |f(t, s) - \phi(t, s)|^2 \, dt \, ds < \frac{\varepsilon}{2}. \qquad (9.2.6)$$

We can sum up the integrals in Equations (9.2.5) and (9.2.6) to get

$$\int_a^b \int_a^b |f(t, s) - \phi(t, s)|^2 \, dt \, ds < \varepsilon.$$

Note that the function ϕ vanishes on the set D_δ. Hence the function ϕ is an off-diagonal step function, as pointed out in the first paragraph following Definition 9.2.1. Thus we have proved the next approximation lemma.

Lemma 9.2.4. *Let f be a function in $L^2([a, b]^2)$. Then there exists a sequence $\{f_n\}$ of off-diagonal step functions such that*

$$\lim_{n \to \infty} \int_a^b \int_a^b |f(t, s) - f_n(t, s)|^2 \, dt \, ds = 0. \qquad (9.2.7)$$

Now, we are ready to extend the mapping I_2 to the space $L^2([a, b]^2)$. Let $f \in L^2([a, b]^2)$. Choose a sequence $\{f_n\}$ of off-diagonal step functions converging to f in $L^2([a, b]^2)$. The existence of such a sequence is guaranteed by Lemma 9.2.4. Then by the linearity of I_2 and Lemma 9.2.3,

$$E\left\{\left(I_2(f_n) - I_2(f_m)\right)^2\right\} = 2\|\widehat{f}_n - \widehat{f}_m\|^2 \leq 2\|f_n - f_m\|^2 \to 0,$$

as $n, m \to \infty$. Hence the sequence $\{I_2(f_n)\}$ is Cauchy in $L^2(\Omega)$. Define

$$I_2(f) = \lim_{n \to \infty} I_2(f_n), \quad \text{in } L^2(\Omega). \qquad (9.2.8)$$

It is easily seen that $I_2(f)$ is well-defined, namely, it does not depend on the choice of the sequence $\{f_n\}$ used in Equation (9.2.8).

Definition 9.2.5. *Let $f \in L^2([a,b]^2)$. The limit $I_2(f)$ in Equation (9.2.8) is called the* double Wiener–Itô integral *of f.*

We will also use $\int_a^b \int_a^b f(t,s) \, dB(t) \, dB(s)$ to denote the double Wiener–Itô integral $I_2(f)$ of f.

Example 9.2.6. Note that the function $f \equiv 1$ on $[0,1]^2$ is not an off-diagonal step function. Hence its double Wiener–Itô integral is not defined in Step 1. Define f_n by Equation (9.1.8). Then we have $I_2(f_n) = S_n$ by Equation (9.1.5). As shown in Equation (9.1.9), f_n converges to f in $L^2([0,1]^2)$. Therefore, by Definition 9.2.5 and Equation (9.1.6),

$$I_2(f) = \lim_{n \to \infty} I_2(f_n) = \lim_{n \to \infty} S_n = B(1)^2 - 1, \quad \text{in } L^2(\Omega).$$

Hence we have $\int_0^1 \int_0^1 1 \, dB(t) \, dB(s) = B(1)^2 - 1$.

Lemmas 9.2.2 and 9.2.3 can be extended to functions in $L^2([a,b]^2)$ using the approximation in Lemma 9.2.4 and the definition of a double Wiener–Itô integral. We state the extension as the next theorem without proof since the verification is routine.

Theorem 9.2.7. *Let $f(t,s) \in L^2([a,b]^2)$. Then we have*

(1) $I_2(f) = I_2(\widehat{f})$. *Here $\widehat{f}$ is the symmetrization of f.*
(2) $E[I_2(f)] = 0$.
(3) $E[I_2(f)^2] = 2\|\widehat{f}\|^2$. *Here $\| \cdot \|$ is the norm on $L^2([a,b]^2)$.*

There is an important relationship between double Wiener–Itô integrals and iterated Itô integrals. It is given by the next theorem.

Theorem 9.2.8. *Let $f(t,s) \in L^2([a,b]^2)$. Then*

$$\int_a^b \int_a^b f(t,s) \, dB(t) \, dB(s) = 2 \int_a^b \left[\int_a^t \widehat{f}(t,s) \, dB(s) \right] dB(t), \qquad (9.2.9)$$

where $\widehat{f}$ is the symmetrization of f.

Remark 9.2.9. The inner integral $X_t = \int_a^t \widehat{f}(t,s) \, dB(s)$ is a Wiener integral, defined in Section 2.3. By Theorem 2.3.4, $E(X_t^2) = \int_a^t \widehat{f}(t,s)^2 \, ds$. Hence

$$\int_a^b E(X_t^2) \, dt = \int_a^b \left[\int_a^t \widehat{f}(t,s)^2 \, ds \right] dt = \frac{1}{2}\|\widehat{f}\|^2 \leq \frac{1}{2}\|f\|^2 < \infty.$$

This shows that the stochastic process X_t belongs to $L^2_{\text{ad}}([a,b] \times \Omega)$ and the integral $\int_a^b X_t \, dB(t)$ is an Itô integral as defined in Section 4.3.

Proof. First consider the case $f = 1_{[t_1,t_2) \times [s_1,s_2)}$ with $[t_1, t_2) \cap [s_1, s_2) = \emptyset$. The symmetrization of f is given by

$$\widehat{f} = \frac{1}{2} \left(1_{[t_1,t_2) \times [s_1,s_2)} + 1_{[s_1,s_2) \times [t_1,t_2)} \right).$$

We may assume that $s_1 < s_2 < t_1 < t_2$. For the other case $t_1 < t_2 < s_1 < s_2$, just interchange $[s_1, s_2)$ with $[t_1, t_2)$. By the definition of $I_2(f)$ in Step 1,

$$\int_a^b \int_a^b f(t, s) \, dB(t) \, dB(s) = \big(B(t_2) - B(t_1)\big)\big(B(s_2) - B(s_1)\big). \qquad (9.2.10)$$

On the other hand, we have

$$\int_a^b \left[\int_a^t \widehat{f}(t, s) \, dB(s) \right] dB(t) = \frac{1}{2} \int_{t_1}^{t_2} \left[\int_{s_1}^{s_2} 1 \, dB(s) \right] dB(t)$$

$$= \frac{1}{2} \int_{t_1}^{t_2} \big(B(s_2) - B(s_1)\big) \, dB(t)$$

$$= \frac{1}{2} \big(B(s_2) - B(s_1)\big)\big(B(t_2) - B(t_1)\big). \quad (9.2.11)$$

Hence Equation (9.2.9) follows from Equations (9.2.10) and (9.2.11).

Next, by the linearity of the mapping I_2 and the symmetrization operation, Equation (9.2.9) is also valid for any off-diagonal step function. Finally, use the approximation to extend Equation (9.2.9) to functions $f \in L^2([a, b]^2)$. □

The generalization of the double Wiener–Itô integral to higher order is almost straightforward except for complicated notation. This will be done in Section 9.6 after we study homogeneous chaos in Sections 9.4 and 9.5.

9.3 Hermite Polynomials

In this section we will explain Hermite polynomials and their properties to be used in Sections 9.4 and 9.5.

Let ν be the Gaussian measure with mean 0 and variance $\rho > 0$, i.e.,

$$d\nu(x) = \frac{1}{\sqrt{2\pi\rho}} e^{-\frac{1}{2\rho} x^2} \, dx.$$

Consider the sequence $1, x, x^2, \ldots, x^n, \ldots$ of monomials in the real Hilbert space $L^2(\nu)$. Apply the Gram–Schmidt orthogonalization procedure to this sequence (in order of increasing powers) to obtain orthogonal polynomials $P_0(x), P_1(x), \ldots, P_n(x), \ldots$ in the Hilbert space $L^2(\nu)$, where $P_0(x) = 1$ and $P_n(x)$ is a polynomial of degree $n \geq 1$ with leading coefficient 1.

Question 9.3.1. Is it possible to define the polynomials $P_n(x)$ directly so that they are computable?

Take the function $\theta(t, x) = e^{tx}$. The expectation of $\theta(t, \cdot)$ with respect to the measure ν is given by

$$E_\nu[\theta(t, \cdot)] = \int_{-\infty}^{\infty} e^{tx} \frac{1}{\sqrt{2\pi\rho}} e^{-\frac{1}{2\rho}x^2} \, dx = e^{\frac{1}{2}\rho t^2}. \tag{9.3.1}$$

The multiplicative renormalization $\psi(t, x)$ of $\theta(t, x)$ is defined by

$$\psi(t, x) = \frac{\theta(t, x)}{E_\nu[\theta(t, \cdot)]} = e^{tx - \frac{1}{2}\rho t^2}.$$

We can expand the function $\psi(t, x)$ as a power series in t. Note that the coefficient of t^n in the series expansion depends on n, x, and ρ. Let $H_n(x; \rho)/n!$ denote this coefficient. Then we have the equality

$$e^{tx - \frac{1}{2}\rho t^2} = \sum_{n=0}^{\infty} \frac{H_n(x; \rho)}{n!} t^n. \tag{9.3.2}$$

On the other hand, we can use the power series $e^z = \sum_{n=0}^{\infty} z^n/n!$ to find the coefficient of t^n in another way. Write $e^{tx - \frac{1}{2}\rho t^2}$ as $e^{tx} e^{-\frac{1}{2}\rho t^2}$ to get

$$e^{tx - \frac{1}{2}\rho t^2} = \left[1 + xt + \cdots + \frac{x^n}{n!} t^n + \cdots \right]\left[1 + \frac{-\rho}{2} t^2 + \cdots + \frac{(-\rho)^m}{m!2^m} t^{2m} + \cdots \right],$$

which can be multiplied out to find the coefficient of t^n:

$$\frac{x^n}{n!} \cdot 1 + \frac{x^{n-2}}{(n-2)!} \frac{(-\rho)}{2} + \cdots + \frac{x^{n-2k}}{(n-2k)!} \frac{(-\rho)^k}{k!2^k} + \cdots,$$

with the last term being an x-term (when n is odd) or a constant (when n is even). But the coefficient of t^n is also given by $H_n(x; \rho)/n!$ in Equation (9.3.2). Therefore,

$$H_n(x; \rho) = n!\left(\frac{x^n}{n!} \cdot 1 + \frac{x^{n-2}}{(n-2)!} \frac{(-\rho)}{2} + \cdots + \frac{x^{n-2k}}{(n-2k)!} \frac{(-\rho)^k}{k!2^k} + \cdots \right)$$

$$= x^n + \frac{n!}{(n-2)!2}(-\rho)x^{n-2} + \cdots + \frac{n!}{(n-2k)!k!2^k}(-\rho)^k x^{n-2k} + \cdots.$$

Observe that

$$\frac{n!}{(n-2k)!k!2^k} = \binom{n}{2k}(2k-1)!!, \tag{9.3.3}$$

where $(2k-1)!! = (2k-1)(2k-3)\cdots 3 \cdot 1$ and by convention $(-1)!! = 1$. The number in Equation (9.3.3) represents the number of ways of choosing k pairs from a set of n distinct objects.

By using Equation (9.3.3), we can express $H_n(x; \rho)$ as follows:

$$H_n(x; \rho) = \sum_{k=0}^{[n/2]} \binom{n}{2k}(2k-1)!!\,(-\rho)^k x^{n-2k}, \tag{9.3.4}$$

where $[n/2]$ is the integer part of $n/2$. Note that $H_n(x; \rho)$ is a polynomial in x of degree n with leading coefficient 1. For example, the first few of these polynomials are given by

$$
\begin{aligned}
H_0(x; \rho) &= 1, \\
H_1(x; \rho) &= x, \\
H_2(x; \rho) &= x^2 - \rho, \\
H_3(x; \rho) &= x^3 - 3\rho x, \\
H_4(x; \rho) &= x^4 - 6\rho x^2 + 3\rho^2, \\
H_5(x; \rho) &= x^5 - 10\rho x^3 + 15\rho^2 x.
\end{aligned}
$$

Definition 9.3.2. *The polynomial $H_n(x; \rho)$ given by Equation (9.3.4) is called the* Hermite polynomial *of degree n with parameter ρ.*

Theorem 9.3.3. *Let $H_n(x; \rho)$ be the Hermite polynomial of degree n defined by Equation (9.3.4). Then*

$$
H_n(x; \rho) = (-\rho)^n \, e^{x^2/2\rho} \, D_x^n \, e^{-x^2/2\rho}, \tag{9.3.5}
$$

where D_x is the differentiation operator in the x-variable.

Remark 9.3.4. Most books, if not all, define the Hermite polynomial $H_n(x; \rho)$ by Equation (9.3.5). In that case, Equation (9.3.4) becomes a theorem.

Proof. The exponent in the left-hand side of Equation (9.3.2) can be rewritten as $-\frac{1}{2}\rho(t - \frac{x}{\rho})^2 + \frac{x^2}{2\rho}$. Hence we have

$$
e^{x^2/2\rho} e^{-\frac{1}{2}\rho(t-\frac{x}{\rho})^2} = \sum_{n=0}^{\infty} \frac{H_n(x; \rho)}{n!} \, t^n.
$$

Differentiate both sides n times in the t-variable and then put $t = 0$ to get

$$
H_n(x; \rho) = e^{x^2/2\rho} \left(D_t^n \, e^{-\frac{1}{2}\rho(t-\frac{x}{\rho})^2} \right) \Big|_{t=0}.
$$

Note that $D_t^n \, e^{-\frac{1}{2}\rho(t-\frac{x}{\rho})^2} = (-\rho)^n D_x^n \, e^{-\frac{1}{2}\rho(t-\frac{x}{\rho})^2}$. Therefore,

$$
\begin{aligned}
H_n(x; \rho) &= e^{x^2/2\rho} \left((-\rho)^n D_x^n \, e^{-\frac{1}{2}\rho(t-\frac{x}{\rho})^2} \right) \Big|_{t=0} \\
&= (-\rho)^n \, e^{x^2/2\rho} \, D_x^n \, e^{-x^2/2\rho},
\end{aligned}
$$

which proves Equation (9.3.5). $\qquad\square$

Now, what is the relationship between the Hermite polynomial $H_n(x; \rho)$ and the polynomial $P_n(x)$ in Question 9.3.1? If we can show that the Hermite polynomials $H_n(x; \rho), n = 0, 1, \ldots$, are orthogonal, then we can conclude that $H_n(x; \rho) = P_n(x)$. This is indeed the case.

Theorem 9.3.5. *Let ν be the Gaussian measure with mean 0 and variance ρ. Then the Hermite polynomials $H_n(x;\rho)$, $n \geq 0$, are orthogonal in $L^2(\nu)$. Moreover, we have*

$$\int_{-\infty}^{\infty} H_n(x;\rho)^2 \, d\nu(x) = n!\rho^n, \quad n \geq 0. \tag{9.3.6}$$

Proof. For any $t, s \in \mathbb{R}$, use Equation (9.3.2) to get

$$e^{(t+s)x - \frac{1}{2}\rho(t^2+s^2)} = \sum_{n,m=0}^{\infty} \frac{t^n s^m}{n!m!} H_n(x;\rho) H_m(x;\rho). \tag{9.3.7}$$

By Equation (9.3.1),

$$\int_{-\infty}^{\infty} e^{(t+s)x - \frac{1}{2}\rho(t^2+s^2)} \, d\nu(x) = e^{-\frac{1}{2}\rho(t^2+s^2)} \, e^{\frac{1}{2}\rho(t+s)^2} = e^{\rho ts}.$$

Therefore, upon integrating both sides of Equation (9.3.7), we obtain

$$e^{\rho ts} = \sum_{n,m=0}^{\infty} \frac{t^n s^m}{n!m!} \int_{-\infty}^{\infty} H_n(x;\rho) H_m(x;\rho) \, d\nu(x). \tag{9.3.8}$$

Since the left-hand side is a function of the product ts, the coefficient of $t^n s^m$ in the right-hand side must be 0 for any $n \neq m$, namely,

$$\int_{-\infty}^{\infty} H_n(x;\rho) H_m(x;\rho) \, d\nu(x) = 0, \quad \forall \, n \neq m. \tag{9.3.9}$$

Hence the Hermite polynomials are orthogonal in $L^2(\nu)$. And then Equation (9.3.8) becomes

$$e^{\rho ts} = \sum_{n=0}^{\infty} \frac{(ts)^n}{(n!)^2} \int_{-\infty}^{\infty} H_n(x;\rho)^2 \, d\nu(x).$$

But we also have the power series expansion

$$e^{\rho ts} = \sum_{n=0}^{\infty} \frac{\rho^n}{n!} (ts)^n.$$

Obviously, we get Equation (9.3.6) upon comparing the coefficients of $(ts)^n$ in the last two power series.   □

By Theorem 9.3.5, the collection $\{H_n(x;\rho)/\sqrt{n!\rho^n} \, ; \, n = 0, 1, 2, \dots\}$ is an orthonormal system. Its orthogonal complement can be shown to be $\{0\}$. Hence this collection forms an orthonormal basis for $L^2(\nu)$. We state this fact as the next theorem.

Theorem 9.3.6. *Let ν be the Gaussian measure with mean 0 and variance ρ. Then every function f in $L^2(\nu)$ has a unique series expansion*

$$f(x) = \sum_{n=0}^{\infty} a_n \frac{H_n(x; \rho)}{\sqrt{n! \rho^n}},\tag{9.3.10}$$

where the coefficients a_n are given by

$$a_n = \frac{1}{\sqrt{n! \rho^n}} \int_{-\infty}^{\infty} f(x) H_n(x; \rho)\, d\nu(x), \quad n \geq 0.$$

Moreover, we have

$$\|f\|^2 = \sum_{n=0}^{\infty} a_n^2.$$

We list below several identities for the Hermite polynomials.

1. *Generating function:* $e^{tx - \frac{1}{2} \rho t^2} = \sum_{n=0}^{\infty} \frac{t^n}{n!} H_n(x; \rho)$.

2. *Monomials:* $x^n = \sum_{k=0}^{[n/2]} \binom{n}{2k} (2k-1)!! \rho^k H_{n-2k}(x; \rho)$.

3. *Recursion formula:* $H_{n+1}(x; \rho) = x H_n(x; \rho) - \rho n H_{n-1}(x; \rho)$.

4. *Derivative:* $D_x H_n(x; \rho) = n H_{n-1}(x; \rho)$.

5. *Eigenfunctions:* $\left(-\rho D_x^2 + x D_x \right) H_n(x; \rho) = n H_n(x; \rho)$.

6. *Product:* $H_n(x; \rho) H_m(x; \rho) = \sum_{k=0}^{n \wedge m} k! \binom{n}{k} \binom{m}{k} \rho^k H_{n+m-2k}(x; \rho)$.

7. *Partial derivatives:* $\frac{\partial}{\partial \rho} H_n(x; \rho) = -\frac{1}{2} \frac{\partial^2}{\partial x^2} H_n(x; \rho)$.

9.4 Homogeneous Chaos

Let ν be the Gaussian measure on $\mathbb{R}$ with mean 0 and variance ρ. By Theorem 9.3.6 in the previous section, every function in the Hilbert space $L^2(\nu)$ has a unique series expansion by the Hermite polynomials $H_n(x; \rho)$, $n \geq 0$.

Recall the Wiener space in Section 3.1. Let C denote the Banach space of real-valued continuous functions on $[0, 1]$ vanishing at 0 and let μ be the Wiener measure on C. The Wiener space (C, μ) is an infinite-dimensional analogue of the one-dimensional probability space $(\mathbb{R}, \nu)$.

Question 9.4.1. What is the analogue of Equation (9.3.10) for the Hilbert space $L^2(\mu)$ of square integrable functions on the Wiener space (C, μ)?

By Theorem 3.1.2, the stochastic process $B(t, \omega) = \omega(t)$, $0 \leq t \leq 1$, $\omega \in C$, is a Brownian motion. To prevent us from being restricted to the interval $[0, 1]$, we will consider a little bit more general setup for Question 9.4.1.

Let $(\Omega, \mathcal{F}, P)$ be a fixed probability space and let $B(t)$ be a Brownian motion with respect to P. We will fix a finite interval $[a, b]$. Recall that for a function $f \in L^2[a, b]$, the Wiener integral $\int_a^b f(t)\, dB(t)$ is measurable with respect to a smaller σ-field than $\mathcal{F}$, namely, the Brownian σ-field

$$\mathcal{F}^B \equiv \sigma\{B(t)\,; \, a \leq t \leq b\}.$$

In general, $\mathcal{F}^B \neq \mathcal{F}$. An $\mathcal{F}^B$-measurable function on Ω can be referred to as a Brownian function. Let $L_B^2(\Omega)$ denote the Hilbert space of P-square integrable functions on Ω that are measurable with respect to $\mathcal{F}^B$. Note that $L_B^2(\Omega)$ is a subspace of $L^2(\Omega)$.

Question 9.4.2. What is the analogue of Equation (9.3.10) for the space $L_B^2(\Omega)$ of square integrable functions on the probability space $(\Omega, \mathcal{F}^B, P)$?

Note that when $\Omega = C$ and $P = \mu$, we have $\mathcal{F} = \mathcal{F}^B = \mathcal{B}(C)$, the Borel σ-field of C. Hence the above question reduces to Question 9.4.1.

First we prove a lemma, which we will need below.

Lemma 9.4.3. *Let $(\Omega, \mathcal{G}, P)$ be a probability space and let $\{\mathcal{G}_n\}$ be a filtration such that $\sigma\{\cup_n \mathcal{G}_n\} = \mathcal{G}$. Suppose $X \in L^1(\Omega)$. Then $E[X \,|\, \mathcal{G}_n]$ converges to X in $L^1(\Omega)$ as $n \to \infty$.*

Remark 9.4.4. Actually, it is also true that $E[X \,|\, \mathcal{G}_n]$ converges to X almost surely. In general, the *martingale convergence theorem* says that if $\{X_n\}$ is an $L^1(\Omega)$-bounded martingale, then X_n converges almost surely to some random variable. But the proof is much more involved than this lemma.

Proof. Let $\varepsilon > 0$ be given. Since $\sigma\{\cup_n \mathcal{G}_n\} = \mathcal{G}$, there exist $S_\varepsilon = \sum_{i=1}^k a_i 1_{A_i}$, $a_i \in \mathbb{R}, A_i \in \cup_n \mathcal{G}_n$, such that

$$\|X - S_\varepsilon\|_1 < \varepsilon/2, \tag{9.4.1}$$

where $\| \cdot \|_1$ is the norm on $L^1(\Omega)$. For simplicity, let $X_n = E[X \,|\, \mathcal{G}_n]$. Then

$$\|X - X_n\|_1 = \|\{X - S_\varepsilon\} + \{S_\varepsilon - E[S_\varepsilon | \mathcal{G}_n]\} + \{E[S_\varepsilon - X | \mathcal{G}_n]\}\|_1$$
$$\leq \|X - S_\varepsilon\|_1 + \|S_\varepsilon - E[S_\varepsilon | \mathcal{G}_n]\|_1 + \|E[S_\varepsilon - X | \mathcal{G}_n]\|_1. \tag{9.4.2}$$

Note that $|E[S_\varepsilon - X \,|\, \mathcal{G}_n]| \leq E[|S_\varepsilon - X| \,|\, \mathcal{G}_n]$ by the conditional Jensen's inequality. Then take the expectation to get

$$\|E[S_\varepsilon - X \,|\, \mathcal{G}_n]\|_1 \leq E|S_\varepsilon - X| = \|S_\varepsilon - X\|_1 < \varepsilon/2. \tag{9.4.3}$$

Moreover, observe that $A_i \in \cup_n \mathcal{G}_n$ for $i = 1, 2, \ldots, k$. Hence there exists N such that $A_i \in \mathcal{G}_N$ for all $i = 1, 2, \ldots, k$. This implies that

$$E[S_\varepsilon | \mathcal{G}_n] = S_\varepsilon, \quad \forall\, n \geq N. \tag{9.4.4}$$

By putting Equations (9.4.1), (9.4.3), and (9.4.4) into Equation (9.4.2), we immediately see that $\|X - X_n\|_1 \leq \varepsilon$ for all $n \geq N$. Hence X_n converges to X in $L^1(\Omega)$ as $n \to \infty$. $\qquad \square$

Now we return to the Brownian motion $B(t)$ on the probability space $(\Omega, \mathcal{F}^B, P)$. Let $I(f) = \int_a^b f(t)\, dB(t)$ be the Wiener integral of $f \in L^2[a, b]$. A product $I(f_1)I(f_2)\cdots I(f_k)$ with $f_1, f_2, \ldots, f_k \in L^2[a, b]$ is called a *polynomial chaos* of order k; cf. Equation (9.1.2).

Let $J_0 = \mathbb{R}$ and for $n \geq 1$, define J_n to be the $L_B^2(\Omega)$-closure of the linear space spanned by constant functions and polynomial chaos of degree $\leq n$. Then we have the inclusions

$$J_0 \subset J_1 \subset \cdots \subset J_n \subset \cdots \subset L_B^2(\Omega). \tag{9.4.5}$$

Theorem 9.4.5. *The union $\bigcup_{n=0}^{\infty} J_n$ is dense in $L_B^2(\Omega)$.*

Proof. Let $\{e_n\}_{n=1}^{\infty}$ be an orthonormal basis for $L^2[a, b]$ and let $\mathcal{G}_n$ be the σ-field generated by $I(e_1), I(e_2), \ldots, I(e_n)$. Then $\{\mathcal{G}_n\}$ is a filtration and we have $\sigma\{\cup_n \mathcal{G}_n\} = \mathcal{F}^B$.

Let $\phi \in L_B^2(\Omega)$ be orthogonal to $\bigcup_{n=0}^{\infty} J_n$. Then for any fixed n,

$$E\big\{\phi \cdot I(e_1)^{k_1} I(e_2)^{k_2} \cdots I(e_n)^{k_n}\big\} = 0, \quad \forall\, k_1, k_2, \ldots, k_n \geq 0.$$

Observe that $I(e_1)^{k_1} I(e_2)^{k_2} \cdots I(e_n)^{k_n}$ is $\mathcal{G}_n$-measurable. Hence

$$E\big\{\phi \cdot I(e_1)^{k_1} I(e_2)^{k_2} \cdots I(e_n)^{k_n}\big\}$$
$$= E\big\{I(e_1)^{k_1} I(e_2)^{k_2} \cdots I(e_n)^{k_n} E[\phi \,|\, \mathcal{G}_n]\big\}.$$

Therefore, for all integers $k_1, k_2, \ldots, k_n \geq 0$,

$$E\big\{I(e_1)^{k_1} I(e_2)^{k_2} \cdots I(e_n)^{k_n} E[\phi \,|\, \mathcal{G}_n]\big\} = 0.$$

Note that the random variables $I(e_1), I(e_2), \ldots, I(e_n)$ are independent with the same standard normal distribution. Moreover,

$$E[\phi \,|\, \mathcal{G}_n] = \theta_n\big(I(e_1), I(e_2), \ldots, I(e_n)\big) \tag{9.4.6}$$

for some measurable function θ_n on $\mathbb{R}^n$ (this fact can be shown by the same arguments as those in Exercise 13 of Chapter 2). Thus for all $k_1, k_2, \ldots, k_n \geq 0$,

$$\int_{\mathbb{R}^n} x_1^{k_1} x_2^{k_2} \cdots x_n^{k_n} \theta_n(x_1, x_2, \ldots, x_n)\, d\mu(x) = 0,$$

where μ is the standard Gaussian measure on $\mathbb{R}^n$. In view of Equation (9.3.4), this equality implies that for any integers $k_1, k_2, \ldots, k_n \geq 0$,

$$\int_{\mathbb{R}^n} H_{k_1}(x_1; 1) H_{k_2}(x_2; 1) \cdots H_{k_n}(x_n; 1) \theta_n(x_1, x_2, \ldots, x_n)\, d\mu(x) = 0.$$

But it follows from Theorem 9.3.6 that the collection

$$\big\{H_{k_1}(x_1; 1) H_{k_2}(x_2; 1) \cdots H_{k_n}(x_n; 1)\, ;\, k_1, k_2, \ldots, k_n \geq 0\big\}$$

is an orthogonal basis for $L^2(\mathbb{R}^n, \mu)$. Hence $\theta_n = 0$ almost everywhere with respect to μ. Then by Equation (9.4.6), we have

$$E[\phi \,|\, \mathcal{G}_n] = 0, \quad \text{almost surely}$$

for any $n \geq 1$. On the other hand, by Lemma 9.4.3, $E[\phi \,|\, \mathcal{G}_n]$ converges to ϕ in $L^2_B(\Omega)$ as $n \to \infty$. Hence $\phi = 0$ almost surely. This proves the assertion that the union $\bigcup_{n=0}^{\infty} J_n$ is dense in $L^2_B(\Omega)$. $\qquad\qquad\square$

Now recall the sequence $\{J_n\}$ of subspaces of $L^2_B(\Omega)$ in Equation (9.4.5). Let $K_0 = \mathbb{R}$ and for each $n \geq 1$, define K_n to be the orthogonal complement of J_{n-1} in J_n, namely,

$$J_n = J_{n-1} \oplus K_n \quad \text{(orthogonal direct sum)}.$$

Then we have the following sequence of orthogonal subspaces of the real Hilbert space $L^2_B(\Omega)$:

$$K_0, \;\; K_1, \;\; K_2, \;\; \ldots, \;\; K_n, \;\; \ldots .$$

Definition 9.4.6. *Let n be a nonnegative integer. The elements of the Hilbert space K_n are called* homogeneous chaoses *of order n.*

The spaces K_n, $n \geq 1$, are all infinite-dimensional. The homogeneous chaoses of order 1 are Gaussian random variables. The next theorem follows from Theorem 9.4.5 and the construction of the homogeneous chaos.

Theorem 9.4.7. *The space $L^2_B(\Omega)$ is the orthogonal direct sum of the spaces K_n of homogeneous chaoses of order $n \geq 0$, namely,*

$$L^2_B(\Omega) = K_0 \oplus K_1 \oplus K_2 \oplus \cdots \oplus K_n \oplus \cdots .$$

Each function ϕ in $L^2_B(\Omega)$ has a unique homogeneous chaos expansion

$$\phi = \sum_{n=0}^{\infty} \phi_n. \tag{9.4.7}$$

Moreover,

$$\|\phi\|^2 = \sum_{n=0}^{\infty} \|\phi_n\|^2, \tag{9.4.8}$$

where $\|\cdot\|$ is the norm on $L^2_B(\Omega)$.

Example 9.4.8. Let $f \in L^2[a,b]$ and $I(f) = \int_a^b f(t)\,dB(t)$. Obviously,

$$I(f)^2 = \|f\|^2 + \left(I(f)^2 - \|f\|^2\right).$$

Direct computation shows that $I(f)^2 - \|f\|^2$ is orthogonal to K_0 and K_1. Hence $I(f)^2 - \|f\|^2$ belongs to K_2. On the other hand, consider the double Wiener–Itô integral $\int_a^b \int_a^b f(t)f(s)\,dB(t)\,dB(s)$. By Theorem 9.2.8,

$$\int_a^b \int_a^b f(t)f(s)\,dB(t)\,dB(s) = 2\int_a^b f(t)\left[\int_a^t f(s)\,dB(s)\right]dB(t)$$

$$= 2\int_a^b f(t)X_t\,dB(t), \qquad (9.4.9)$$

where $X_t = \int_a^t f(s)\,dB(s)$. Then $dX_t = f(t)\,dB(t)$. Use Itô's formula to get

$$d(X_t^2) = 2X_t\,dX_t + (dX_t)^2 = 2f(t)X_t\,dB(t) + f(t)^2\,dt,$$

which gives the stochastic integral

$$2\int_a^b f(t)X_t\,dB(t) = X_b^2 - X_a^2 - \int_a^b f(t)^2\,dt = I(f)^2 - \|f\|^2. \qquad (9.4.10)$$

From Equations (9.4.9) and (9.4.10) we get the double Wiener–Itô integral

$$\int_a^b \int_a^b f(t)f(s)\,dB(t)\,dB(s) = I(f)^2 - \|f\|^2. \qquad (9.4.11)$$

This equality shows that the homogeneous chaos $I(f)^2 - \|f\|^2$ is a double Wiener–Itô integral.

Theorem 9.4.9. *Let P_n denote the orthogonal projection of $L_B^2(\Omega)$ onto K_n. If $f_1,\ldots,f_k$ are nonzero orthogonal functions in $L^2[a,b]$ and $n_1,\ldots,n_k$ are nonnegative integers, then*

$$P_n\big(I(f_1)^{n_1}\cdots I(f_k)^{n_k}\big) = H_{n_1}\big(I(f_1)\,;\|f_1\|^2\big)\cdots H_{n_k}\big(I(f_k)\,;\|f_k\|^2\big),$$

where $n = n_1 + \cdots + n_k$. In particular, we have

$$P_n\big(I(f)^n\big) = H_n(I(f)\,;\|f\|^2) \qquad (9.4.12)$$

for any nonzero function f in $L^2[a,b]$.

Remark 9.4.10. It follows from the theorem that $H_n(I(f)\,;\|f\|^2)$, and more generally, $H_{n_1}\big(I(f_1)\,;\|f_1\|^2\big)\cdots H_{n_k}\big(I(f_k)\,;\|f_k\|^2\big)$ with $n_1 + \cdots + n_k = n$ are all homogeneous chaoses of order n.

Proof. In order to avoid complicated notation, we will prove only Equation (9.4.12). First use the second identity in the list of identities for the Hermite polynomials at the end of Section 9.3 to get

$$I(f)^n = H_n(I(f)\,;\|f\|^2) + \frac{n(n-1)}{2}\|f\|^2 H_{n-2}(I(f)\,;\|f\|^2) + \cdots. \qquad (9.4.13)$$

Observe that all terms, except for the first one, in the right-hand side are orthogonal to K_n because they have degree less than n. Therefore, to prove Equation (9.4.12), it suffices to prove that

$$H_n(I(f); \|f\|^2) \perp J_{n-1},$$

or equivalently, for any nonzero $g_1, \ldots, g_m \in L^2[a, b]$ and $0 \leq m \leq n - 1$,

$$E\{H_n(I(f); \|f\|^2) \cdot I(g_1) \cdots I(g_m)\} = 0. \qquad (9.4.14)$$

Obviously, we may assume that $g_1, \ldots, g_m$ are orthogonal. Moreover, for each g_i, we can write it as $g_i = c_i f + h_i$ with $c_i \in \mathbb{R}$ and $h_i \perp f$. Hence in order to prove Equation (9.4.14), it suffices to prove that for any nonzero $h_i, \ldots, h_m$ orthogonal to f and $p + q_1 + \cdots + q_m \leq n - 1$,

$$E\{H_n(I(f); \|f\|^2) \cdot I(f)^p I(h_1)^{q_1} \cdots I(h_m)^{q_m}\} = 0.$$

Finally, using Equation (9.4.13), we see that it suffices to prove that for any $0 \leq r + r_1 + \cdots + r_m \leq n - 1$,

$$E\left\{H_n(I(f); \|f\|^2) \cdot H_r(I(f); \|f\|^2) \prod_{i=1}^{m} H_{r_i}(I(h_i); \|h_i\|^2)\right\} = 0.$$

Note that the random variables $I(f), I(h_1), \ldots, I(h_m)$ are independent. Hence it suffices to show that for any $r \leq n - 1$,

$$E\{H_n(I(f); \|f\|^2) H_r(I(f); \|f\|^2)\} = 0.$$

But this is true by Theorem 9.3.6. Hence the theorem is proved.   □

Example 9.4.11. Let $f \in L^2[a, b]$ be a nonzero function. In the generating function in Equation (9.3.2), put $x = I(f)$ and $\rho = \|f\|^2$ to get

$$e^{tI(f) - \frac{1}{2}\|f\|^2 t^2} = \sum_{n=0}^{\infty} \frac{t^n}{n!} H_n(I(f); \|f\|^2). \qquad (9.4.15)$$

By Theorem 9.4.9, $H_n(I(f); \|f\|^2) \in K_n$ for each $n \geq 0$. Hence Equation (9.4.15) gives the homogeneous chaos expansion in Equation (9.4.7) for the function $\phi = e^{tI(f) - \frac{1}{2}\|f\|^2 t^2}$.

9.5 Orthonormal Basis for Homogeneous Chaos

In this section we will give an answer to Question 9.4.2 raised in the previous section, namely, what is the analogue of Equation (9.3.10) for the Hilbert space $L_B^2(\Omega)$ of square integrable functions on $(\Omega, \mathcal{F}^B, P)$?

By Theorem 9.4.7, $L_B^2(\Omega)$ is the orthogonal direct sum of homogeneous chaoses $L_B^2(\Omega) = K_0 \oplus K_1 \oplus K_2 \oplus \cdots \oplus K_n \oplus \cdots$. What is the analogue of this orthogonal direct sum for the standard Gaussian measure on the real line? Observe that in view of Equation (9.3.10), the corresponding space for each K_n is the one-dimensional space spanned by the Hermite polynomial

$H_n(x; 1)$. This observation gives us a clue for finding an orthonormal basis for $L_B^2(\Omega)$ that can be used for series expansion of functions in $L_B^2(\Omega)$. Then we have the answer to Question 9.4.2. However, in the case of $L_B^2(\Omega)$, every space $K_n, n \geq 1$, is an infinite-dimensional Hilbert space.

Let $f \in L^2[a, b]$. For convenience, we will also use the notation $\widetilde{f}$ to denote the Wiener integral of f, namely,

$$\widetilde{f} = I(f) = \int_a^b f(t) \, dB(t).$$

In this section we fix an orthonormal basis $\{e_k\}_{k=1}^\infty$ for the space $L^2[a, b]$. For a sequence $\{n_k\}_{k=1}^\infty$ of nonnegative integers with finite sum, define

$$\mathcal{H}_{n_1, n_2, \ldots} = \prod_k \frac{1}{\sqrt{n_k!}} H_{n_k}(\widetilde{e}_k; 1)$$

$$= \frac{1}{\sqrt{n_1! n_2! \cdots}} H_{n_1}(\widetilde{e}_1; 1) H_{n_2}(\widetilde{e}_2; 1) \cdots. \tag{9.5.1}$$

Note that $H_0(x; 1) = 1$ and there are only finitely many nonzero n_k's. Hence the product in this equation is a product of only finitely many factors.

Example 9.5.1. Suppose $n_1 + n_2 + \cdots = 1$. This equation has infinitely many solutions given by $1, 0, 0, \ldots$; $0, 1, 0, \ldots$; $0, 0, 1, \ldots$; $\ldots$. Hence we get the corresponding functions $\widetilde{e}_n$, $n = 1, 2, 3, \ldots$, which are all in the space K_1 of homogeneous chaoses of order 1.

Example 9.5.2. Consider the case $n_1 + n_2 + \cdots = 2$. The solutions of this equation are given by either $n_k = 2$ for a single index $1 \leq k < \infty$ and 0 for other indices or $n_i = n_j = 1$ for two indices $1 \leq i < j < \infty$ and 0 for other indices. Then we get the functions

$$\frac{1}{\sqrt{2}}\big(I(e_k)^2 - 1\big), \ 1 \leq k < \infty; \quad I(e_i)I(e_j), \ 1 \leq i < j < \infty,$$

all of which belong to the space K_2 of homogeneous chaoses of order 2.

Lemma 9.5.3. *For any fixed integer $n \geq 1$, the collection of functions*

$$\big\{\mathcal{H}_{n_1, n_2, \ldots}; \ n_1 + n_2 + \cdots = n\big\}$$

is a subset of K_n. Moreover, the linear space spanned by this collection of functions is dense in K_n.

Proof. The first assertion follows from Theorem 9.4.9. To prove the second assertion, notice that the same arguments as those in the proof of Theorem 9.4.5 show the following implication:

$$E\big[\phi \cdot \mathcal{H}_{n_1, n_2, \ldots}\big] = 0, \ \forall n_1, n_2, \ldots \geq 0 \quad \Longrightarrow \quad \phi = 0. \tag{9.5.2}$$

Now, suppose $\phi \in K_n$ is orthogonal to $\mathcal{H}_{n_1,n_2,\dots}$ for all $n_1, n_2, \dots$ with $n_1 + n_2 + \cdots = n$. Then $P_n \phi = \phi$ and $E(\phi \cdot \mathcal{H}_{n_1,n_2,\dots}) = 0$ for all such n_k's. Here P_n is the orthogonal projection of $L_B^2(\Omega)$ onto K_n. Moreover, observe that for any $n_1, n_2, \dots$,

$$P_n\left(\mathcal{H}_{n_1,n_2,\dots}\right) = \begin{cases} \mathcal{H}_{n_1,n_2,\dots}, & \text{if } n_1 + n_2 + \cdots = n; \\ 0, & \text{otherwise.} \end{cases} \qquad (9.5.3)$$

Then for any $n_1, n_2, \dots \geq 0$, we can use the assumption on ϕ and Equation (9.5.3) to show that

$$\begin{aligned} E\left[\phi \cdot \mathcal{H}_{n_1,n_2,\dots}\right] &= \langle \phi, \mathcal{H}_{n_1,n_2,\dots} \rangle = \langle P_n \phi, \mathcal{H}_{n_1,n_2,\dots} \rangle \\ &= \langle \phi, P_n \mathcal{H}_{n_1,n_2,\dots} \rangle = E\left[\phi \cdot (P_n \mathcal{H}_{n_1,n_2,\dots})\right] \\ &= 0, \end{aligned}$$

where $\langle \cdot, \cdot \rangle$ is the inner product on the Hilbert space $L_B^2(\Omega)$. Therefore, by the implication in Equation (9.5.2), we conclude that $\phi = 0$. This finishes the proof for the second assertion of the lemma. □

Theorem 9.5.4. *For any fixed integer $n \geq 1$, the collection of functions*

$$\left\{ \mathcal{H}_{n_1,n_2,\dots} \, ; \, n_1 + n_2 + \cdots = n \right\} \qquad (9.5.4)$$

is an orthonormal basis for the space K_n of homogeneous chaoses of order n.

Proof. In view of Lemma 9.5.3, we only need to show that the collection of functions in Equation (9.5.4) is an orthonormal system. For each k, we use Equation (9.3.6) to show that

$$\begin{aligned} \|H_{n_k}(\widetilde{e}_k; 1)\|^2 &= \int_{\Omega} H_{n_k}(\widetilde{e}_k; 1)^2 \, dP \\ &= \int_{-\infty}^{\infty} H_{n_k}(x; 1)^2 \frac{1}{\sqrt{2\pi}} e^{-x^2/2} \, dx \\ &= n!. \end{aligned}$$

Moreover, note that the random variables $\widetilde{e}_1, \widetilde{e}_2, \dots$ are independent. Hence we see easily from Equation (9.5.1) that $\|\mathcal{H}_{n_1,n_2,\dots}\|^2 = 1$. Thus all functions in Equation (9.5.4) have norm 1.

Next we show that the functions in Equation (9.5.4) are orthogonal. Let $\{n_k\}$ and $\{j_k\}$ be two distinct sequences of nonnegative integers, regardless of whether they have the same sum. Then there exists some k_0 such that $n_{k_0} \neq j_{k_0}$. For simplicity, let

$$\phi_k = \frac{1}{\sqrt{n_k!}} H_{n_k}(\widetilde{e}_k; 1), \quad \theta_k = \frac{1}{\sqrt{j_k!}} H_{j_k}(\widetilde{e}_k; 1).$$

Use the independence of $\tilde{e}_1, \tilde{e}_2, \ldots$ and Equation (9.3.9) to check that

$$E\left[\mathcal{H}_{n_1,n_2,\ldots}\mathcal{H}_{j_1,j_2,\ldots}\right] = E\left[\phi_{k_0}\theta_{k_0}\prod_{k\neq k_0}\phi_k\theta_k\right]$$

$$= E(\phi_{k_0}\theta_{k_0})\,E\left[\prod_{k\neq k_0}\phi_k\theta_k\right]$$

$$= 0.$$

This shows that $\mathcal{H}_{n_1,n_2,\ldots}$ and $\mathcal{H}_{j_1,j_2,\ldots}$ are orthogonal. $\qquad\square$

Example 9.5.5. Suppose $f \in L^2[a,b]$ has the expansion $f = \sum_{n=1}^{\infty} a_n e_n$, where $a_n = \langle f, e_n \rangle$. Then the expansion of $I(f) \in K_1$ in $L_B^2(\Omega)$ is given by

$$I(f) = \sum_{n=1}^{\infty} a_n I(e_n).$$

This expansion is a special case of Theorem 9.5.4 when $n = 1$. But it also follows from the fact that the mapping $I : L^2[a,b] \to L_B^2(\Omega)$ is an isometry (see Theorem 2.3.4).

Example 9.5.6. Let $f = \sum_{n=1}^{\infty} a_n e_n \in L^2[a,b]$ be as in the previous example. The orthonormal basis in Theorem 9.5.4 for the case $n = 2$ is given by

$$\xi_k \equiv \frac{1}{\sqrt{2}}\left(\tilde{e}_k^2 - 1\right),\ 1 \leq k < \infty;\quad \eta_{ij} = \tilde{e}_i\tilde{e}_j,\ 1 \leq i < j < \infty.$$

By Theorem 9.4.9, $P_2(I(f)^2) = I(f)^2 - \|f\|^2$ belongs to the space K_2. Note that $\langle P_2(I(f)^2), \xi_k \rangle = \langle I(f)^2, P_2\,\xi_k \rangle = \langle I(f)^2, \xi_k \rangle$. Hence for each $k \geq 1$,

$$\langle P_2(I(f)^2), \xi_k \rangle = E\left[I(f)^2\xi_k\right] = E\left[\sum_{i,j}^{\infty} a_i a_j \tilde{e}_i \tilde{e}_j \frac{1}{\sqrt{2}}\left(\tilde{e}_k^2 - 1\right)\right].$$

We can in fact bring the expectation inside the summation sign. Observe that the expectation is zero except possibly when $i = j = k$. Hence,

$$\langle P_2(I(f)^2), \xi_k \rangle = \frac{1}{\sqrt{2}}a_k^2 E\left[\tilde{e}_k^2\left(\tilde{e}_k^2 - 1\right)\right] = \sqrt{2}\,a_k^2.$$

Similarly, we can check that for any $1 \leq i < j < \infty$,

$$\langle P_2(I(f)^2), \eta_{ij} \rangle = 2a_i a_j.$$

Therefore, the expansion of $P_2(I(f)^2)$ is given by

$$P_2(I(f)^2) = \sum_{k=1}^{\infty} \sqrt{2}\,a_k^2 \frac{1}{\sqrt{2}}\left(I(e_k)^2 - 1\right) + \sum_{i<j} 2a_i a_j I(e_i)I(e_j).$$

Next we give a concrete example for the above expansions of the functions f and $P_2(I(f)^2)$ in the spaces $L^2[a, b]$ and K_2, respectively. An orthonormal basis for $L^2[0, 1]$ is given by

$$e_n(t) = \sqrt{2} \sin\left((n - \tfrac{1}{2})\pi t\right), \quad n = 1, 2, 3, \ldots$$

(see page 50 of the book [55]). Consider a simple function $f(t) = t$. Use the integration by parts formula to get

$$a_n = \int_0^1 t\sqrt{2} \sin\left((n - \tfrac{1}{2})\pi t\right) dt = (-1)^n \sqrt{2}\left((n - \tfrac{1}{2})\pi\right)^{-2}.$$

Therefore, $f(t)$ has the expansion

$$f(t) = \sum_{n=1}^{\infty} (-1)^n \sqrt{2}\left((n - \tfrac{1}{2})\pi\right)^{-2} \sqrt{2} \sin\left((n - \tfrac{1}{2})\pi t\right)$$

and the expansion of $P_2(I(f)^2)$ is given by

$$2\sum_{k=1}^{\infty} \left((k - \tfrac{1}{2})\pi\right)^{-4}\left(\tilde{e}_k^2 - 1\right) + 4\sum_{i<j}(-1)^{i+j}\left((i - \tfrac{1}{2})\pi\right)^{-2}\left((j - \tfrac{1}{2})\pi\right)^{-2}\tilde{e}_i\tilde{e}_j.$$

Now we can put together the orthonormal bases in Theorem 9.5.4 for $n \geq 1$ to give an answer to Question 9.4.2.

Theorem 9.5.7. *The collection of functions*

$$\left\{\mathcal{H}_{n_1,n_2,\ldots}; \; n_1 + n_2 + \cdots = n, \; n = 0, 1, 2, \ldots\right\} \tag{9.5.5}$$

is an orthonormal basis for the Hilbert space $L_B^2(\Omega)$. Every ϕ in $L_B^2(\Omega)$ has a unique series expansion

$$\phi = \sum_{n=0}^{\infty} \sum_{n_1+n_2+\cdots=n} a_{n_1,n_2,\ldots} \mathcal{H}_{n_1,n_2,\ldots}, \tag{9.5.6}$$

where $a_{n_1,n_2,\ldots} = E\left(\phi\mathcal{H}_{n_1,n_2,\ldots}\right) = \int_{\Omega} \phi\mathcal{H}_{n_1,n_2,\ldots} \, dP$.

9.6 Multiple Wiener–Itô Integrals

Recall that $B(t)$ is a fixed Brownian motion and $L_B^2(\Omega)$ is the Hilbert space of square integrable functions on the probability space $(\Omega, \mathcal{F}^B, P)$. Here $\mathcal{F}^B$ is the σ-field $\mathcal{F}^B = \sigma\{B(t); a \leq t \leq b\}$. By Theorem 9.4.7, the Hilbert space $L_B^2(\Omega)$ is the orthogonal direct sum of the spaces K_n, $n \geq 0$. In Example 9.4.8 the homogeneous chaos $I(f)^2 - \|f\|^2$ for $f \in L^2[a, b]$ is shown to be given by a double Wiener–Itô integral. It is reasonable to expect that there is a one-to-one correspondence between the homogeneous chaoses of order 2 and double Wiener–Itô integrals. This leads to the next question.

Question 9.6.1. Are all homogeneous chaoses of order $n \geq 2$ given by some kind of integrals?

This was the question considered by K. Itô in 1951. We will follow his original idea in the paper [34] to define the multiple Wiener–Itô integrals. In Section 9.7 we will show that there is a one-to-one correspondence between the homogeneous chaoses and multiple Wiener–Itô integrals.

For simplicity, let $T \equiv [a, b]$. The first aim in this section is to define the multiple Wiener–Itô integral

$$\int_{T^n} f(t_1, t_2, \ldots, t_n) \, dB(t_1) \, dB(t_2) \cdots dB(t_n)$$

for $f \in L^2(T^n)$. The essential idea is already given in the case $n = 2$ for the double Wiener–Itô integral. We will simply modify the arguments and notation in Section 9.2 to the case $n \geq 3$.

Let $D = \{(t_1, t_2, \ldots, t_n) \in T^n; \exists i \neq j \text{ such that } t_i = t_j\}$ be the "diagonal set" of T^n. A subset of T^n of the form $[t_1^{(1)}, t_1^{(2)}) \times [t_2^{(1)}, t_2^{(2)}) \times \cdots \times [t_n^{(1)}, t_n^{(2)})$ is called a rectangle.

Step 1. *Off-diagonal step functions*

A step function on T^n is a function of the form

$$f = \sum_{1 \leq i_1, i_2, \ldots, i_n \leq k} a_{i_1, i_2, \ldots, i_n} 1_{[\tau_{i_1-1}, \tau_{i_1}) \times [\tau_{i_2-1}, \tau_{i_2}) \times \cdots \times [\tau_{i_n-1}, \tau_{i_n})}, \qquad (9.6.1)$$

where $a = \tau_0 < \tau_1 < \tau_2 < \cdots < \tau_k = b$. An *off-diagonal step function* is a step function with coefficients satisfying the condition

$$a_{i_1, i_2, \ldots, i_n} = 0 \quad \text{if } i_p = i_q \text{ for some } p \neq q. \qquad (9.6.2)$$

The condition in Equation (9.6.2) means that the function f vanishes on the set D. The collection of off-diagonal step functions is a vector space.

For an off-diagonal step function f given by Equation (9.6.1), define

$$I_n(f) = \sum_{1 \leq i_1, i_2, \ldots, i_n \leq k} a_{i_1, i_2, \ldots, i_n} \xi_{i_1} \xi_{i_2} \cdots \xi_{i_n}, \qquad (9.6.3)$$

where $\xi_{i_p} = B(\tau_{i_p}) - B(\tau_{i_p-1})$, $1 \leq p \leq n$. The value $I_n(f)$ is well-defined, i.e., it does not depend on how f is represented by Equation (9.6.1). Moreover, the mapping I_n is linear on the vector space of off-diagonal step functions.

The symmetrization $\widehat{f}(t_1, \ldots, t_n)$ of a function $f(t_1, \ldots, t_n)$ is defined by

$$\widehat{f}(t_1, t_2, \ldots, t_n) = \frac{1}{n!} \sum_{\sigma} f(t_{\sigma(1)}, t_{\sigma(2)}, \ldots, t_{\sigma(n)}),$$

where the summation is over all permutations σ of the set $\{1, 2, \ldots, n\}$. Since the Lebesgue measure is symmetric, we have

$$\int_{T^n} |f(t_{\sigma(1)}, \ldots, t_{\sigma(n)})|^2 \, dt_1 \cdots dt_n = \int_{T^n} |f(t_1, \ldots, t_n)|^2 \, dt_1 \cdots dt_n.$$

for any permutation σ. Therefore, by the triangle inequality,

$$\|\widehat{f}\| \le \frac{1}{n!} \sum_{\sigma} \|f\| = \frac{1}{n!} \, n! \, \|f\| = \|f\|.$$

Thus we have the inequality $\|\widehat{f}\| \le \|f\|$. In general, strict inequality can occur as shown in Section 9.2. If f is an off-diagonal step function, then $\widehat{f}$ is also an off-diagonal step function.

Lemma 9.6.2. *If f is an off-diagonal step function, then $I_n(f) = I_n(\widehat{f})$.*

Proof. Note that I_n and the symmetrization operator are linear. Hence it suffices to prove the lemma for the case

$$f = 1_{[t_1^{(1)}, t_1^{(2)}) \times [t_2^{(1)}, t_2^{(2)}) \times \cdots \times [t_n^{(1)}, t_n^{(2)})},$$

where the intervals $[t_i^{(1)}, t_i^{(2)})$, $1 \le i \le n$, are disjoint. Then we have

$$I_n(f) = \prod_{i=1}^{n} \left(B(t_i^{(2)}) - B(t_i^{(1)}) \right). \tag{9.6.4}$$

On the other hand, the symmetrization $\widehat{f}$ of f is given by

$$\widehat{f} = \frac{1}{n!} \sum_{\sigma} 1_{[t_{\sigma(1)}^{(1)}, t_{\sigma(1)}^{(2)}) \times [t_{\sigma(2)}^{(1)}, t_{\sigma(2)}^{(2)}) \times \cdots \times [t_{\sigma(n)}^{(1)}, t_{\sigma(n)}^{(2)})}.$$

Therefore,

$$I_n(\widehat{f}) = \frac{1}{n!} \sum_{\sigma} \prod_{i=1}^{n} \left(B(t_{\sigma(i)}^{(2)}) - B(t_{\sigma(i)}^{(1)}) \right).$$

Observe that $\prod_{i=1}^{n} \left(B(t_{\sigma(i)}^{(2)}) - B(t_{\sigma(i)}^{(1)}) \right) = \prod_{i=1}^{n} \left(B(t_i^{(2)}) - B(t_i^{(1)}) \right)$ for any permutation σ. Moreover, there are $n!$ permutations of the set $\{1, 2, \ldots, n\}$. It follows that

$$I_n(\widehat{f}) = \frac{1}{n!} \sum_{\sigma} \prod_{i=1}^{n} \left(B(t_i^{(2)}) - B(t_i^{(1)}) \right) = \prod_{i=1}^{n} \left(B(t_i^{(2)}) - B(t_i^{(1)}) \right). \tag{9.6.5}$$

Equations (9.6.4) and (9.6.5) prove the assertion of the lemma.   □

Lemma 9.6.3. *If f is an off-diagonal step function, then $E[I_n(f)] = 0$ and*

$$E\left[I_n(f)^2 \right] = n! \int_{T^n} |\widehat{f}(t_1, t_2, \ldots, t_n)|^2 \, dt_1 dt_2 \cdots dt_n. \tag{9.6.6}$$

Proof. Let f be an off-diagonal step function given by Equation (9.6.1). Then $I_n(f)$ is given by Equation (9.6.3). Since the function f satisfies the condition in Equation (9.6.2), the coefficient $a_{i_1,i_2,\ldots,i_n}$ must be 0 whenever the intervals $[\tau_{i_1-1}, \tau_{i_1}), [\tau_{i_2-1}, \tau_{i_2}), \ldots, [\tau_{i_n-1}, \tau_{i_n})$ are not disjoint. On the other hand, when these intervals are disjoint, the corresponding product $\xi_{i_1}\xi_{i_2}\cdots\xi_{i_n}$ has expectation 0. Hence we have $E[I_n(f)] = 0$.

Note that $I_n(f) = I_n(\widehat{f})$ by Lemma 9.6.2. Hence we may assume that f is symmetric in proving Equation (9.6.6). In that case,

$$a_{i_{\sigma(1)}, i_{\sigma(2)}, \ldots, i_{\sigma(n)}} = a_{i_1, i_2, \ldots, i_n}$$

for any permutation σ. Thus $I_n(f)$ in Equation (9.6.3) can be rewritten as

$$I_n(f) = n! \sum_{1 \leq i_1 < i_2 < \cdots < i_n \leq k} a_{i_1, i_2, \ldots, i_n} \xi_{i_1}\xi_{i_2}\cdots\xi_{i_n}.$$

Therefore,

$$E\big[I_n(f)^2\big] = (n!)^2 \sum_{i_1 < \cdots < i_n} \sum_{j_1 < \cdots < j_n} a_{i_1, \ldots, i_n} a_{j_1, \ldots, j_n} E\big[\xi_{i_1}\cdots\xi_{i_n}\,\xi_{j_1}\cdots\xi_{j_n}\big],$$

where, for simplicity of notation, we have omitted the indices such as i_2, j_2, etc. Observe that for a fixed set of indices $i_1 < \cdots < i_n$, we have

$$E\big[\xi_{i_1}\cdots\xi_{i_n}\,\xi_{j_1}\cdots\xi_{j_n}\big] = \begin{cases} \prod_{p=1}^n (\tau_{i_p} - \tau_{i_p-1}), & \text{if } j_1 = i_1, \ldots, j_n = i_n; \\ 0, & \text{otherwise.} \end{cases}$$

It follows that

$$E\big[I_n(f)^2\big] = (n!)^2 \sum_{i_1 < \cdots < i_n} a_{i_1, \ldots, i_n}^2 \prod_{p=1}^n (\tau_{i_p} - \tau_{i_p-1})$$

$$= n! \sum_{i_1, \ldots, i_n} a_{i_1, \ldots, i_n}^2 \prod_{p=1}^n (\tau_{i_p} - \tau_{i_p-1})$$

$$= n! \int_{T^n} f(t_1, \ldots, t_n)^2 \, dt_1 \cdots dt_n,$$

which proves Equation (9.6.6) since f is assumed to be symmetric. $\qquad\square$

Step 2. *Approximation by off-diagonal step functions*

Recall the set D defined earlier in this section. The set D can be rewritten as $D = \cup_{i \neq j}[\{t_i = t_j\} \cap D]$, which means that D is a finite union of the intersections of $(n-1)$-dimensional hyperplanes with D. Hence the Lebesgue measure of D is zero. This fact allows us to adopt the same arguments as those in the derivation of Equations (9.2.5), (9.2.6), and (9.2.7) to prove the next approximation lemma.

Lemma 9.6.4. *Let f be a function in $L^2(T^n)$. Then there exists a sequence $\{f_k\}$ of off-diagonal step functions such that*

$$\lim_{k \to \infty} \int_{T^n} \left| f(t_1, t_2, \ldots, t_n) - f_k(t_1, t_2, \ldots, t_n) \right|^2 dt_1 dt_2 \cdots dt_n = 0. \quad (9.6.7)$$

Now suppose $f \in L^2(T^n)$. Choose a sequence $\{f_k\}$ of off-diagonal step functions converging to f in $L^2(T^n)$. By Lemma 9.6.4 such a sequence exists. Then by the linearity of I_n and Lemma 9.6.3,

$$E\left[\left(I_n(f_k) - I_n(f_\ell)\right)^2\right] = n! \|\widehat{f}_k - \widehat{f}_\ell\|^2 \le n! \|f_k - f_\ell\|^2 \to 0,$$

as $k, \ell \to \infty$. Hence the sequence $\{I_n(f_k)\}_{k=1}^{\infty}$ is Cauchy in $L^2(\Omega)$. Define

$$I_n(f) = \lim_{k \to \infty} I_n(f_k), \quad \text{in } L^2(\Omega). \quad (9.6.8)$$

The value $I_n(f)$ is well-defined, namely, it does not depend on the choice of the sequence $\{f_k\}$ used in Equation (9.6.8).

Definition 9.6.5. *Let $f \in L^2(T^n)$. The limit $I_n(f)$ in Equation (9.6.8) is called the* multiple Wiener–Itô integral *of f and is denoted by*

$$\int_{T^n} f(t_1, t_2, \ldots, t_n) \, dB(t_1) \, dB(t_2) \cdots dB(t_n).$$

Note that $I_1(f)$ is simply the Wiener integral $I(f)$ of f defined in Section 2.3 and $I_2(f)$ is the double Wiener–Itô integral of f defined in Section 9.2.

Obviously, Lemmas 9.6.2 and 9.6.3 can be extended to functions in $L^2(T^n)$ using the approximation in Lemma 9.6.4 and the definition of the multiple Wiener–Itô integral. We state this fact as the next theorem.

Theorem 9.6.6. *Let $f \in L^2(T^n)$, $n \ge 1$. Then we have*

(1) $I_n(f) = I_n(\widehat{f})$. *Here $\widehat{f}$ is the symmetrization of f.*
(2) $E[I_n(f)] = 0$.
(3) $E\left[I_n(f)^2\right] = n! \|\widehat{f}\|^2$. *Here $\|\cdot\|$ is the norm on $L^2(T^n)$.*

The next theorem gives an equality to write a multiple Wiener–Itô integral as an iterated Itô integral. It is useful for computation.

Theorem 9.6.7. *Let $f \in L^2(T^n)$, $n \ge 2$. Then*

$$\int_{T^n} f(t_1, t_2, \ldots, t_n) \, dB(t_1) \, dB(t_2) \cdots dB(t_n)$$

$$= n! \int_a^b \cdots \int_a^{t_{n-2}} \left[\int_a^{t_{n-1}} \widehat{f}(t_1, t_2, \ldots, t_n) \, dB(t_n) \right] dB(t_{n-1}) \cdots dB(t_1),$$

where $\widehat{f}$ is the symmetrization of f.

Proof. It suffices to prove the theorem for the case that f is the characteristic function of a rectangle that is disjoint from the set D. By Lemma 9.6.2, we may assume that f is of the form

$$f = 1_{[t_1^{(1)}, t_1^{(2)}) \times [t_2^{(1)}, t_2^{(2)}) \times \cdots \times [t_n^{(1)}, t_n^{(2)})},$$

where $t_n^{(1)} < t_n^{(2)} \le t_{n-1}^{(1)} < t_{n-1}^{(2)} \le \cdots \le t_2^{(1)} < t_2^{(2)} \le t_1^{(1)} < t_1^{(2)}$. Then the multiple Wiener–Itô integral of f is given by

$$\int_{T^n} f(t_1, \ldots, t_n) \, dB(t_1) \cdots dB(t_n) = \prod_{i=1}^n \left(B(t_i^{(2)}) - B(t_i^{(1)}) \right). \tag{9.6.9}$$

On the other hand, note that $\widehat{f} = \frac{1}{n!} f$ on the region $t_n < t_{n-1} < \cdots < t_1$. Hence we get

$$\int_a^{t_{n-1}} \widehat{f}(t_1, t_2, \ldots, t_n) \, dB(t_n)$$

$$= \frac{1}{n!} 1_{[t_1^{(1)}, t_1^{(2)}) \times [t_2^{(1)}, t_2^{(2)}) \times \cdots \times [t_{n-1}^{(1)}, t_{n-1}^{(2)})} \left(B(t_n^{(2)}) - B(t_n^{(1)}) \right),$$

which is $\mathcal{F}_{t_{n-1}^{(1)}}$-measurable and can be regarded as a "constant" stochastic process for integration on the interval $[t_{n-1}^{(1)}, t_{n-1}^{(2)}]$ with respect to $dB(t_{n-1})$. Hence we can repeat the above arguments to get

$$\int_a^b \cdots \int_a^{t_{n-2}} \left[\int_a^{t_{n-1}} \widehat{f}(t_1, \ldots, t_n) \, dB(t_n) \right] dB(t_{n-1}) \cdots dB(t_1),$$

$$= \frac{1}{n!} \prod_{i=1}^n \left(B(t_i^{(2)}) - B(t_i^{(1)}) \right). \tag{9.6.10}$$

The theorem follows from Equations (9.6.9) and (9.6.10). $\qquad\square$

Definition 9.6.8. *Let $g_1, \ldots, g_n \in L^2[a, b]$. The* tensor product *$g_1 \otimes \cdots \otimes g_n$ is defined to be the function*

$$g_1 \otimes \cdots \otimes g_n(t_1, \ldots, t_n) = g_1(t_1) \cdots g_n(t_n).$$

The tensor product *$f_1^{\otimes n_1} \otimes \cdots \otimes f_k^{\otimes n_k}$ means that f_j is repeated n_j times, $1 \le j \le k$.*

Theorem 9.6.9. *Let $f_1, f_2, \ldots, f_k$ be nonzero orthogonal functions in $L^2[a, b]$ and let $n_1, \ldots, n_k$ be positive integers. Then*

$$I_n \left(f_1^{\otimes n_1} \otimes \cdots \otimes f_k^{\otimes n_k} \right) = \prod_{j=1}^k H_{n_j} \left(I(f_j); \|f_j\|^2 \right), \tag{9.6.11}$$

where $n = n_1 + \cdots + n_k$. In particular, for any nonzero $f \in L^2[a, b]$,

$$I_n \left(f^{\otimes n} \right) = H_n \left(I(f); \|f\|^2 \right). \tag{9.6.12}$$

Proof. We first prove Equation (9.6.12). The case $n = 1$ is obvious. The case $n = 2$ has already been proved by Equation (9.4.11). For general integer n, we use mathematical induction. Suppose Equation (9.6.12) is valid for n. Then by Theorem 9.6.7,

$$\int_{T^{n+1}} f(t_1) \cdots f(t_{n+1}) \, dB(t_1) \cdots dB(t_{n+1}) = (n+1)! \int_a^b f(t_1) X_{t_1} \, dB(t_1),$$

where X_t is given by

$$X_t = \int_a^t \cdots \left[\int_a^{t_n} f(t_2) \cdots f(t_{n+1}) \, dB(t_{n+1}) \right] \cdots dB(t_2).$$

By Theorem 9.6.7 and the induction on n,

$$X_t = \frac{1}{n!} \int_{[a,t]^n} f(t_2) \cdots f(t_{n+1}) \, dB(t_2) \cdots dB(t_{n+1})$$

$$= \frac{1}{n!} H_n \left(\int_a^t f(s) \, dB(s) \, ; \int_a^t f(s)^2 \, ds \right).$$

Therefore, we have the equality

$$\int_{T^{n+1}} f(t_1) \cdots f(t_{n+1}) \, dB(t_1) \cdots dB(t_{n+1})$$

$$= (n+1) \int_a^b f(t_1) H_n \left(\int_a^{t_1} f(s) \, dB(s) \, ; \int_a^{t_1} f(s)^2 \, ds \right) dB(t_1). \quad (9.6.13)$$

On the other hand, we can apply Itô's formula to $H_{n+1}(x; \rho)$ to get

$$dH_{n+1} \left(\int_a^t f(s) \, dB(s) \, ; \int_a^t f(s)^2 \, ds \right)$$

$$= \left(\frac{\partial}{\partial x} H_{n+1} \right) f(t) \, dB(t) + \frac{1}{2} \left(\frac{\partial^2}{\partial x^2} H_{n+1} \right) f(t)^2 \, dt + \left(\frac{\partial}{\partial \rho} H_{n+1} \right) f(t)^2 \, dt.$$

Recall the following identities from Section 9.3:

$$\frac{\partial}{\partial x} H_{n+1}(x; \rho) = (n+1) H_n(x; \rho),$$

$$\frac{\partial}{\partial \rho} H_{n+1}(x; \rho) = -\frac{1}{2} \frac{\partial^2}{\partial x^2} H_{n+1}(x; \rho).$$

Thus we obtain

$$dH_{n+1} \left(\int_a^t f(s) \, dB(s) \, ; \int_a^t f(s)^2 \, ds \right)$$

$$= (n+1) f(t) H_n \left(\int_a^t f(s) \, dB(s) \, ; \int_a^t f(s)^2 \, ds \right) dB(t),$$

which, upon integration over $[a, b]$, gives the equality

$$H_{n+1}\big(I(f); \|f\|^2\big)$$
$$= (n+1) \int_a^b f(t) H_n\bigg(\int_a^t f(s)\, dB(s) \,;\, \int_a^t f(s)^2\, ds \bigg)\, dB(t). \quad (9.6.14)$$

Equations (9.6.13) and (9.6.14) show that Equation (9.6.12) is valid for $n+1$. Hence by induction, Equation (9.6.12) holds for any positive integer n.

Now we prove Equation (9.6.11). Let $\widetilde{f} = I(f)$. For any real numbers $r_1, r_2, \ldots, r_k$, apply Equation (9.4.15) to get

$$\exp\bigg[\sum_{i=1}^k r_i \widetilde{f_i} - \frac{1}{2} \sum_{i=1}^k r_i^2 \|f_i\|^2 \bigg] = \prod_{i=1}^k e^{r_i \widetilde{f_i} - \frac{1}{2} r_i^2 \|f_i\|^2}$$
$$= \prod_{i=1}^k \sum_{n_i=0}^\infty \frac{r_i^{n_i}}{n_i!} H_{n_i}\big(\widetilde{f_i}; \|f_i\|^2\big). \quad (9.6.15)$$

On the other hand, by Equation (9.4.15) with $t = 1$ and $f = \sum_{i=1}^k r_i \widetilde{f_i}$,

$$\exp\bigg[\sum_{i=1}^k r_i \widetilde{f_i} - \frac{1}{2} \sum_{i=1}^k r_i^2 \|f_i\|^2 \bigg] = \sum_{m=0}^\infty \frac{1}{m!} H_m\bigg(\sum_{i=1}^k r_i \widetilde{f_i} \,;\, \frac{1}{2} \sum_{i=1}^k r_i^2 \|f_i\|^2 \bigg).$$

Then we apply what we have already proved in Equation (9.6.12) to H_m in the right-hand side to get

$$\exp\bigg[\sum_{i=1}^k r_i \widetilde{f_i} - \frac{1}{2} \sum_{i=1}^k r_i^2 \|f_i\|^2 \bigg]$$
$$= \sum_{m=0}^\infty \frac{1}{m!} \int_{T^m} \prod_{j=1}^m \bigg[\sum_{i=1}^k r_i f_i(t_j) \bigg]\, dB(t_1) \cdots dB(t_m). \quad (9.6.16)$$

Equation (9.6.11) follows by comparing the coefficient of $r_1^{n_1} r_2^{n_2} \cdots r_k^{n_k}$ in the right-hand sides of Equations (9.6.15) and (9.6.16). $\qquad \square$

Finally we prove a theorem on the orthogonality of $I_n(f)$ and $I_m(g)$ in the Hilbert space $L_B^2(\Omega)$ when $n \neq m$.

Theorem 9.6.10. *Let $n \neq m$. Then $E\big(I_n(f) I_m(g)\big) = 0$ for any $f \in L^2(T^n)$ and $g \in L^2(T^m)$.*

Proof. It suffices to prove the theorem for f and g of the following form:

$$f = 1_{[t_1^{(1)}, t_1^{(2)}) \times [t_2^{(1)}, t_2^{(2)}) \times \cdots \times [t_n^{(1)}, t_n^{(2)})},$$
$$g = 1_{[s_1^{(1)}, s_1^{(2)}) \times [s_2^{(1)}, s_2^{(2)}) \times \cdots \times [s_m^{(1)}, s_m^{(2)})},$$

where the intervals satisfy the condition

$$t_n^{(1)} < t_n^{(2)} \le t_{n-1}^{(1)} < t_{n-1}^{(2)} \le \cdots \le t_2^{(1)} < t_2^{(2)} \le t_1^{(1)} < t_1^{(2)},$$

$$s_m^{(1)} < s_m^{(2)} \le s_{m-1}^{(1)} < s_{m-1}^{(2)} \le \cdots \le s_2^{(1)} < s_2^{(2)} \le s_1^{(1)} < s_1^{(2)}.$$

Then $I_n(f)I_m(g)$ is given by

$$I_n(f)I_m(g) = \left[\prod_{i=1}^{n} \left(B(t_i^{(2)}) - B(t_i^{(1)})\right)\right]\left[\prod_{j=1}^{m} \left(B(s_j^{(2)}) - B(s_j^{(1)})\right)\right]. \quad (9.6.17)$$

Now, put these points t and s together to form an increasing set of points $\tau_1 < \tau_2 < \cdots < \tau_r$ with $r \le n + m$. Then each factor in the first product of Equation (9.6.17) can be rewritten as a sum of increments of $B(t)$ on the τ-intervals. Hence, upon multiplying out the n factors, each term in the first product $\prod_{i=1}^{n}$ must be of the form

$$\left(B(\tau_{i_1}) - B(\tau_{i_1-1})\right)\cdots\left(B(\tau_{i_n}) - B(\tau_{i_n-1})\right), \quad (9.6.18)$$

where $\tau_{i_1} < \cdots < \tau_{i_n}$. Similarly, each term in the second product $\prod_{j=1}^{m}$ in Equation (9.6.17) must be of the form

$$\left(B(\tau_{j_1}) - B(\tau_{j_1-1})\right)\cdots\left(B(\tau_{j_m}) - B(\tau_{j_m-1})\right), \quad (9.6.19)$$

where $\tau_{j_1} < \cdots < \tau_{j_m}$.

It is easy to see that the product of Equations (9.6.18) and (9.6.19) has expectation 0 because $n \ne m$. Thus we can conclude from Equation (9.6.17) that $E\big(I_n(f)I_m(g)\big) = 0$. □

9.7 Wiener–Itô Theorem

Let $L^2_{\mathrm{sym}}(T^n)$ denote the real Hilbert space of symmetric square integrable functions on T^n. Recall the following facts from the previous sections:

1. By Theorem 9.4.7, we have a decomposition of $L^2_B(\Omega)$ as an orthogonal direct sum $L^2_B(\Omega) = K_0 \oplus K_1 \oplus K_2 \oplus \cdots \oplus K_n \oplus \cdots$.

2. By Theorem 9.5.4, the collection $\{\mathcal{H}_{n_1,n_2,\dots}; n_1 + n_2 + \cdots = n\}$ is an orthonormal basis for K_n.

3. By Theorem 9.6.6, $E\big[I_n(f)^2\big] = n!\|f\|^2$ for any $f \in L^2_{\mathrm{sym}}(T^n)$. Hence the mapping $\frac{1}{\sqrt{n!}}I_n : L^2_{\mathrm{sym}}(T^n) \to L^2_B(\Omega)$ is an isometry.

4. By Theorem 9.6.9, multiple Wiener–Itô integrals are related to Hermite polynomials of Wiener integrals.

Observe from the second and fourth facts that homogeneous chaoses must be related to multiple Wiener–Itô integrals. In fact, the homogeneous chaoses of order n are exactly multiple Wiener–Itô integrals of order n.

Theorem 9.7.1. *Let $n \geq 1$. If $f \in L^2(T^n)$, then $I_n(f) \in K_n$. Conversely, if $\phi \in K_n$, then there exists a unique function $f \in L^2_{\mathrm{sym}}(T^n)$ such that $\phi = I_n(f)$.*

Remark 9.7.2. It follows from this theorem that $K_n = \{I_n(f); f \in L^2_{\mathrm{sym}}(T^n)\}$. Moreover, $\|I_n(f)\| = \sqrt{n!}\,\|f\|$ for all $f \in L^2_{\mathrm{sym}}(T^n)$ by Theorem 9.6.6.

Proof. Let $f_1^{\otimes n_1} \widehat{\otimes} f_2^{\otimes n_2} \widehat{\otimes} \cdots \widehat{\otimes} f_k^{\otimes n_k}$ denote the symmetrization of the function $f_1^{\otimes n_1} \otimes f_2^{\otimes n_2} \otimes \cdots \otimes f_k^{\otimes n_k}$ defined in Definition 9.6.8. Let $\{e_k\}_{k=1}^{\infty}$ be an orthonormal basis for $L^2(T)$, $T = [a,b]$. It is a fact that the collection of functions

$$\left\{ \frac{\sqrt{n!}}{\sqrt{n_1! n_2! \cdots}}\, e_1^{\otimes n_1} \widehat{\otimes} e_2^{\otimes n_2} \widehat{\otimes} \cdots\, ;\ n_1 + n_2 + \cdots = n \right\} \qquad (9.7.1)$$

forms an orthonormal basis for $L^2_{\mathrm{sym}}(T^n)$.

To prove the first assertion, we may assume that $f \in L^2(T^n)$ is a symmetric function since $I_n(f) = I_n(\widehat{f})$ by Theorem 9.6.6. Then f can be written as

$$f = \sum_{n_1 + n_2 + \cdots = n} a_{n_1, n_2, \ldots} \frac{\sqrt{n!}}{\sqrt{n_1! n_2! \cdots}}\, e_1^{\otimes n_1} \widehat{\otimes} e_2^{\otimes n_2} \widehat{\otimes} \cdots,$$

where the coefficients satisfy the condition

$$\|f\|^2 = \sum_{n_1 + n_2 + \cdots = n} a_{n_1, n_2, \ldots}^2 < \infty.$$

Then use Equations (9.5.1) and (9.6.11) to get

$$I_n(f) = \sum_{n_1 + n_2 + \cdots = n} a_{n_1, n_2, \ldots} \frac{\sqrt{n!}}{\sqrt{n_1! n_2! \cdots}} \prod_j H_{n_j}(\widetilde{e}_j; 1)$$

$$= \sqrt{n!} \sum_{n_1 + n_2 + \cdots = n} a_{n_1, n_2, \ldots} \mathcal{H}_{n_1, n_2, \ldots}.$$

Hence by Theorem 9.5.4 we have

$$\|I_n(f)\|^2 = n! \sum_{n_1 + n_2 + \cdots = n} a_{n_1, n_2, \ldots}^2 = n! \, \|f\|^2 < \infty.$$

If follows that $I_n(f) \in K_n$.

Conversely, suppose $\phi \in K_n$. By Theorem 9.5.4, ϕ can be expanded as

$$\phi = \sum_{n_1 + n_2 + \cdots = n} b_{n_1, n_2, \ldots} \mathcal{H}_{n_1, n_2, \ldots}.$$

Define a function f on T^n by

$$f = \sum_{n_1 + n_2 + \cdots = n} b_{n_1, n_2, \ldots} \frac{1}{\sqrt{n_1! n_2! \cdots}}\, e_1^{\otimes n_1} \widehat{\otimes} e_2^{\otimes n_2} \widehat{\otimes} \cdots.$$

Then by the above arguments,

$$\|f\|^2 = \frac{1}{n!} \sum_{n_1+n_2+\cdots=n} b_{n_1,n_2,\ldots}^2 = \frac{1}{n!}\|\phi\|^2 < \infty.$$

This shows that $f \in L^2_{\mathrm{sym}}(T^n)$. Moreover, we see from the above arguments that $I_n(f) = \phi$. To show the uniqueness part, suppose $f, g \in L^2_{\mathrm{sym}}(T^n)$ and $\phi = I_n(f) = I_n(g)$. Then

$$\|f - g\| = \frac{1}{\sqrt{n!}}\|I_n(f-g)\| = \frac{1}{\sqrt{n!}}\|I_n(f) - I_n(g)\| = 0,$$

which implies that $f = g$. $\qquad\qquad\qquad\qquad\qquad\qquad\qquad\qquad\qquad\square$

The next theorem follows from Theorems 9.4.7 and 9.7.1.

Theorem 9.7.3. (Wiener–Itô theorem) *The space $L^2_B(\Omega)$ can be decomposed into the orthogonal direct sum*

$$L^2_B(\Omega) = K_0 \oplus K_1 \oplus K_2 \oplus \cdots \oplus K_n \oplus \cdots,$$

where K_n consists of multiple Wiener–Itô integrals of order n. Each function ϕ in $L^2_B(\Omega)$ can be uniquely represented by

$$\phi = \sum_{n=0}^{\infty} I_n(f_n), \quad f_n \in L^2_{\mathrm{sym}}(T^n), \tag{9.7.2}$$

and the following equality holds:

$$\|\phi\|^2 = \sum_{n=0}^{\infty} n!\, \|f_n\|^2.$$

Now we ask the following question.

Question 9.7.4. Given a function $\phi \in L^2_B(\Omega)$, how can one derive the functions $f_n, n \geq 0$, in Equation (9.7.2)?

We will give an answer to this question using the concept of derivative from white noise theory. We state this concept in an informal way. For a rigorous treatment, see the books [27] and [56].

Definition 9.7.5. *Let $\phi = I_n(f), f \in L^2_{\mathrm{sym}}(T^n)$. The* variational derivative *of ϕ is defined to be*

$$\frac{\delta}{\delta t}\phi = nI_{n-1}\big(f_n(t,\cdot)\big), \tag{9.7.3}$$

where the right-hand side is understood to be 0 when $n = 0$. In particular,

$$\frac{\delta}{\delta t}I(f) = f(t). \tag{9.7.4}$$

Suppose $\phi \in L_B^2(\Omega)$ is represented by Equation (9.7.2). By applying the operator $\delta/\delta t$ informally n times, we will get

$$\frac{\delta^n}{\delta t_1 \delta t_2 \cdots \delta t_n} \phi = n! f_n(t_1, t_2, \ldots, t_n)$$
$$+ (n+1)! I_1\big(f_{n+1}(t_1, t_2, \ldots, t_n, \cdot)\big) + \cdots. \qquad (9.7.5)$$

Recall that $E(I_n(f)) = 0$ for all $n \geq 1$. Hence by taking the expectation in both sides of Equation (9.7.5), we obtain the next theorem.

Theorem 9.7.6. *Let $\phi \in L_B^2(\Omega)$. Assume that all variational derivatives exist and have expectation. Put*

$$f_n(t_1, t_2, \ldots, t_n) = \frac{1}{n!} E\Big[\frac{\delta^n}{\delta t_1 \delta t_2 \cdots \delta t_n} \phi\Big].$$

Then the Wiener–Itô expansion of ϕ is given by $\phi = \sum_{n=0}^{\infty} I_n(f_n)$.

Example 9.7.7. Consider $\phi = e^{-B(1)^2}$. Obviously, $\phi \in L_B^2(\Omega)$. Let us find the first three nonzero terms in the Wiener–Itô expansion of ϕ. First we have

$$Ee^{-B(1)^2} = \frac{1}{\sqrt{2\pi}} \int_{-\infty}^{\infty} e^{-x^2} e^{-x^2/2} \, dx = \frac{1}{\sqrt{3}}.$$

Write ϕ as $\phi = \exp\big[-\big(\int_0^1 1 \, dB(s)\big)^2 \big]$ and use Equation (9.7.4) to obtain

$$\frac{\delta}{\delta t_1} \phi = \phi \cdot \Big\{ -2 \int_0^1 1 \, dB(s) \cdot 1 \Big\} = -2B(1) e^{-B(1)^2}.$$

Therefore,

$$f_1(t_1) = E\Big[\frac{\delta}{\delta t_1} \phi\Big] = 0, \quad \forall 0 \leq t_1 \leq 1.$$

Differentiate one more time to get

$$\frac{\delta^2}{\delta t_1 \delta t_2} \phi = -2\big(1 - 2B(1)^2\big) e^{-B(1)^2},$$

which has the following expectation:

$$f_2(t_1, t_2) = E\Big[\frac{\delta^2}{\delta t_1 \delta t_2} \phi\Big] = -\frac{2}{3\sqrt{3}}.$$

The next two variational derivatives of ϕ are given by

$$\frac{\delta^3}{\delta t_1 \delta t_2 \delta t_3} \phi = 4\big(3B(1) - 2B(1)^3\big) e^{-B(1)^2},$$
$$\frac{\delta^4}{\delta t_1 \delta t_2 \delta t_3 \delta t_4} \phi = 4\big(3 - 12B(1)^2 + 4B(1)^4\big) e^{-B(1)^2}.$$

Obviously, $f_3 = 0$. By direct computation, we have

$$f(t_1, t_2, t_3, t_4) = E\left[\frac{\delta^4}{\delta t_1 \delta t_2 \delta t_3 \delta t_4}\phi\right] = \frac{4}{3\sqrt{3}}.$$

Thus we get the first three nonzero terms in the Wiener–Itô expansion:

$$e^{-B(1)^2} = \frac{1}{\sqrt{3}} - \frac{2}{3\sqrt{3}} I_2(1) + \frac{4}{3\sqrt{3}} I_4(1) + \cdots.$$

9.8 Representation of Brownian Martingales

Let $B(t)$ be a Brownian motion and let $T = [a, b]$ be a fixed interval. As before, let $\mathcal{F}^B = \sigma\{B(t); a \le t \le b\}$ and $\mathcal{F}_t^B = \sigma\{B(s); a \le s \le t\}$. Suppose $f(t)$ is a stochastic process in $L_{\mathrm{ad}}^2([a, b] \times \Omega)$. Then by Theorems 4.6.1 and 4.6.2, the stochastic process $X_t = \int_a^t f(s)\, dB(s), a \le t \le b$, is a continuous martingale with respect to the filtration $\{\mathcal{F}_t^B\}$. Moreover, by Theorem 4.3.5 we have $E(X_t^2) = \int_a^t E\big(f(s)^2\big)\, dt < \infty$ for all $t \in [a, b]$. In this section we will show that every square integrable $\{\mathcal{F}_t^B\}$-martingale occurs this way. This fact is a simple application of the Wiener–Itô theorem.

Let $n \ge 1$ and consider a multiple Wiener–Itô integral

$$X = \int_{T^n} f(t_1, t_2, \ldots, t_n)\, dB(t_1) dB(t_2) \cdots dB(t_n),$$

where $f \in L_{\mathrm{sym}}^2(T^n)$. By Theorem 9.6.7, X can be rewritten as

$$X = n! \int_a^b \int_a^{t_1} \cdots \left[\int_a^{t_{n-1}} f(t_1, \ldots, t_n)\, dB(t_n)\right] \cdots dB(t_2)\, dB(t_1).$$

Define a stochastic process $\theta(t)$ by

$$\theta(t) = n! \int_a^t \cdots \left[\int_a^{t_{n-1}} f(t, t_2, \ldots, t_n)\, dB(t_n)\right] \cdots dB(t_2), \quad a \le t \le b.$$

Then we can write X as a stochastic integral

$$X = \int_a^b \theta(t)\, dB(t). \tag{9.8.1}$$

Note that the stochastic process $\theta(t)$ is adapted with respect to the filtration $\{\mathcal{F}_t^B\}$. Moreover, we can again use Theorem 9.6.7 to rewrite $\theta(t)$ as a multiple Wiener–Itô integral

$$\theta(t) = n \cdot (n-1)! \int_a^t \cdots \left[\int_a^{t_{n-1}} f(t, t_2, \ldots, t_n)\, dB(t_n)\right] \cdots dB(t_2)$$

$$= n \int_{[a,t]^{n-1}} f(t, t_2, \ldots, t_n)\, dB(t_2) \cdots dB(t_n). \tag{9.8.2}$$

Then apply Theorem 9.6.6 to find that

$$E\big(\theta(t)^2\big) = n^2 \cdot (n-1)! \int_{[a,t]^{n-1}} f(t,t_2,\ldots,t_n)^2 \, dt_2 \cdots dt_n$$

$$\leq n \cdot n! \int_{[a,b]^{n-1}} f(t,t_2,\ldots,t_n)^2 \, dt_2 \cdots dt_n,$$

which implies that

$$\int_a^b E\big(\theta(t)^2\big) \, dt \leq n \cdot n! \int_{[a,b]^n} f(t_1,t_2,\ldots,t_n)^2 \, dt_1 \, dt_2 \cdots dt_n$$

$$= nE(X^2) < \infty.$$

Thus $\theta \in L^2_{\mathrm{ad}}([a,b] \times \Omega)$ and the stochastic integral in Equation (9.8.1) is an Itô integral as defined in Chapter 4.

Now we use the above discussion to prove the next theorem.

Theorem 9.8.1. *Let $X \in L^2_B(\Omega)$ and $EX = 0$. Then there exists a stochastic process $\theta \in L^2_{\mathrm{ad}}([a,b] \times \Omega)$ such that*

$$X = \int_a^b \theta(t) \, dB(t).$$

Remark 9.8.2. Suppose X belongs to the domain of $\delta/\delta t$. Then by the proof below, we have the following formula:

$$\theta(t) = E\Big[\frac{\delta}{\delta t} X \,\Big|\, \mathcal{F}^B_t\Big].$$

Proof. By the Wiener–Itô theorem, X has the expansion $X = \sum_{n=0}^\infty I_n(f_n)$ with $f_n \in L^2_{\mathrm{sym}}(T^n)$. But $f_0 = EX = 0$. Hence we have

$$X = \sum_{n=1}^\infty I_n(f_n), \quad \|X\|^2 = \sum_{n=1}^\infty n! \, \|f_n\|^2. \tag{9.8.3}$$

For each $n \geq 1$, define a stochastic process

$$\theta_n(t) = n! \int_a^t \cdots \Big[\int_a^{t_{n-1}} f_n(t,t_2,\ldots,t_n) \, dB(t_n)\Big] \cdots dB(t_2).$$

Then by Equation (9.8.1) we get the equality

$$X = \sum_{n=1}^\infty \int_a^b \theta_n(t) \, dB(t). \tag{9.8.4}$$

By Equation (9.8.2) we can rewrite $\theta_n(t)$ as a multiple Wiener–Itô integral

$$\theta_n(t) = n \int_{[a,t]^{n-1}} f_n(t, t_2, \ldots, t_n) \, dB(t_2) \cdots dB(t_n).$$

Hence by Theorem 9.6.10, the θ_n's are orthogonal. Define

$$S_n(t) = \sum_{k=1}^{n} \theta_k(t), \quad a \le t \le b.$$

Note that $I_n(f_n) = \int_a^b \theta_n(t) \, dB(t)$. Hence $E\big(I_n(f_n)^2\big) = \int_a^b E\big(\theta_n(t)^2\big) \, dt$. Then for any $n > m$,

$$\int_a^b E\big(|S_n(t) - S_m(t)|^2\big) \, dt = \sum_{k=m+1}^{n} \int_a^b E\big(\theta_k(t)^2\big) \, dt$$

$$= \sum_{k=m+1}^{n} E\big(I_k(f_k)^2\big),$$

which converges to 0 as $n > m \to \infty$. This shows that $\theta = \lim_{n\to\infty} S_n$ exists in the space $L^2_{\mathrm{ad}}([a,b] \times \Omega)$. Then by Equation (9.8.4) we have

$$X = \lim_{n\to\infty} \sum_{k=1}^{n} \int_a^b \theta_k(t) \, dB(t) = \lim_{n\to\infty} \int_a^b S_n(t) \, dB(t) = \int_a^b \theta(t) \, dB(t).$$

This completes the proof of the theorem.   □

Recall from Definition 3.3.5 that a stochastic process $\widetilde{X}(t)$ is a version of $X(t)$ if $P\{\widetilde{X}(t) = X(t)\} = 1$ for each t.

Theorem 9.8.3. (Martingale representation theorem) *Let $M_t, a \le t \le b$, be a square integrable martingale with respect to $\{\mathcal{F}^B_t; a \le t \le b\}$ and $M_a = 0$. Then M_t has a continuous version $\widetilde{M}(t)$ given by*

$$\widetilde{M}(t) = \int_a^t \theta(s) \, dB(s), \quad a \le t \le b,$$

where $\theta \in L^2_{\mathrm{ad}}([a,b] \times \Omega)$.

Proof. By assumption $M_b \in L^2_B(\Omega)$ and $EM_b = EM_a = 0$. Hence we can apply Theorem 9.8.1 to the random variable M_b to get a stochastic process $\theta(t)$ in $L^2_{\mathrm{ad}}([a,b] \times \Omega)$ such that

$$M_b = \int_a^b \theta(s) \, dB(s). \tag{9.8.5}$$

Define a stochastic process $\widetilde{M}(t)$ by

$$\widetilde{M}(t) = \int_a^t \theta(s)\,dB(s), \quad a \le t \le b.$$

By Theorems 4.6.1 and 4.6.2, $\widetilde{M}(t)$ is a continuous martingale with respect to the filtration $\{\mathcal{F}_t^B\}$. Moreover, for each $t \in [a, b]$,

$$M_t = E\big[M_b \,|\, \mathcal{F}_t^B\big] = E\big[\widetilde{M}(b) \,|\, \mathcal{F}_t^B\big] = \widetilde{M}(t)$$

almost surely. Hence $\widetilde{M}(t)$ is a version of M_t and the theorem is proved. $\square$

Exercises

1. Consider the opposite diagonal squares $\cup_{i=1}^n [t_{i-1}, t_i) \times [t_{n-i}, t_{n-i+1})$ in Equation (9.1.3). The corresponding Riemann sum is given by

$$T_n = \sum_{i=1}^n \big(B(t_i) - B(t_{i-1})\big)\big(B(t_{n-i+1}) - B(t_{n-i})\big).$$

   Show that $T_n \to 0$ in $L^2(\Omega)$ as $\|\Delta_n\| \to 0$.

2. Prove the identities listed at the end of Section 9.3.

3. The Hermite polynomial of degree n is often defined by

$$H_n(x) = (-1)^n e^{x^2} D_x^n\, e^{-x^2}, \quad n \ge 0.$$

   Show that $H_n(x) = 2^n H_n\big(x; \tfrac{1}{2}\big)$ and $H_n(x; \rho) = 2^{-n/2} \rho^{n/2} H_n(x/\sqrt{2\rho})$.

4. Suppose $f, g \in L^2[a, b]$. Show that the double Wiener–Itô integral of the function $f(t)g(s)$ is given by

$$\int_a^b \int_a^b f(t)g(s)\,dB(t)\,dB(s) = I(f)I(g) - \int_a^b f(t)g(t)\,dt,$$

   where $I(f)$ and $I(g)$ are the Wiener integrals of f and g, respectively.

5. Evaluate the double Wiener–Itô integral $\int_0^1 \int_0^1 t\,dB(s)\,dB(t)$.

6. Let f be a nonzero function in $L^2[a, b]$ and let $I(f)$ be the Wiener integral of f. Find the homogeneous chaos expansion for the function $\phi = e^{I(f)}$.

7. Find the homogeneous chaos expansion for the function $B(t)^4 + 2B(t)^3$.

8. Find the homogeneous chaos expansion for the function $B(t)e^{B(t)}$.

9. Find the homogeneous chaos expansion for the functions $\sinh(B(t))$ and $\cosh(B(t))$.

10. Find the homogeneous chaos expansion for the functions $\sin(B(t))$ and $\cos(B(t))$.

11. List the orthonormal basis for the space K_3 of homogeneous chaoses of order 3 as given by Theorem 9.5.4.

12. Let μ be the standard Gaussian measure on $\mathbb{R}^2$, i.e., its density function is given by $f(x,y) = (2\pi)^{-1} e^{-(x^2+y^2)/2}$. Find an orthonormal basis for the Hilbert space $L^2(\mu)$.

13. Let $f \in L^2(T)$ and $g \in L^2(T^n)$. The *tensor product* $f \otimes g$ is defined to be the following function on T^{n+1}:

$$f \otimes g(t_1, t_2, \ldots, t_{n+1}) = f(t_1)g(t_2, \ldots, t_n).$$

The *contraction* $f \odot_k g$ of f with the kth variable of g is defined to be the following function on T^{n-1}:

$$f \odot_k g(t_1, \ldots, \widehat{t_k}, \ldots, t_n) = \int_a^b f(t_k)g(t_1, \ldots, t_k, \ldots, t_n) \, dt_k,$$

where $\widehat{t_k}$ means that t_k is deleted. Prove the equality

$$I_{n+1}(f \otimes g) = I(f)I_n(g) - \sum_{k=1}^n I_{n-1}(f \odot_k g).$$

14. Let $0 \le a < c < d < b \le 1$. Find the Wiener–Itô decomposition of the function $\phi = \big(B(b) - B(a)\big)\big(B(d) - B(c)\big)$.

15. (a) Suppose $\theta \in L^2[a, b]$ has the expansion

$$\theta(x) = \sum_{n=0}^{\infty} a_n H_n(x; 1).$$

Let f be a unit vector in $L^2[a, b]$. Show that the Wiener–Itô expansion of $\theta(I(f))$ is given by

$$\theta(I(f)) = \sum_{n=0}^{\infty} a_n H_n\big(I(f); 1\big).$$

(b) Use the result in part (a) to find the Wiener–Itô expansion of $|B(1)|$.

16. Let $X = I_n(f)$ and $f \in L^2_{\text{sym}}(T^n)$. Prove that

$$E\big[X \,|\, \mathcal{F}_t^B\big] = \int_{[a,t]^n} f(t_1, t_2, \ldots, t_n) \, dB(t_1) \, dB(t_2) \cdots dB(t_n).$$

10

Stochastic Differential Equations

A differential equation $\frac{dx}{dt} = f(t, x)$ in the Leibniz–Newton calculus can also be interpreted as an integral equation $x(t) = x_a + \int_a^t f(s, x(s)) \, ds$. When perturbed by the informal derivative $\dot{B}(t)$ of a Brownian motion $B(t)$, this equation takes the form $\frac{dX}{dt} = \sigma(t, X) \dot{B}(t) + f(t, X)$. In the Itô calculus, $\dot{B}(t)$ and dt are combined together to form the Brownian differential $dB(t)$. The stochastic differential equation $dX = \sigma(t, X) \, dB(t) + f(t, X) \, dt$ is just a symbolic expression and is interpreted as meaning the stochastic integral equation $X_t = X_a + \int_a^t \sigma(s, X_s) \, dB(s) + \int_a^t f(s, X_s) \, ds$, $a \le t \le b$. The original motivation for K. Itô to develop the theory of stochastic integration was to construct diffusion processes by solving stochastic differential equations.

10.1 Some Examples

First we recall several examples of stochastic differential equations that arose in the previous chapters.

Example 10.1.1. From Example 7.4.5, we have the Langevin equation

$$dX_t = \alpha \, dB(t) - \beta X_t \, dt, \quad X_0 = x_0.$$

The solution is an Ornstein–Uhlenbeck process

$$X_t = e^{-\beta t} x_0 + \alpha \int_0^t e^{-\beta(t-u)} \, dB(u).$$

Example 10.1.2. Recall that in Example 7.5.2 we used Itô's formula to derive an $\mathbb{R}^2$-valued stochastic differential equation

$$dV_t = \begin{bmatrix} 0 & -1 \\ 1 & 0 \end{bmatrix} V_t \, dB(t) - \frac{1}{2} V_t \, dt, \quad V_0 = \begin{bmatrix} 1 \\ 0 \end{bmatrix}.$$

The solution V_t is the transpose of $[\cos B(t), \sin B(t)]$.

Example 10.1.3. In Example 7.5.3, we again used Itô's formula to derive an $\mathbb{R}^3$-valued stochastic differential equation

$$dV_t = (KV_t + q)\, dB(t) + LV_t\, dt, \quad V_0 = v_0,$$

where K, q, L, and v_0 are given by

$$K = \begin{bmatrix} 0 & -1 & 0 \\ 1 & 0 & 0 \\ 0 & 0 & 0 \end{bmatrix}, \quad q = \begin{bmatrix} 0 \\ 0 \\ 1 \end{bmatrix}, \quad L = \begin{bmatrix} -\frac{1}{2} & 0 & 0 \\ 0 & -\frac{1}{2} & 0 \\ 0 & 0 & 0 \end{bmatrix}, \quad v_0 = \begin{bmatrix} 1 \\ 0 \\ 0 \end{bmatrix}.$$

The solution V_t is the transpose of $[\cos B(t),\ \sin B(t),\ B(t)]$.

Example 10.1.4. Let $B_1(t)$ and $B_2(t)$ be independent Brownian motions. In Example 7.5.4, we derived an $\mathbb{R}^3$-valued stochastic differential equation

$$dV_t = KV_t\, dB_1(t) + CV_t\, dB_2(t) + LV_t\, dt, \quad V_0 = v_0,$$

where K, C, L, and v_0 are given by

$$K = \begin{bmatrix} 1 & 0 & 0 \\ 0 & 1 & 0 \\ 0 & 0 & 1 \end{bmatrix}, \quad C = \begin{bmatrix} 0 & -1 & 0 \\ 1 & 0 & 0 \\ 0 & 0 & 0 \end{bmatrix}, \quad L = \begin{bmatrix} 0 & 0 & 0 \\ 0 & 0 & 0 \\ 0 & 0 & \frac{1}{2} \end{bmatrix}, \quad v_0 = \begin{bmatrix} 1 \\ 0 \\ 1 \end{bmatrix}.$$

The solution V_t is the transpose of $\left[e^{B_1(t)} \cos B_2(t),\ e^{B_1(t)} \sin B_2(t),\ e^{B_1(t)} \right]$.

Example 10.1.5. In Example 8.3.6, we have stochastic differential equations

$$dX_t = X_t \circ dB(t),\ X_0 = 1; \quad dY_t = Y_t\, dB(t),\ Y_0 = 1,$$

where $\circ$ denotes the Stratonovich integral. The solutions are given by

$$X_t = e^{B(t)}; \quad Y_t = e^{B(t) - \frac{1}{2}t}.$$

Example 10.1.6. The Bessel process $L_{n-1}(t)$ defined in Definition 8.5.3 is shown in Theorem 8.5.4 to satisfy the stochastic differential equation

$$dX_t = dW(t) + \frac{n-1}{2} \frac{1}{X_t}\, dt, \quad X_0 = 0.$$

Example 10.1.7. The exponential process $\mathcal{E}_h(t)$ defined by Equation (8.7.3) is shown in the proof of Theorem 8.9.4 to satisfy the stochastic differential equation in Equation (8.9.8), namely,

$$dX_t = h(t)X_t\, dB(t), \quad X_0 = 1.$$

Notice that all of the above examples except for Example 10.1.6 are linear equations. Moreover, almost all sample paths of the solutions are functions defined on the whole half-line $[0, \infty)$.

We give two more examples to show that the solution of a stochastic differential equation can explode in finite time almost surely and that there may exist several solutions of a stochastic differential equation.

Example 10.1.8. Consider the stochastic differential equation

$$dX_t = X_t^2 \, dB(t) + X_t^3 \, dt, \quad X_0 = 1, \tag{10.1.1}$$

which means the following stochastic integral equation:

$$X_t = 1 + \int_0^t X_s^2 \, dB(s) + \int_0^t X_s^3 \, ds.$$

To solve this equation, apply Itô's formula to find that

$$
\begin{aligned}
d\left(\frac{1}{X_t}\right) &= -\frac{1}{X_t^2} \, dX_t + \frac{1}{2}\frac{2}{X_t^3} \, (dX_t)^2 \\
&= -\frac{1}{X_t^2}\left(X_t^2 \, dB(t) + X_t^3 \, dt\right) + X_t \, dt \\
&= -dB(t).
\end{aligned}
$$

Therefore, $\frac{1}{X_t} = -B(t) + C$. The initial condition $X_0 = 1$ implies that $C = 1$. Hence the solution of Equation (10.1.1) is given by

$$X_t = \frac{1}{1 - B(t)}.$$

Note that the solution X_t explodes at the first exit time of the Brownian motion $B(t)$ from the interval $(-\infty, 1)$.

Example 10.1.9. Consider the stochastic differential equation

$$dX_t = 3X_t^{2/3} \, dB(t) + 3X_t^{1/3} \, dt, \quad X_0 = 0, \tag{10.1.2}$$

which means the following stochastic integral equation:

$$X_t = 3\int_0^t X_s^{2/3} \, dB(s) + 3\int_0^t X_s^{1/3} \, ds.$$

For any fixed constant $a > 0$, define a function $\theta_a(x) = (x - a)^3 \, 1_{\{x \geq a\}}$. It is easily verified that the first two derivatives of θ_a are given by

$$\theta_a'(x) = 3\theta_a(x)^{2/3}, \quad \theta_a''(x) = 6\theta_a(x)^{1/3}.$$

By Itô's formula, we have

$$d\bigl(\theta_a(B(t))\bigr) = 3\theta_a(B(t))^{2/3} \, dB(t) + 3\theta_a(B(t))^{1/3} \, dt.$$

Moreover, $\theta_a(B(0)) = 0$. Hence $\theta_a(B(t))$ is a solution of Equation (10.1.2) for any $a > 0$. This shows that Equation (10.1.2) has infinitely many solutions.

Observe that the last two examples are in fact simple modifications of well-known examples in ordinary differential equations for the Leibniz–Newton calculus. Hence we would expect to encounter similar phenomena in stochastic differential equations for the Itô calculus. This means that in order to ensure the existence and uniqueness of a globally defined solution, we need to impose the Lipschitz and growth conditions.

10.2 Bellman–Gronwall Inequality

In this section we will prove two inequalities to be used in the next section. The first one, called the Bellman–Gronwall inequality, covers the following situation. Suppose we have a function $\phi \in L^1[a, b]$ satisfying the inequality

$$\phi(t) \leq f(t) + \beta \int_a^t \phi(s) \, ds, \quad \forall t \in [a, b], \tag{10.2.1}$$

where $f \in L^1[a, b]$ and β is a positive constant.

Question 10.2.1. Given Equation (10.2.1), how can we estimate $\phi(t)$ in terms of $f(t)$ and β?

Define a new function $g(t)$ by

$$g(t) = \beta \int_a^t \phi(s) \, ds, \quad a \leq t \leq b.$$

Then by the fundamental theorem of calculus and Equation (10.2.1),

$$g'(t) = \beta\phi(t) \leq \beta f(t) + \beta g(t), \quad \text{almost everywhere.}$$

Hence we have

$$g'(t) - \beta g(t) \leq \beta f(t).$$

Multiply both sides of this inequality by the integrating factor $e^{-\beta t}$ to get

$$\frac{d}{dt}\left(e^{-\beta t}g(t)\right) = e^{-\beta t}\left(g'(t) - \beta g(t)\right) \leq \beta f(t)e^{-\beta t},$$

which, after we integrate from a to t, becomes

$$e^{-\beta t}g(t) \leq \beta \int_a^t f(s)e^{-\beta s} \, ds.$$

Therefore,

$$g(t) \leq \beta \int_a^t f(s)e^{\beta(t-s)} \, ds.$$

Hence by Equation (10.2.1),

$$\phi(t) \leq f(t) + g(t) \leq f(t) + \beta \int_a^t f(s)e^{\beta(t-s)} \, ds.$$

Lemma 10.2.2. (Bellman–Gronwall inequality) *Suppose* $\phi \in L^1[a, b]$ *satisfies Equation* (10.2.1). *Then*

$$\phi(t) \leq f(t) + \beta \int_a^t f(s)e^{\beta(t-s)} \, ds.$$

In particular, when $f(t)$ *is a constant* α, *we have*

$$\phi(t) \leq \alpha e^{\beta(t-a)}, \quad \forall a \leq t \leq b. \tag{10.2.2}$$

For the second inequality needed in the next section, let $\{\theta_n\}_{n=1}^\infty$ be a sequence of functions in $L^1[a,b]$ satisfying the inequality

$$\theta_{n+1}(t) \le f(t) + \beta \int_a^t \theta_n(s)\,ds, \quad \forall t \in [a,b], \tag{10.2.3}$$

where $f \in L^1[a,b]$ and β is a positive constant.

Question 10.2.3. Given Equation (10.2.3), how can we estimate $\theta_{n+1}(t)$ in terms of $f(t)$, β, n, and $\theta_1(t)$?

The above trick in proving Lemma 10.2.2 does not work for this case. Instead, we need to iterate the given inequality and apply induction, which in fact can also be used to prove Lemma 10.2.2.

Put $n = 1$ in Equation (10.2.3) to get

$$\theta_2(t) \le f(t) + \beta \int_a^t \theta_1(s)\,ds. \tag{10.2.4}$$

Put $n = 2$ in Equation (10.2.3) and then use Equation (10.2.4) to see that

$$\theta_3(t) \le f(t) + \beta \int_a^t \theta_2(s)\,ds$$

$$\le f(t) + \beta \int_a^t f(s)\,ds + \beta^2 \int_a^t \left(\int_a^s \theta_1(u)\,du \right) ds.$$

Change the order of integration in the above iterated integral to obtain

$$\theta_3(t) \le f(t) + \beta \int_a^t f(s)\,ds + \beta^2 \int_a^t (t-u)\theta_1(u)\,du.$$

Similarly, we have

$$\theta_4(t) \le f(t) + \beta \int_a^t f(s)\,ds + \beta^2 \int_a^t (t-u)f(u)\,du + \beta^3 \int_a^t \frac{(t-u)^2}{2}\theta_1(u)\,du.$$

In general, we can use induction to show that for any $n \ge 1$,

$$\theta_{n+1}(t) \le f(t) + \beta \int_a^t f(s)\,ds + \beta^2 \int_a^t (t-u)f(u)\,du$$

$$+ \cdots + \beta^{n-1} \int_a^t \frac{(t-u)^{n-2}}{(n-2)!} f(u)\,du$$

$$+ \beta^n \int_a^t \frac{(t-u)^{n-1}}{(n-1)!} \theta_1(u)\,du.$$

Note that $\sum_{k=0}^{n-2} \frac{\beta^k(t-u)^k}{k!} \le e^{\beta(t-u)}$. Hence we can simplify the estimate of $\theta_{n+1}(t)$ by

$$\theta_{n+1}(t) \leq f(t) + \beta \int_a^t f(u)e^{\beta(t-u)}\,du + \beta^n \int_a^t \frac{(t-u)^{n-1}}{(n-1)!}\theta_1(u)\,du. \quad (10.2.5)$$

Thus we have proved the next lemma.

Lemma 10.2.4. *Let $\{\theta_n\}_{n=1}^\infty$ be a sequence of functions in $L^1[a,b]$ satisfying Equation (10.2.3). Then Equation (10.2.5) holds for any $n \geq 1$. In particular, when $f(t) \equiv \alpha$ and $\theta_1(t) \equiv c$ are constants, then the following inequality holds for any $n \geq 1$:*

$$\theta_{n+1}(t) \leq \alpha e^{\beta(t-a)} + c\,\frac{\beta^n(t-a)^n}{n!}. \quad (10.2.6)$$

10.3 Existence and Uniqueness Theorem

Let $B(t)$ be a Brownian motion and $\{\mathcal{F}_t; a \leq t \leq b\}$ a filtration satisfying the conditions given in the beginning of Section 4.3, namely, $B(t)$ is $\mathcal{F}_t$-measurable for each t and $B(t) - B(s)$ is independent of $\mathcal{F}_s$ for any $s \leq t$.

Let $\sigma(t,x)$ and $f(t,x)$ be measurable functions of $t \in [a,b]$ and $x \in \mathbb{R}$. Consider the stochastic differential equation (SDE)

$$dX_t = \sigma(t,X_t)\,dB(t) + f(t,X_t)\,dt, \quad X_a = \xi,$$

which must be interpreted as the stochastic integral equation (SIE)

$$X_t = \xi + \int_a^t \sigma(s,X_s)\,dB(s) + \int_a^t f(s,X_s)\,ds, \quad a \leq t \leq b. \quad (10.3.1)$$

First we need to explain what it means that a stochastic process X_t is a solution of the SIE in Equation (10.3.1).

Definition 10.3.1. *A jointly measurable stochastic process X_t, $a \leq t \leq b$, is called a* solution *of the SIE in Equation (10.3.1) if it satisfies the following conditions:*

(1) *The stochastic process $\sigma(t,X_t)$ belongs to $\mathcal{L}_{\mathrm{ad}}(\Omega, L^2[a,b])$ (see Notation 5.1.1), so that $\int_a^t \sigma(s,X_s)\,dB(s)$ is an Itô integral for each $t \in [a,b]$;*
(2) *Almost all sample paths of the stochastic process $f(t,X_t)$ belong to $L^1[a,b]$;*
(3) *For each $t \in [a,b]$, Equation (10.3.1) holds almost surely.*

As pointed out by examples in Section 10.1, we need to impose conditions on the functions $\sigma(t,x)$ and $f(t,x)$ in order to ensure the existence of a unique nonexplosive solution of the SIE in Equation (10.3.1). We state the conditions in the next two definitions.

Definition 10.3.2. *A measurable function $g(t,x)$ on $[a,b]\times\mathbb{R}$ is said to satisfy the* Lipschitz condition *in x if there exists a constant $K > 0$ such that*

$$|g(t,x) - g(t,y)| \leq K|x-y|, \quad \forall\, a \leq t \leq b,\ x,y \in \mathbb{R}.$$

Definition 10.3.3. *A measurable function $g(t, x)$ on $[a, b] \times \mathbb{R}$ is said to satisfy the* linear growth condition *in x if there exists a constant $K > 0$ such that*

$$|g(t, x)| \leq K(1 + |x|), \quad \forall\, a \leq t \leq b, \ x \in \mathbb{R}. \tag{10.3.2}$$

Note the following inequalities for all $x \geq 0$:

$$1 + x^2 \leq (1 + x)^2 \leq 2(1 + x^2).$$

Hence the condition in Equation (10.3.2) is equivalent to the existence of a constant $C > 0$ such that

$$|g(t, x)|^2 \leq C(1 + x^2), \quad \forall\, a \leq t \leq b, \ x \in \mathbb{R}.$$

Lemma 10.3.4. *Let $\sigma(t, x)$ and $f(t, x)$ be measurable functions on $[a, b] \times \mathbb{R}$ satisfying the Lipschitz condition in x. Suppose ξ is an $\mathcal{F}_a$-measurable random variable with $E(\xi^2) < \infty$. Then the stochastic integral equation in Equation (10.3.1) has at most one continuous solution X_t.*

Proof. Let X_t and Y_t be two continuous solutions of the SIE in Equation (10.3.1). Put $Z_t = X_t - Y_t$. Then Z_t is a continuous stochastic process and

$$Z_t = \int_a^t \big(\sigma(s, X_s) - \sigma(s, Y_s)\big)\, dB(s) + \int_a^t \big(f(s, X_s) - f(s, Y_s)\big)\, ds.$$

Use the inequality $(a + b)^2 \leq 2(a^2 + b^2)$ to get

$$Z_t^2 \leq 2\bigg[\bigg(\int_a^t \big(\sigma(s, X_s) - \sigma(s, Y_s)\big)\, dB(s) \bigg)^2$$
$$+ \bigg(\int_a^t \big(f(s, X_s) - f(s, Y_s)\big)\, ds \bigg)^2 \bigg]. \tag{10.3.3}$$

By the Lipschitz condition of the function $\sigma(t, x)$, we have

$$E\bigg(\int_a^t \big(\sigma(s, X_s) - \sigma(s, Y_s)\big)\, dB(s) \bigg)^2 = \int_a^t E\Big[\big(\sigma(s, X_s) - \sigma(s, Y_s)\big)^2\Big]\, ds$$
$$\leq K^2 \int_a^t E(Z_s^2)\, ds. \tag{10.3.4}$$

On the other hand, by the Lipschitz condition of the function $f(t, x)$,

$$\bigg(\int_a^t \big(f(s, X_s) - f(s, Y_s)\big)\, ds \bigg)^2 \leq (t - a) \int_a^t \big(f(s, X_s) - f(s, Y_s)\big)^2\, ds$$
$$\leq (b - a)K^2 \int_a^t Z_s^2\, ds. \tag{10.3.5}$$

Put Equations (10.3.4) and (10.3.5) into Equation (10.3.3) to see that

$$E(Z_t^2) \leq 2K^2(1 + b - a) \int_a^t E(Z_s^2)\,ds.$$

By the Bellman–Gronwall inequality in Lemma 10.2.2, we have $E(Z_t^2) = 0$ for all $t \in [a, b]$. Hence $Z_t = 0$ almost surely for each $t \in [a, b]$. Let $\{r_1, r_2, \ldots\}$ be a counting of the rational numbers in the interval $[a, b]$. Then for each r_n, there exists Ω_n such that $P(\Omega_n) = 1$ and $Z_{r_n}(\omega) = 0$ for all $\omega \in \Omega_n$. Let $\Omega' = \cap_{n=1}^\infty \Omega_n$. Then $P(\Omega') = 1$ and for each $\omega \in \Omega'$, we have $Z_{r_n}(\omega) = 0$ for all n. Since Z_t is a continuous stochastic process, there exists Ω'' such that $P(\Omega'') = 1$ and for each $\omega \in \Omega''$, the function $Z_t(\omega)$ is a continuous function of t. Finally, let $\Omega_0 = \Omega' \cap \Omega''$. Then $P(\Omega_0) = 1$ and for each $\omega \in \Omega_0$, the function $Z_t(\omega)$ is a continuous function that vanishes at all rational numbers in $[a, b]$. It follows that for each $\omega \in \Omega_0$, the function $Z_t(\omega)$ is 0 for all $t \in [a, b]$. Therefore, X_t and Y_t are the same continuous stochastic process. □

Theorem 10.3.5. *Let $\sigma(t, x)$ and $f(t, x)$ be measurable functions on $[a, b] \times \mathbb{R}$ satisfying the Lipschitz and linear growth conditions in x. Suppose ξ is an $\mathcal{F}_a$-measurable random variable with $E(\xi^2) < \infty$. Then the stochastic integral equation in Equation (10.3.1) has a unique continuous solution X_t.*

Remark 10.3.6. We point out that the above theorem covers Examples 10.1.1 and 10.1.5. It also covers Example 10.1.7 when h is a bounded deterministic function. However, it does not cover Examples 10.1.6, 10.1.8, and 10.1.9.

Proof. The uniqueness of a continuous solution follows from Lemma 10.3.4. We proceed to prove the existence of a solution.

By assumption, there exists a constant $C > 0$ such that the following inequalities hold for all $t \in [a, b]$ and $x, y \in \mathbb{R}$:

$$|\sigma(t, x) - \sigma(t, y)| \leq C|x - y|, \quad |f(t, x) - f(t, y)| \leq C|x - y|; \qquad (10.3.6)$$

$$|\sigma(t, x)|^2 \leq C(1 + x^2), \quad |f(t, x)|^2 \leq C(1 + x^2). \qquad (10.3.7)$$

We use a similar iteration procedure as the one for ordinary differential equations to produce a solution of the SIE in Equation (10.3.1). Define a sequence $\{X_t^{(n)}\}_{n=1}^\infty$ of continuous stochastic processes inductively by setting $X_t^{(1)} \equiv \xi$ and for $n \geq 1$,

$$X_t^{(n+1)} = \xi + \int_a^t \sigma(s, X_s^{(n)})\,dB(s) + \int_a^t f(s, X_s^{(n)})\,ds. \qquad (10.3.8)$$

Obviously, $X_t^{(1)}$ belongs to $L_{ad}^2([a, b] \times \Omega)$ (see Notation 4.3.1.) Assume by induction that the stochastic process $X_t^{(n)}$ belongs to $L_{ad}^2([a, b] \times \Omega)$. Then by the linear growth condition in Equation (10.3.7),

$$E \int_a^b \sigma(t, X_t^{(n)})^2\,dt \leq C(b - a) + CE \int_a^b |X_t^{(n)}|^2\,dt < \infty;$$

$$\int_a^t |f(s, X_s^{(n)})|\,ds \leq \sqrt{C(b - a)}\left(\int_a^b (1 + |X_t^{(n)}|^2)\,dt\right)^{1/2} < \infty, \quad \text{a.s.}$$

Hence the first integral in Equation (10.3.8) is an Itô integral as defined in Chapter 4, while the second integral is a Lebesgue integral in t for almost all $\omega \in \Omega$. Thus $X_t^{(n+1)}$ is a continuous stochastic process and is adapted to the filtration $\{\mathcal{F}_t\}$. Moreover, since $|a + b + c|^2 \leq 3(a^2 + b^2 + c^2)$, we have

$$|X_t^{(n+1)}|^2 \leq 3\left[\xi^2 + \left(\int_a^t \sigma(s, X_s^{(n)})\, dB(s)\right)^2 + \left(\int_a^t f(s, X_s^{(n)})\, ds\right)^2\right], \quad (10.3.9)$$

which together with the linear growth condition implies that

$$E \int_a^b |X_t^{(n+1)}|^2\, dt < \infty.$$

This shows that the stochastic process $X_t^{(n+1)}$ belongs to $L_{\mathrm{ad}}^2([a, b] \times \Omega)$. Thus by induction we have a sequence $\{X_t^{(n)}\}_{n=1}^{\infty}$ of continuous stochastic processes in the space $L_{\mathrm{ad}}^2([a, b] \times \Omega)$.

Next, we estimate $E(|X_t^{(n+1)} - X_t^{(n)}|^2)$, which will be stated by Equation (10.3.13) below. For convenience, let

$$Y_t^{(n+1)} = \int_a^t \sigma(s, X_s^{(n)})\, dB(s), \quad Z_t^{(n+1)} = \int_a^t f(s, X_s^{(n)})\, ds.$$

Then $X_t^{(n+1)} = \xi + Y_t^{(n+1)} + Z_t^{(n+1)}$. Since $(a + b)^2 \leq 2(a^2 + b^2)$, we have

$$E(|X_t^{(n+1)} - X_t^{(n)}|^2)$$
$$\leq 2\left\{E(|Y_t^{(n+1)} - Y_t^{(n)}|^2) + E(|Z_t^{(n+1)} - Z_t^{(n)}|^2)\right\}. \quad (10.3.10)$$

By the Lipschitz condition in Equation (10.3.6),

$$E(|Y_t^{(n+1)} - Y_t^{(n)}|^2) = \int_a^t E(|\sigma(s, X_s^{(n)}) - \sigma(s, X_s^{(n-1)}|^2)\, ds$$
$$\leq C^2 \int_a^t E(|X_s^{(n)} - X_s^{(n-1)}|^2)\, ds. \quad (10.3.11)$$

Similarly, the Lipschitz condition in Equation (10.3.6) yields that

$$|Z_t^{(n+1)} - Z_t^{(n)}|^2 \leq (b - a)C^2 \int_a^t |X_s^{(n)} - X_s^{(n-1)}|^2\, ds. \quad (10.3.12)$$

Equations (10.3.10), (10.3.11), and (10.3.12) imply that for any $n \geq 2$,

$$E(|X_t^{(n+1)} - X_t^{(n)}|^2) \leq 2C^2(1 + b - a) \int_a^t E(|X_s^{(n)} - X_s^{(n-1)}|^2)\, ds.$$

On the other hand, by the growth condition in Equation (10.3.7),

$$E\big(\big|X_t^{(2)} - X_t^{(1)}\big|^2\big) \leq 2C^2(1 + b - a) \int_a^t \big(1 + E(\xi^2)\big)\, ds.$$

Then by Lemma 10.2.4,

$$E\big(\big|X_t^{(n+1)} - X_t^{(n)}\big|^2\big) \leq \rho\, \frac{\beta^n (t - a)^n}{n!}, \tag{10.3.13}$$

where $\rho = 1 + E(\xi^2)$ and $\beta = 2C^2(1 + b - a)$.

Now note that for any $t \in [a, b]$,

$$\big|X_t^{(n+1)} - X_t^{(n)}\big| \leq \big|Y_t^{(n+1)} - Y_t^{(n)}\big| + \big|Z_t^{(n+1)} - Z_t^{(n)}\big|.$$

Hence we have

$$\sup_{a \leq t \leq b} \big|X_t^{(n+1)} - X_t^{(n)}\big| \leq \sup_{a \leq t \leq b} \big|Y_t^{(n+1)} - Y_t^{(n)}\big| + \sup_{a \leq t \leq b} \big|Z_t^{(n+1)} - Z_t^{(n)}\big|,$$

which implies that

$$\left\{ \sup_{a \leq t \leq b} \big|X_t^{(n+1)} - X_t^{(n)}\big| > \frac{1}{n^2} \right\}$$

$$\subset \left\{ \sup_{a \leq t \leq b} \big|Y_t^{(n+1)} - Y_t^{(n)}\big| > \frac{1}{2n^2} \right\} \cup \left\{ \sup_{a \leq t \leq b} \big|Z_t^{(n+1)} - Z_t^{(n)}\big| > \frac{1}{2n^2} \right\}.$$

Therefore,

$$P\left\{ \sup_{a \leq t \leq b} \big|X_t^{(n+1)} - X_t^{(n)}\big| > \frac{1}{n^2} \right\} \leq P\left\{ \sup_{a \leq t \leq b} \big|Y_t^{(n+1)} - Y_t^{(n)}\big| > \frac{1}{2n^2} \right\}$$

$$+ P\left\{ \sup_{a \leq t \leq b} \big|Z_t^{(n+1)} - Z_t^{(n)}\big| > \frac{1}{2n^2} \right\}. \tag{10.3.14}$$

Apply the Doob submartingale inequality in Theorem 4.5.1 and then use Equations (10.3.6) and (10.3.13) to get

$$P\left\{ \sup_{a \leq t \leq b} \big|Y_t^{(n+1)} - Y_t^{(n)}\big| > \frac{1}{2n^2} \right\} \leq 4n^4 E\big(\big|Y_b^{(n+1)} - Y_b^{(n)}\big|^2\big)$$

$$\leq 4n^4 C^2 \int_a^b E\big(\big|X_t^{(n)} - X_t^{(n-1)}\big|^2\big)\, dt$$

$$\leq 4n^4 C^2 \rho\, \frac{\beta^{n-1}(b - a)^n}{n!}. \tag{10.3.15}$$

On the other hand, by Equation (10.3.6) using the function f,

$$\big|Z_t^{(n+1)} - Z_t^{(n)}\big|^2 \leq C^2(b - a) \int_a^t \big|X_s^{(n)} - X_s^{(n-1)}\big|^2\, ds,$$

which immediately implies that

$$\sup_{a \le t \le b} \left| Z_t^{(n+1)} - Z_t^{(n)} \right|^2 \le C^2 (b - a) \int_a^b \left| X_t^{(n)} - X_t^{(n-1)} \right|^2 ds.$$

From this inequality and Equation (10.3.13) we see that

$$P\left\{ \sup_{a \le t \le b} \left| Z_t^{(n+1)} - Z_t^{(n)} \right| > \frac{1}{2n^2} \right\} \le 4n^4 E\left[\left\{ \sup_{a \le t \le b} \left| Z_t^{(n+1)} - Z_t^{(n)} \right| \right\}^2 \right]$$

$$\le 4n^4 C^2 (b - a)\rho \, \frac{\beta^{n-1}(b-a)^n}{n!}. \quad (10.3.16)$$

It follows from Equations (10.3.14), (10.3.15), and (10.3.16) that

$$P\left\{ \sup_{a \le t \le b} \left| X_t^{(n+1)} - X_t^{(n)} \right| > \frac{1}{n^2} \right\} \le 2\rho \, \frac{n^4 \beta^n (b-a)^n}{n!}.$$

It is easy to check that the series $\sum_n \frac{n^4 \beta^n (b-a)^n}{n!}$ is convergent. Hence by the Borel–Cantelli lemma in Theorem 3.2.1, we have

$$P\left\{ \sup_{a \le t \le b} \left| X_t^{(n+1)} - X_t^{(n)} \right| > \frac{1}{n^2} \text{ i.o.} \right\} = 0.$$

This implies that the series $\xi + \sum_{n=1}^\infty \left(X_t^{(n+1)} - X_t^{(n)} \right)$ converges uniformly on $[a, b]$ with probability 1 to, say, X_t. Note that the nth partial sum of this series is $X_t^{(n)}$. Hence with probability 1,

$$\lim_{n \to \infty} X_t^{(n)} = X_t, \quad \text{uniformly for } t \in [a, b].$$

Obviously, the stochastic process X_t is continuous and adapted to the filtration $\{\mathcal{F}_t ; a \le t \le b\}$. Moreover, Equation (10.3.13) implies that

$$\left\| X_t^{(n+1)} - X_t^{(n)} \right\| \le \sqrt{\rho} \, \frac{\beta^{n/2}(b-a)^{n/2}}{\sqrt{n!}},$$

where $\| \cdot \|$ is the $L^2(\Omega)$-norm. This inequality implies that for each t, the series $\xi + \sum_{n=1}^\infty \left(X_t^{(n+1)} - X_t^{(n)} \right)$ converges in $L^2(\Omega)$ and

$$\| X_t \| \le \| \xi \| + \sum_{n=1}^\infty \sqrt{\rho} \, \frac{\beta^{n/2}(b-a)^{n/2}}{\sqrt{n!}}.$$

It follows that $E \int_a^b |X_t|^2 \, dt < \infty$. Hence the stochastic process X_t belongs to the space $L_{\text{ad}}^2([a, b] \times \Omega) \subset \mathcal{L}_{\text{ad}}(\Omega, L^2[a, b])$. We can easily check that X_t satisfies the conditions (1) and (2) in Definition 10.3.1. Moreover, we can verify that the limit as $n \to \infty$ can be brought inside the integral signs in Equation (10.3.8) to get

$$X_t = \xi + \int_a^t \sigma(s, X_s) \, dB(s) + \int_a^t f(s, X_s) \, ds.$$

Hence X_t is a solution of Equation (10.3.1) and the proof is complete.   □

10.4 Systems of Stochastic Differential Equations

Let $B_t^{(1)}, B_t^{(2)}, \ldots, B_t^{(m)}$ be m independent Brownian motions. For $1 \leq i \leq n$ and $1 \leq j \leq m$, let $\sigma_{ij}(t, x)$ and $f_i(t, x)$ be functions of $t \in [a, b]$ and $x \in \mathbb{R}^n$. Consider the following system of stochastic differential equations for $1 \leq i \leq n$:

$$dX_t^{(i)} = \sum_{j=1}^{m} \sigma_{ij}(t, X_t^{(1)}, \ldots, X_t^{(n)}) \, dB_t^{(j)} + f_i(t, X_t^{(1)}, \ldots, X_t^{(n)}) \, dt, \quad (10.4.1)$$

with the initial condition $X_a^{(i)} = \xi_i$. If we use the matrix notation

$$X_t = \begin{bmatrix} X_t^{(1)} \\ \vdots \\ X_t^{(n)} \end{bmatrix}, \quad \sigma = \begin{bmatrix} \sigma_{11} & \cdots & \sigma_{1m} \\ \vdots & \ddots & \vdots \\ \sigma_{n1} & \cdots & \sigma_{nm} \end{bmatrix}, \quad B_t = \begin{bmatrix} B_t^{(1)} \\ \vdots \\ B_t^{(m)} \end{bmatrix}, \quad f = \begin{bmatrix} f_1 \\ \vdots \\ f_n \end{bmatrix}, \quad \xi = \begin{bmatrix} \xi_1 \\ \vdots \\ \xi_n \end{bmatrix},$$

then Equation (10.4.1) can be rewritten as a matrix SDE by

$$dX_t = \sigma(t, X_t) \, dB(t) + f(t, X_t) \, dt, \quad X_a = \xi. \quad (10.4.2)$$

For an $n \times 1$ column vector v in $\mathbb{R}^n$, let $|v|$ denote the Euclidean norm of v. The *Hilbert–Schmidt norm* of an $n \times m$ matrix $\sigma = [\sigma_{ij}]$ is defined by

$$\|\sigma\| = \left(\sum_{i=1}^{n} \sum_{j=1}^{m} \sigma_{ij}^2 \right)^{1/2}. \quad (10.4.3)$$

The multidimensional generalization of Theorem 10.3.5 is given in the next theorem.

Theorem 10.4.1. *Let the matrix-valued function $\sigma(t, x)$ and the vector-valued function $f(t, x)$ in Equation (10.4.2) be measurable. Assume that there exists a constant $K > 0$ such that for all $t \in [a, b]$ and $x, y \in \mathbb{R}^n$,*

(1) *(Lipschitz condition)*

$$\|\sigma(t, x) - \sigma(t, y)\| \leq K|x - y|, \quad |f(t, x) - f(t, y)| \leq K|x - y|;$$

(2) *(Linear growth condition)*

$$\|\sigma(t, x)\|^2 \leq K(1 + |x|^2), \quad |f(t, x)|^2 \leq K(1 + |x|^2).$$

Then for any $\mathcal{F}_a$-measurable random vector ξ with $E(|\xi|^2) < \infty$, the stochastic integral equation

$$X_t = \xi + \int_a^t \sigma(s, X_s) \, dB_s + \int_a^t f(s, X_s) \, ds, \quad a \leq t \leq b,$$

has a unique continuous solution.

This theorem can be proved by exactly the same arguments as those in the proof of Theorem 10.3.5 with simple modifications of notation. We point out that Examples 10.1.2, 10.1.3, and 10.1.4 are covered by this theorem.

10.5 Markov Property

In this section we explain the Markov property, an important concept that will be needed in the next two sections.

Let $X, Y_1, Y_2, \ldots, Y_n$ be random variables. Let $\sigma\{Y_1, Y_2, \ldots, Y_n\}$ denote the σ-field generated by $Y_1, Y_2, \ldots, Y_n$. The conditional expectation of X given $\sigma\{Y_1, Y_2, \ldots, Y_n\}$ will also be denoted by $E[X \mid Y_1, Y_2, \ldots, Y_n]$, namely,

$$E[X \mid Y_1, Y_2, \ldots, Y_n] = E[X \mid \sigma\{Y_1, Y_2, \ldots, Y_n\}].$$

Then there exists a Borel measurable function θ on $\mathbb{R}^n$ such that

$$E[X \mid Y_1, Y_2, \ldots, Y_n] = \theta(Y_1, Y_2, \ldots, Y_n).$$

Notation 10.5.1. We will use $E[X \mid Y_1 = y_1, Y_2 = y_2, \ldots, Y_n = y_n]$ to denote $\theta(y_1, y_2, \ldots, y_n)$, i.e.,

$$E[X \mid Y_1 = y_1, Y_2 = y_2, \ldots, Y_n = y_n] = \theta(y_1, y_2, \ldots, y_n)$$
$$= E[X \mid Y_1, Y_2, \ldots, Y_n]\big|_{Y_1 = y_1, Y_2 = y_2, \ldots, Y_n = y_n}.$$

The function θ is characterized as the unique, up to the joint distribution of $Y_1, Y_2, \ldots, Y_n$, Borel measurable function such that

$$E\{X g(Y_1, Y_2, \ldots, Y_n)\} = E\{\theta(Y_1, Y_2, \ldots, Y_n) \, g(Y_1, Y_2, \ldots, Y_n)\} \quad (10.5.1)$$

for all bounded measurable functions g on $\mathbb{R}^n$.

Definition 10.5.2. *The conditional probability of an event A given a σ-field $\mathcal{G}$ is defined by*

$$P(A \mid \mathcal{G}) = E[1_A \mid \mathcal{G}],$$

where 1_A is the characteristic function of A, i.e., $1_A(x) = 1$ if $x \in A$ and $= 0$ if $x \notin A$. In particular, we have

$$P(X \le x \mid Y_1, Y_2, \ldots, Y_n) = E\{1_{\{X \le x\}} \mid Y_1, Y_2 \ldots, Y_n\},$$

$$P(X \le x \mid Y_1 = y_1, \ldots, Y_n = y_n) = E[\{1_{\{X \le x\}} \mid Y_1 = y_1, \ldots, Y_n = y_n\}.$$

Example 10.5.3. Let $f(x, y_1, y_2, \ldots, y_n)$ be the joint density function of $X, Y_1, Y_2, \ldots, Y_n$. Then we have

$$P(X \le x \mid Y_1 = y_1, Y_2 = y_2, \ldots, Y_n = y_n)$$
$$= \frac{1}{f_{Y_1, Y_2, \ldots, Y_n}(y_1, y_2, \ldots, y_n)} \int_{-\infty}^{x} f(u, y_1, y_2, \ldots, y_n) \, du, \quad (10.5.2)$$

where $f_{Y_1, Y_2, \ldots, Y_n}$ is the marginal density function of $Y_1, Y_2, \ldots, Y_n$. This fact can be easily verified using Equation (10.5.1).

Consider a Brownian motion $B(t)$. Let $0 < t_1 < t_2 < \cdots < t_n < t$. Using Equation (10.5.2) and the marginal density function of $B(t)$ in Equation (3.1.2), we immediately see that

$$P\big(B(t) \le x \,|\, B(t_1) = y_1, B(t_2) = y_2, \ldots, B(t_n) = y_n\big)$$

$$= \frac{1}{\sqrt{2\pi(t - t_n)}} \int_{-\infty}^{x} \exp\left[-\frac{(u - y_n)^2}{2(t - t_n)} \right] du.$$

Observe that the right-hand side does not depend on $y_1, y_2, \ldots, y_{n-1}$. In fact, we also have

$$P\big(B(t) \le x \,|\, B(t_n) = y_n\big) = \frac{1}{\sqrt{2\pi(t - t_n)}} \int_{-\infty}^{x} \exp\left[-\frac{(u - y_n)^2}{2(t - t_n)} \right] du.$$

Therefore, a Brownian motion $B(t)$ satisfies the following equality:

$$P\big(B(t) \le x \,|\, B(t_i) = y_i, i = 1, 2, \ldots, n\big) = P\big(B(t) \le x \,|\, B(t_n) = y_n\big).$$

Now, if we use this property of a Brownian motion as a condition, then we come to a new concept and an important class of stochastic processes in the next definition.

Definition 10.5.4. *A stochastic process X_t, $a \le t \le b$, is said to satisfy the Markov property if for any $a \le t_1 < t_2 < \cdots < t_n < t \le b$, the equality*

$$P\big(X_t \le x \,|\, X_{t_1}, X_{t_2}, \ldots, X_{t_n}\big) = P\big(X_t \le x \,|\, X_{t_n}\big)$$

holds for any $x \in \mathbb{R}$, or equivalently, the equality

$$P\big(X_t \le x \,|\, X_{t_i} = y_i, i = 1, 2, \ldots, n\big) = P\big(X_t \le x \,|\, X_{t_n} = y_n\big) \qquad (10.5.3)$$

holds for any $x, y_1, y_2, \ldots, y_n \in \mathbb{R}$. By a Markov process, *we mean a stochastic process X_t, $a \le t \le b$, satisfying the Markov property.*

Remark 10.5.5. We can interpret t_n as the present time, $t_1, t_2, \ldots, t_{n-1}$ as the past, and t as the future. Then the Markov property in Equation (10.5.3) means that given the past and the present, the future depends only on the present. Equivalently, it means that given the present, the future and the past are independent.

Obviously, a Brownian motion $B(t)$ is a Markov process. Below we give more examples to illustrate the Markov property.

Example 10.5.6. Let X_t be a stochastic process with independent increments and $X_0 = 0$. Then for any $t_1 < t_2 < \cdots < t_n < t$ and $x \in \mathbb{R}$,

$$P\big(X_t \le x \,|\, X_{t_1}, X_{t_2}, \ldots, X_{t_n}\big) = P\big((X_t - X_{t_n}) + X_{t_n} \le x \,|\, X_{t_1}, X_{t_2}, \ldots, X_{t_n}\big)$$

$$= P\big((X_t - X_{t_n}) + X_{t_n} \le x \,|\, X_{t_n}\big)$$

$$= P\big(X_t \le x \,|\, X_{t_n}\big).$$

This shows that X_t is a Markov process.

Example 10.5.7. Let ϕ be an increasing function on $\mathbb{R}$. Consider the stochastic process $X_t = \phi(B(t))$. For any $t_1 < t_2 < \cdots < t_n < t$ and $x \in \mathbb{R}$, we have

$$P\big(X_t \le x \,|\, X_{t_i} = y_i, i = 1, 2, \ldots, n\big)$$
$$= P\big(B(t) \le \phi^{-1}(x) \,|\, B(t_i) = \phi^{-1}(y_i), i = 1, 2, \ldots, n\big)$$
$$= P\big(B(t) \le \phi^{-1}(x) \,|\, B(t_n) = \phi^{-1}(y_n)\big)$$
$$= P\big(X_t \le x \,|\, X_{t_n} = y_n\big).$$

Therefore, $X_t = \phi(B(t))$ is a Markov process.

Example 10.5.8. Let $f \in L^2[a, b]$ and $X_t = \int_a^t f(s)\,dB(s), a \le t \le b$. Then for any $t_1 < t_2 < \cdots < t_n < t$, the random variable $\int_{t_n}^t f(s)\,dB(s)$ is independent of X_{t_i} for $i = 1, 2, \ldots, n$. Hence for any $x \in \mathbb{R}$,

$$P\big(X_t \le x \,|\, X_{t_1}, X_{t_2}, \ldots, X_{t_n}\big) = P\big(X_{t_n} + \textstyle\int_{t_n}^t f(s)\,dB(s) \,\big|\, X_{t_1}, X_{t_2}, \ldots, X_{t_n}\big)$$
$$= P\big(X_{t_n} + \textstyle\int_{t_n}^t f(s)\,dB(s) \,\big|\, X_{t_n}\big)$$
$$= P\big(X_t \le x \,|\, X_{t_n}\big).$$

This shows that the stochastic process $X_t = \int_a^t f(s)\,dB(s), a \le t \le b$, is a Markov process for any $f \in L^2[a, b]$.

In the next section we will need the following lemma.

Lemma 10.5.9. *Suppose a stochastic process $X_t, a \le t \le b$, is adapted to a filtration $\{\mathcal{F}_t \,;\, a \le t \le b\}$ and satisfies the condition*

$$P\big(X_t \le x \,|\, \mathcal{F}_s\big) = P\big(X_t \le x \,|\, X_s\big), \quad \forall s < t, \ x \in \mathbb{R}. \tag{10.5.4}$$

Then X_t is a Markov process.

Proof. Let $t_1 < t_2 < \cdots < t_n < t$ and $x \in \mathbb{R}$. Use Property 5 of conditional expectation in Section 2.4 and the assumption in Equation (10.5.4) to get

$$P\big(X_t \le x \,|\, X_{t_1}, X_{t_2}, \ldots, X_{t_n}\big) = E\big[P\big(X_t \le x \,|\, \mathcal{F}_{t_n}\big) | X_{t_1}, X_{t_2}, \ldots, X_{t_n}\big]$$
$$= E\big[P\big(X_t \le x \,|\, X_{t_n}\big) \,|\, X_{t_1}, X_{t_2}, \ldots, X_{t_n}\big]$$
$$= P\big(X_t \le x \,|\, X_{t_n}\big).$$

Hence X_t satisfies the Markov property. Thus X_t is a Markov process.   □

Recall that by Kolmogorov's extension theorem, a stochastic process can be defined as a collection of marginal distributions satisfying the consistency condition. This leads to the following question for Markov processes.

Question 10.5.10. How can we describe the Markov property by the marginal distributions of a Markov process?

Let X_t, $a \le t \le b$, be a Markov process. Let $a = t_1 < t_2$ and $c_1, c_2 \in \mathbb{R}$. Then using the conditional probability, we get

$$P(X_{t_1} \le c_1, X_{t_2} \le c_2) = \int_{-\infty}^{c_1} P(X_{t_2} \le c_2 \,|\, X_{t_1} = x_1) \, P_{X_{t_1}}(dx_1)$$

$$\bullet \qquad = \int_{-\infty}^{c_1} \int_{-\infty}^{c_2} P(X_{t_2} \in dx_2 \,|\, X_{t_1} = x_1) \, \nu(dx_1),$$

where P_X denotes the distribution of a random variable X and $\nu = P_{X_a}$ is the *initial distribution* of the Markov process X_t.

Next, consider $a = t_1 < t_2 < t_3$. In this case, we need to apply the Markov property to get

$$P(X_{t_1} \le c_1, X_{t_2} \le c_2, X_{t_3} \le c_3)$$

$$= \int_{-\infty}^{c_1} \int_{-\infty}^{c_2} P(X_{t_3} \le c_3 \,|\, X_{t_1} = x_1, X_{t_2} = x_2) \, P_{X_{t_1}, X_{t_2}}(dx_1 dx_2)$$

$$= \int_{-\infty}^{c_1} \int_{-\infty}^{c_2} P(X_{t_3} \le c_3 \,|\, X_{t_2} = x_2) \, P_{X_{t_1}, X_{t_2}}(dx_1 dx_2)$$

$$= \int_{-\infty}^{c_1} \int_{-\infty}^{c_2} \int_{-\infty}^{c_3} P(X_{t_3} \in dx_3 \,|\, X_{t_2} = x_2) P(X_{t_2} \in dx_2 \,|\, X_{t_1} = x_1) \, \nu(dx_1),$$

where $P_{X,Y}$ denotes the joint distribution of X and Y.

Now, observe from the above marginal distributions that the conditional probability $P(X_t \in dy \,|\, X_s = x)$ for $s < t$ is a very special quantity. Thus we give it a name in the following definition.

Definition 10.5.11. *The conditional probability $P(X_t \in dy \,|\, X_s = x)$ is called a transition probability of a Markov process X_t.*

Notation 10.5.12. For convenience, we will use $P_{s,x}(t, dy)$ to denote the transition probability $P(X_t \in dy \,|\, X_s = x)$.

If X_t is interpreted as the position of an object at time t, then the quantity $P_{s,x}(t, A)$ represents the probability that this object, being at x at time s, will be in the set A at time t.

Using the transition probabilities, we can find the marginal distributions of X_t. Let $a = t_1 < t_2 < \cdots < t_n$. Use the same arguments as those in the above derivation to get

$$P(X_{t_1} \le c_1, X_{t_2} \le c_2, \ldots, X_{t_n} \le c_n)$$

$$= \int_{-\infty}^{c_1} \int_{-\infty}^{c_2} \cdots \int_{-\infty}^{c_n} P_{t_{n-1}, x_{n-1}}(t_n, dx_n)$$

$$\times P_{t_{n-2}, x_{n-2}}(t_{n-1}, dx_{n-1}) \cdots P_{t_1, x_1}(t_2, dx_2) \, \nu(dx_1). \qquad (10.5.5)$$

Conversely, suppose $\{P_{s,x}(t, \cdot)\,; \, a \le s < t \le b, \, x \in \mathbb{R}\}$ is a collection of probability measures and ν is a probability measure. Does there exist a

Markov process X_t with transition probabilities given by this collection of probability measures and having the initial distribution ν?

First of all, note that this collection of probability measures must satisfy some condition. For any $s < u < t$, we have

$$P(X_s \leq \alpha, X_t \leq \beta) = P(X_s \leq \alpha, -\infty < X_u < \infty, X_t \leq \beta).$$

Therefore, by Equation (10.5.5),

$$\int_{-\infty}^{\alpha} \int_{-\infty}^{\beta} P_{s,x}(t, dy) \, P_{X_s}(dx) = \int_{-\infty}^{\alpha} \int_{-\infty}^{\infty} \int_{-\infty}^{\beta} P_{u,z}(t, dy) \, P_{s,x}(u, dz) \, P_{X_s}(dx),$$

which implies the equality

$$P_{s,x}(t, A) = \int_{-\infty}^{\infty} P_{u,z}(t, A) \, P_{s,x}(u, dz) \tag{10.5.6}$$

for all $s < u < t$, $x \in \mathbb{R}$, and $A \in \mathcal{B}(\mathbb{R})$, the Borel field of $\mathbb{R}$.

Definition 10.5.13. *The equality in Equation* (10.5.6) *is referred to as the* Chapman–Kolmogorov equation.

Now, from the given $\{P_{s,x}(t, \cdot)\}$ and ν, we can use the right-hand side of Equation (10.5.5) to define a family of probability measures for Kolmogorov's extension theorem. Then using the Chapman–Kolmogorov equation, we can easily check that the consistency condition in Theorem 3.3.2 is satisfied. Hence by Kolmogorov's extension theorem, there is a stochastic process X_t such that for any $a \leq s < t \leq b$,

$$P(X_t \in A \,|\, X_s = x) = P_{s,x}(t, A), \quad \forall x \in \mathbb{R}, \, A \in \mathcal{B}(\mathbb{R}).$$

The resulting stochastic process X_t is in fact a Markov process. This can be seen as follows. Use conditional probability and Equation (10.5.5) to get

$$P(X_t \leq x, X_{t_1} \leq c_1, X_{t_2} \leq c_2, \ldots, X_{t_n} \leq c_n)$$
$$= \int_{-\infty}^{c_1} \cdots \int_{-\infty}^{c_n} P(X_t \leq x \,|\, X_{t_1} = x_1, \ldots, X_{t_n} = x_n) \, P_{X_{t_1},\ldots,X_{t_n}}(dx_1 \cdots dx_n)$$
$$= \int_{-\infty}^{c_1} \cdots \int_{-\infty}^{c_n} P(X_t \leq x \,|\, X_{t_1} = x_1, \ldots, X_{t_n} = x_n)$$
$$\times P_{t_{n-1},x_{n-1}}(t_n, dx_n) \cdots P_{t_1,x_1}(t_2, dx_2) \, \nu(dx_1). \tag{10.5.7}$$

On the other hand, from Equation (10.5.5), we have

$$P(X_t \leq x, X_{t_1} \leq c_1, X_{t_2} \leq c_2, \ldots, X_{t_n} \leq c_n)$$
$$= \int_{-\infty}^{c_1} \cdots \int_{-\infty}^{c_n} \int_{-\infty}^{x} P_{t_n,x_n}(t, dy)$$
$$\times P_{t_{n-1},x_{n-1}}(t_n, dx_n) \cdots P_{t_1,x_1}(t_2, dx_2) \, \nu(dx_1). \tag{10.5.8}$$

By comparing Equations (10.5.7) and (10.5.8), we immediately see that

$$P(X_t \leq x \mid X_{t_1} = x_1, \ldots, X_{t_n} = x_n) = \int_{-\infty}^{x} P_{t_n, x_n}(t, dy).$$

Obviously, this implies the Markov property of X_t, namely,

$$P(X_t \leq x \mid X_{t_1} = x_1, \ldots, X_{t_n} = x_n) = P(X_t \leq x \mid X_{t_n} = x_n).$$

Hence X_t is a Markov process. Thus we have proved the next theorem.

Theorem 10.5.14. *If X_t, $a \leq t \leq b$, is a Markov process, then its transition probabilities*

$$P_{s,x}(t, \cdot) = P(X_t \in (\cdot) \mid X_s = x), \quad a \leq s < t \leq b, \ x \in \mathbb{R}, \qquad (10.5.9)$$

satisfy the Chapman–Kolmogorov equation. Conversely, if ν is a probability measure on $\mathbb{R}$ and $\{P_{s,x}(t, \cdot); a \leq s < t \leq b, x \in \mathbb{R}\}$ is a collection of probability measures on $\mathbb{R}$ satisfying the Chapman–Kolmogorov equation, then there exists a Markov process X_t with initial distribution ν such that Equation (10.5.9) holds.

From this theorem, we see that a Markov process is determined by its initial distribution and transition probabilities satisfying the Chapman–Kolmogorov equation. We give one example to illustrate this point.

Example 10.5.15. Let ν be the probability measure on $\mathbb{R}$ given by

$$\nu\{1\} = \nu\{-1\} = \frac{1}{2}.$$

Define a collection $\{P_{s,x}(t, \cdot); s < t, x \in \mathbb{R}\}$ of probability measures as follows. For $x = \pm 1$, and $s < t$, define

$$P_{s,1}(t, \{1\}) = \frac{1}{2}(1 + e^{-\lambda(t-s)}),$$

$$P_{s,1}(t, \{-1\}) = \frac{1}{2}(1 - e^{-\lambda(t-s)}),$$

$$P_{s,-1}(t, \{1\}) = \frac{1}{2}(1 - e^{-\lambda(t-s)}),$$

$$P_{s,-1}(t, \{-1\}) = \frac{1}{2}(1 + e^{-\lambda(t-s)}),$$

where $\lambda > 0$. For $x \neq \pm 1$, define $P_{s,x}(t, \cdot) = \delta_x$, the Dirac delta measure at x. In fact, we can replace δ_x by any probability measure. It turns out that this collection $\{P_{s,x}(t, \cdot)\}$ satisfies the Chapman–Kolmogorov equation. Hence there is a Markov process X_t with initial distribution ν and transition probabilities $\{P_{s,x}(t, \cdot)\}$. This Markov process X_t is called a *random telegraph process* with parameter λ. It can be easily checked that for any t,

$$P(X_t = 1) = P(X_t = -1) = \frac{1}{2}.$$

A Markov process X_t with $t \geq 0$ is said to be *stationary* if its transition probabilities $P_{s,x}(t, \cdot)$ depend only on x and the difference $t-s$, or equivalently, $P_{h,x}(t + h, \cdot)$ does not depend on h. In that case, we will use the notation

$$P_t(x, A) = P_{h,x}(t + h, A), \quad t > 0, \ x \in \mathbb{R}, \ A \in \mathcal{B}(\mathbb{R}).$$

Then the Chapman–Kolmogorov equation takes the form

$$P_{s+t}(x, A) = \int_{-\infty}^{\infty} P_t(z, A) \, P_s(x, dz) \tag{10.5.10}$$

for all $s, t > 0$, $x \in \mathbb{R}$, and $A \in \mathcal{B}(\mathbb{R})$.

For example, a Brownian motion $B(t)$ is a stationary Markov process. The Ornstein–Uhlenbeck process in Example 7.4.5 is also a stationary Markov process since the conditional probability in Equation (7.4.7) depends only on $t - s$ as a function of s and t and the resulting $P_t(x, \cdot)$ satisfies Equation (10.5.10). The above random telegraph process is also stationary. On the other hand, it is easy to check that the stochastic process $X_t = \int_0^t \frac{1}{1+s^2} \, dB(s)$ is a nonstationary Markov process.

10.6 Solutions of Stochastic Differential Equations

In this section we will study some properties of the solution of the stochastic differential (or rather integral) equation in Equation (10.3.1), namely,

$$X_t = \xi + \int_a^t \sigma(s, X_s) \, dB(s) + \int_a^t f(s, X_s) \, ds, \quad a \leq t \leq b,$$

where σ and f satisfy the Lipschitz and linear growth conditions. We first prove that the solution X_t is a Markov process.

Let $\{\mathcal{F}_t; a \leq t \leq b\}$ be the filtration given by the Brownian motion $B(t)$, namely, $\mathcal{F}_t$ is defined by $\mathcal{F}_t = \sigma\{B(s); s \leq t\}$. Obviously, the solution X_t is adapted to this filtration.

Let $s \in [a, b]$ and $x \in \mathbb{R}$ be fixed and consider the following SIE:

$$X_t = x + \int_s^t \sigma(u, X_u) \, dB(u) + \int_s^t f(u, X_u) \, du, \quad s \leq t \leq b. \tag{10.6.1}$$

To avoid confusion with the solution X_t, we use $X_t^{s,x}$ to denote the solution of the SIE in Equation (10.6.1). Since the initial condition x of Equation (10.6.1) is a constant, we see from the approximation procedure in the proof of Theorem 10.3.5 that the solution $X_t^{s,x}$ is independent of the σ-field $\mathcal{F}_s$ for each $t \in [s, b]$. If follows that for any $\mathcal{F}_s$-measurable random variable Z, we have the equality

$$P\big(X_t^{s,Z} \leq y \,\big|\, \mathcal{F}_s\big) = P\big(X_t^{s,x} \leq y\big)\big|_{x=Z}, \quad \forall y \in \mathbb{R}.$$

In particular, put $Z = X_s$ in this equation to get

$$P\big(X_t^{s,X_s} \le y \,\big|\, \mathcal{F}_s\big) = P\big(X_t^{s,x} \le y\big)\big|_{x=X_s}, \quad \forall\, y \in \mathbb{R}. \tag{10.6.2}$$

Now, since X_t is a solution of the SIE in Equation (10.3.1), we have

$$X_t = X_s + \int_s^t \sigma(u, X_u)\, dB(u) + \int_s^t f(u, X_u)\, du, \quad s \le t \le b.$$

But X_t^{s,X_s} is also a solution of this SIE. Hence by the uniqueness of a solution, we must have $X_t = X_t^{s,X_s}$. Thus Equation (10.6.2) can be rewritten as

$$P\big(X_t \le y \,\big|\, \mathcal{F}_s\big) = P\big(X_t^{s,x} \le y\big)\big|_{x=X_s}, \quad \forall\, y \in \mathbb{R},$$

which is measurable with respect to the σ-field generated by X_s. Therefore, we can conclude that

$$P\big(X_t \le y \,\big|\, X_s\big) = E\big[P\big(X_t \le y \,\big|\, \mathcal{F}_s\big)\,\big|\, X_s\big]$$
$$= P\big(X_t \le y \,\big|\, \mathcal{F}_s\big), \quad \forall\, s < t,\ y \in \mathbb{R}. \tag{10.6.3}$$

Then by Lemma 10.5.9, the stochastic process X_t is a Markov process. Hence we have proved the next theorem.

Theorem 10.6.1. *Let $\sigma(t, x)$ and $f(t, x)$ be measurable functions on $[a, b] \times \mathbb{R}$ satisfying the Lipschitz and linear growth conditions in x. Suppose ξ is an $\mathcal{F}_a$-measurable random variable with $E(\xi^2) < \infty$. Then the unique continuous solution of the stochastic integral equation*

$$X_t = \xi + \int_a^t \sigma(s, X_s)\, dB(s) + \int_a^t f(s, X_s)\, ds, \quad a \le t \le b,$$

is a Markov process.

Observe that the solution X_t satisfies Equation (10.6.3), which is stronger than the Markov property of X_t in view of Lemma 10.5.9. As a matter of fact, the solution X_t satisfies even a stronger property than Equation (10.6.3), which uses the concept of stopping time introduced in Section 5.4. Let τ be a stopping time with respect to the filtration $\{\mathcal{F}_t; a \le t \le b\}$ given by the Brownian motion $B(t)$. The σ-field $\mathcal{F}_\tau$ associated with τ is defined by

$$\mathcal{F}_\tau = \big\{A\,;\ A \cap \{\tau \le t\} \in \mathcal{F}_t, \forall\, t \in [a, b]\big\}.$$

It can be proved that for any stopping time τ with respect to the filtration $\{\mathcal{F}_t\}$, the stochastic process X_t satisfies the following equality:

$$P\big(X_{\tau+v} \le y \,\big|\, \mathcal{F}_\tau\big) = P\big(X_{\tau+v} \le y \,\big|\, X_\tau\big), \quad \forall\, v \ge 0,\ y \in \mathbb{R}, \tag{10.6.4}$$

where it is understood that $X_{\tau(\omega)+v}(\omega) = X_b(\omega)$ if $\tau(\omega) + v \geq b$. Equation (10.6.4) is called the *strong Markov property* of X_t. Notice that when $\tau = s$ and $v = t - s$, Equation (10.6.4) reduces to Equation (10.6.3). Thus the strong Markov property implies Equation (10.6.3), which in turn implies the Markov property by Lemma 10.5.9.

Next consider a stochastic integral equation with the functions σ and f depending only on the variable x. In this case, observe that the Lipschitz condition $|g(x) - g(y)| \leq K|x - y|$ implies that

$$|g(x)| \leq |g(x) - g(0)| + |g(0)| \leq K|x| + |g(0)| \leq C(1 + |x|),$$

where $C = \max\{K, |g(0)|\}$. Hence the linear growth condition automatically follows from the Lipschitz condition.

Theorem 10.6.2. *Assume that $\sigma(x)$ and $f(x)$ are functions satisfying the Lipschitz condition and let $x_0 \in \mathbb{R}$. Then the unique continuous solution of the stochastic integral equation*

$$X_t = x_0 + \int_0^t \sigma(X_s)\, dB(s) + \int_0^t f(X_s)\, ds, \quad t \geq 0, \tag{10.6.5}$$

is a stationary Markov process.

Proof. Let $s \geq 0$ and $x \in \mathbb{R}$ be fixed and consider the following SIE:

$$X_t = x + \int_s^t \sigma(X_u)\, dB(u) + \int_s^t f(X_u)\, du, \quad t \geq s.$$

Let $X_t^{s,x}$ denote the unique continuous solution of this equation. Then

$$X_t^{s,x} = x + \int_s^t \sigma(X_u^{s,x})\, dB(u) + \int_s^t f(X_u^{s,x})\, du.$$

Make a change of variables $u = s + v$ to get

$$X_t^{s,x} = x + \int_0^{t-s} \sigma(X_{s+v}^{s,x})\, dB(s+v) + \int_0^{t-s} f(X_{s+v}^{s,x})\, dv. \tag{10.6.6}$$

Let $\widetilde{B}_s(t) = B(s+t) - B(s)$. By Proposition 2.2.3, the stochastic process $\widetilde{B}_s(t)$ is a Brownian motion for each fixed s. Moreover, the increments of $\widetilde{B}_s(t)$ and $B(t)$ are related by $\widetilde{B}_s(t_2) - \widetilde{B}_s(t_1) = B(s+t_2) - B(s+t_1)$. Thus Equation (10.6.6) can be rewritten as

$$X_t^{s,x} = x + \int_0^{t-s} \sigma(X_{s+v}^{s,x})\, d\widetilde{B}_s(v) + \int_0^{t-s} f(X_{s+v}^{s,x})\, dv.$$

Make a change of variables $t = s + w$ to get

$$X^{s,x}_{s+w} = x + \int_0^w \sigma(X^{s,x}_{s+v})\, d\widetilde{B}_s(v) + \int_0^w f(X^{s,x}_{s+v})\, dv.$$

Then replace the variable w by t:

$$X^{s,x}_{s+t} = x + \int_0^t \sigma(X^{s,x}_{s+v})\, d\widetilde{B}_s(v) + \int_0^t f(X^{s,x}_{s+v})\, dv, \quad t \geq 0. \tag{10.6.7}$$

Now, Equation (10.6.7) means that for any fixed $s > 0$, the stochastic process $X^{s,x}_{s+t}$, $t \geq 0$, is a solution of the following SIE:

$$Y_t = x + \int_0^t \sigma(Y_u)\, d\widetilde{B}_s(u) + \int_0^t f(Y_u)\, du, \quad t \geq 0.$$

From the approximation procedure in the proof of Theorem 10.3.5, we see that the distribution of its solution $X^{s,x}_{s+t}$ is independent of s. Therefore,

$$P(X^{s,x}_{s+t} \in A) = P(X^{0,x}_t \in A), \quad \forall A \in \mathcal{B}(\mathbb{R}). \tag{10.6.8}$$

Let $P_{s,x}(t, \cdot)$ denote the transition probability of the solution X_t of Equation (10.6.5). Then Equation (10.6.8) can be rewritten as

$$P_{s,x}(s+t, A) = P_{0,x}(t, A), \quad \forall s \geq 0, A \in \mathcal{B}(\mathbb{R}).$$

This means that for any $t \geq 0$ and $x \in \mathbb{R}$, the probability measure $P_{s,x}(s+t, \cdot)$ is independent of $s \geq 0$. Thus the stochastic process X_t is stationary. $\square$

Example 10.6.3. Consider the simple stochastic differential equation

$$dX_t = X_t\, dB(t) + dt, \quad X_0 = x. \tag{10.6.9}$$

By Theorem 10.6.2, the solution of this SDE is a stationary Markov process. On the other hand, we can solve Equation 10.6.9 explicitly as follows. Apply Itô's formula to check that

$$d\big(e^{-B(t)+\frac{1}{2}t} X_t\big) = e^{-B(t)+\frac{1}{2}t}\big(dX_t - X_t\, dB(t)\big) = e^{-B(t)+\frac{1}{2}t}\, dt.$$

Integrate both sides from s to t to get

$$X_t = e^{(B(t)-B(s))-\frac{1}{2}(t-s)} X_s + \int_s^t e^{(B(t)-B(u))-\frac{1}{2}(t-u)}\, du. \tag{10.6.10}$$

In particular, the solution of Equation (10.6.9) is given by

$$X_t = x\, e^{B(t)-\frac{1}{2}t} + \int_0^t e^{(B(t)-B(u))-\frac{1}{2}(t-u)}\, du.$$

Moreover, by Equation (10.6.10),

$$X_{s+t}^{s,x} = x\,e^{B(s+t)-B(s)-\frac{1}{2}t} + \int_s^{s+t} e^{(B(s+t)-B(u))-\frac{1}{2}(s+t-u)}\,du,$$

which, after making a change of variables $u = s + v$, becomes

$$X_{s+t}^{s,x} = x\,e^{B(s+t)-B(s)-\frac{1}{2}t} + \int_0^t e^{(B(s+t)-B(s+v))-\frac{1}{2}(t-v)}\,dv.$$

Observe that $B(s+t) - B(s)$ is a Brownian motion for any fixed $s \geq 0$. Hence the distribution of $X_{s+t}^{s,x}$ is independent of s. This shows that the stochastic process X_t is a stationary Markov process.

Next, we consider the dependence of a solution of a stochastic differential equation on its initial condition. It is given by the next theorem.

Theorem 10.6.4. *Let σ and f be functions satisfying the Lipschitz condition in x, namely, there exists a constant $K > 0$ such that*

$$|\sigma(t,x) - \sigma(t,y)| \leq K|x - y|, \quad |f(t,x) - f(t,y)| \leq K|x - y|,$$

for all $t \in [a,b]$ and $x,y \in \mathbb{R}$. Let $\xi, \eta \in L^2(\Omega)$ be $\mathcal{F}_a$-measurable. Assume that X_t^ξ and X_t^η are two solutions of the stochastic differential equation

$$dX_t = \sigma(t, X_t)\,dB(t) + f(t, X_t)\,dt$$

with initial conditions ξ and η, respectively. Then

$$E\big(|X_t^\xi - X_t^\eta|^2\big) \leq 3E\big(|\xi - \eta|^2\big)\,e^{3K^2(1+b-a)(t-a)}, \quad \forall\, a \leq t \leq b.$$

Remark 10.6.5. There is another part of the Doob submartingale inequality in Lemma 4.5.1, which states that if Y_t, $a \leq t \leq b$, is a right continuous submartingale, then

$$E\left[\left(\sup_{a \leq t \leq b} |Y_t|\right)^p\right] \leq q^p E\big(|Y_b|^p\big),$$

where $p > 1$ and $p^{-1} + q^{-1} = 1$. Using this inequality with $p = 2$ and similar arguments to those in the proof below, we can obtain the following stronger estimate for any $a \leq t \leq b$:

$$E\left[\left(\sup_{a \leq s \leq t} |X_s^\xi - X_s^\eta|\right)^2\right] \leq 3E\big(|\xi - \eta|^2\big)\,e^{3K^2(4+b-a)(t-a)}.$$

Proof. By assumption we have

$$X_t^\xi = \xi + \int_a^t \sigma(s, X_s^\xi)\,dB(s) + \int_a^t f(s, X_s^\xi)\,ds,$$

$$X_t^\eta = \eta + \int_a^t \sigma(s, X_s^\eta)\,dB(s) + \int_a^t f(s, X_s^\eta)\,ds.$$

Therefore,

$$|X_t^\xi - X_t^\eta| \leq |\xi - \eta| + \left| \int_a^t \left(\sigma(s, X_s^\xi) - \sigma(s, X_s^\eta) \right) dB(s) \right|$$

$$+ \left| \int_a^t \left(f(s, X_s^\xi) - f(s, X_s^\eta) \right) ds \right|.$$

Use the inequality $(|a| + |b| + |c|)^2 \leq 3(a^2 + b^2 + c^2)$ to find that

$$|X_t^\xi - X_t^\eta|^2 \leq 3 \left\{ |\xi - \eta|^2 + \left| \int_a^t \left(\sigma(s, X_s^\xi) - \sigma(s, X_s^\eta) \right) dB(s) \right|^2 \right.$$

$$+ \left| \int_a^t \left(f(s, X_s^\xi) - f(s, X_s^\eta) \right) ds \right|^2 \right\}$$

$$\leq 3 \left\{ |\xi - \eta|^2 + \left| \int_a^t \left(\sigma(s, X_s^\xi) - \sigma(s, X_s^\eta) \right) dB(s) \right|^2 \right.$$

$$+ (b - a) \int_a^t \left| f(s, X_s^\xi) - f(s, X_s^\eta) \right|^2 ds \right\}.$$

Then take the expectation and use the Lipschitz condition to get

$$E\left(|X_t^\xi - X_t^\eta|^2\right) \leq 3 \left\{ E\left(|\xi - \eta|^2\right) + E \int_a^t \left| \sigma(s, X_s^\xi) - \sigma(s, X_s^\eta) \right|^2 ds \right.$$

$$+ (b - a) E \int_a^t \left| f(s, X_s^\xi) - f(s, X_s^\eta) \right|^2 ds \right\}$$

$$\leq 3E\left(|\xi - \eta|^2\right) + 3K^2(1 + b - a) \int_a^t E\left(|X_s^\xi - X_s^\eta|^2\right) ds.$$

This inequality and Lemma 10.2.2 yield the assertion of the theorem. □

Finally, we mention that for a system of stochastic differential equations in Section 10.4 we have similar results to those in Theorems 10.6.1, 10.6.2, and 10.6.4. It requires only straightforward modifications to write down the statements of the corresponding theorems.

10.7 Some Estimates for the Solutions

In this section we will prove some estimates for Itô integrals and the solutions of stochastic differential equations. The last estimate will be needed in the next section.

Lemma 10.7.1. *Suppose $h(t)$ is an adapted stochastic process satisfying the condition that $\int_a^b E\left(|h(s)|^4\right) ds < \infty$ and let $Y_t = \int_a^t h(s) \, dB(s)$. Then*

$$E\left(|Y_t|^4\right) \leq 2(17 + 4\sqrt{17})(t - a) \int_a^t E\left(|h(s)|^4\right) ds, \quad a \leq t \leq b.$$

Proof. Apply Itô's formula to get

$$Y_t^2 = 2 \int_a^t Y_s h(s) \, dB(s) + \int_a^t h(s)^2 \, ds.$$

The assumption on $h(t)$ implies that $Y_t h(t)$ belongs to $L_{ad}^2([a, b] \times \Omega)$. By the inequality $(a + b)^2 \le 2a^2 + 2b^2$,

$$Y_t^4 \le 8 \left(\int_a^t Y_s h(s) \, dB(s) \right)^2 + 2 \left(\int_a^t h(s)^2 \, ds \right)^2.$$

Then use Theorem 4.3.5 to show that

$$E(Y_t^4) \le 8 \int_a^t E(Y_s^2 h(s)^2) \, ds + 2E \left(\int_a^t h(s)^2 \, ds \right)^2$$

$$\le 8 \left[\int_a^t E(Y_s^4) \, ds \right]^{\frac{1}{2}} \left[\int_a^t E(h(s)^4) \, ds \right]^{\frac{1}{2}}$$

$$+ 2(t - a) \int_a^t E(h(s)^4) \, ds. \tag{10.7.1}$$

Integrate both sides and use the Schwarz inequality to get

$$\int_a^t E(Y_s^4) \, ds \le 8 \left[\int_a^t \int_a^s E(Y_u^4) \, du \, ds \right]^{\frac{1}{2}} \left[\int_a^t \int_a^s E(h(u)^4) \, du \, ds \right]^{\frac{1}{2}}$$

$$+ 2 \int_a^t (s - a) \int_a^s E(h(u)^4) \, du \, ds. \tag{10.7.2}$$

Change the order of integration to show that

$$\int_a^t \int_a^s E(Y_u^4) \, du \, ds = \int_a^t (t - u) E(Y_u^4) \, du$$

$$\le (t - a) \int_a^t E(Y_u^4) \, du. \tag{10.7.3}$$

Similarly, we have

$$\int_a^t \int_a^s E(h(u)^4) \, du \, ds \le (t - a) \int_a^t E(h(u)^4) \, du \tag{10.7.4}$$

$$\int_a^t (s - a) \int_a^s E(h(u)^4) \, du \, ds \le \frac{1}{2}(t - a)^2 \int_a^t E(h(u)^4) \, du. \tag{10.7.5}$$

Now put Equations (10.7.3), (10.7.4), and (10.7.5) into Equation (10.7.2) and change the variable u to s to obtain the inequality

$$\int_a^t E(Y_s^4) \, ds \le 8(t - a) \left[\int_a^t E(Y_s^4) \, ds \right]^{\frac{1}{2}} \left[\int_a^t E(h(s)^4) \, ds \right]^{\frac{1}{2}}$$

$$+ (t - a)^2 \int_a^t E(h(s)^4) \, ds. \tag{10.7.6}$$

Let $x = \int_a^t E(Y_s^4)\, ds$ and $y = \int_a^t E(h(s)^4)\, ds$. We may assume that $y \neq 0$. Otherwise, we have nothing to prove for the lemma. Let $w = \sqrt{x}/\sqrt{y}$. Then Equation (10.7.6) becomes $w^2 \leq 8(t-a)w + (t-a)^2$, which can be easily solved to yield $w \leq (4 + \sqrt{17})(t-a)$. It follows that

$$\int_a^t E(Y_s^4),\, ds \leq (4 + \sqrt{17})^2 (t-a)^2 \int_a^t E(h(s)^4)\, ds. \tag{10.7.7}$$

Then put Equation (10.7.7) into Equation (10.7.1) to get the lemma.   □

Theorem 10.7.2. *Let $\sigma(t,x)$ and $f(t,x)$ satisfy the Lipschitz condition in x and the following growth condition:*

$$|\sigma(t,x)|^2 \leq C(1 + x^2), \quad |f(t,x)|^2 \leq C(1 + x^2). \tag{10.7.8}$$

Suppose ξ is an $\mathcal{F}_a$-measurable random variable with $E(\xi^4) < \infty$. Then the solution X_t of the stochastic integral equation

$$X_t = \xi + \int_a^t \sigma(s, X_s)\, dB(s) + \int_a^t f(s, X_s)\, ds, \quad a \leq t \leq b,$$

satisfies the inequalities

$$E(|X_t|^4) \leq \{27E(\xi^4) + C_1(b-a)\} e^{C_1(t-a)}, \tag{10.7.9}$$
$$E(|X_t - \xi|^4) \leq C_2\{1 + 27E(\xi^4) + C_1(b-a)\}(t-a)^2 e^{C_1(t-a)}, \tag{10.7.10}$$

where the constants C_1 and C_2 are given by

$$C_1 = 54\{2(17 + 4\sqrt{17}) + (b-a)^2\}(b-a)C^2,$$
$$C_2 = 16\{2(17 + 4\sqrt{17}) + (b-a)^2\}C^2.$$

Proof. Use the inequality $(a + b + c)^4 \leq 27(a^4 + b^4 + c^4)$, Lemma 10.7.1, and the Hölder inequality to get

$$E(|X_t|^4) \leq 27E(\xi^4) + 27\widetilde{C}(t-a)\int_a^t E(\sigma(s, X_s)^4)\, ds$$

$$+ 27(t-a)^3 \int_a^t E(f(s, X_s)^4)\, ds$$

$$\leq 27E(\xi^4) + 27\widetilde{C}(b-a)\int_a^t E(\sigma(s, X_s)^4)\, ds$$

$$+ 27(b-a)^3 \int_a^t E(f(s, X_s)^4)\, ds, \tag{10.7.11}$$

where $\widetilde{C} = 2(17 + 4\sqrt{17})$ is the constant from Lemma 10.7.1. Then use the growth condition in Equation (10.7.8) and the inequality $(1 + x^2)^2 \leq 2(1 + x^4)$ to show that

$$E\big(|X_t|^4\big) \leq 27E(\xi^4) + C_1 \int_a^t \big[1 + E\big(|X_s|^4\big)\big]\, ds,$$

where $C_1 = 54\{\widetilde{C} + (b-a)^2\}(b-a)C^2$. Therefore, we have

$$E\big(|X_t|^4\big) \leq 27E(\xi^4) + C_1(b-a) + C_1 \int_a^t E\big(|X_s|^4\big)\, ds,$$

which implies Equation (10.7.9) by the Bellman–Gronwall inequality.

Next, use the inequality $(a+b)^4 \leq 8(a^4 + b^4)$, Lemma 10.7.1, and the Hölder inequality to show that

$$E\big(|X_t - \xi|^4\big)$$

$$\leq 8\widetilde{C}(t-a) \int_a^t E\big(\sigma(s, X_s)^4\big)\, ds + 8(t-a)^3 \int_a^t E\big(f(s, X_s)^4\big)\, ds$$

$$\leq 8(t-a)\bigg(\widetilde{C} \int_a^t E\big(\sigma(s, X_s)^4\big)\, ds + (b-a)^2 \int_a^t E\big(f(s, X_s)^4\big)\, ds\bigg).$$

Then by Equation (10.7.8) and the inequality $(1+x^2)^2 \leq 2(1+x^4)$,

$$E\big(|X_t - \xi|^4\big) \leq 16\{\widetilde{C} + (b-a)^2\}C^2(t-a) \int_a^t \big(1 + E\big(|X_s|^4\big)\, ds. \quad (10.7.12)$$

Moreover, by Equation (10.7.9) and the inequality $e^x - 1 \leq xe^x$ for $x \geq 0$,

$$\int_a^t E\big(|X_s|^4\big)\, ds \leq \{27E(\xi^4) + C_1(b-a)\}(t-a)\, e^{C_1(t-a)}. \quad (10.7.13)$$

Finally, put Equation (10.7.13) into Equation (10.7.12) to get the inequality in Equation (10.7.10) and the theorem is proved. $\qquad \square$

10.8 Diffusion Processes

Let $x = (x_1, x_2, \ldots, x_n)$ and $y = (y_1, y_2, \ldots, y_n)$ be two vectors in $\mathbb{R}^n$. The notation $y \leq x$ means that $y_i \leq x_i$ for all $1 \leq i \leq n$. By adopting this notation, we can use the same formulation as in Definition 10.5.4 to define the *Markov property* of an $\mathbb{R}^n$-valued stochastic process X_t, $a \leq t \leq b$. Moreover, we can define the *transition probability* $P_{s,x}(t, \cdot)$ of an $\mathbb{R}^n$-valued Markov process in the same way as in Definition 10.5.11, namely,

$$P_{s,x}(t, A) = P\big(X_t \in A \mid X_s = x\big), \quad A \in \mathcal{B}(\mathbb{R}^n),$$

where $a \leq s < t \leq b$ and $x \in \mathbb{R}^n$. The *Chapman–Kolmogorov equation* is the following equality:

$$P_{s,x}(t, A) = \int_{\mathbb{R}^n} P_{u,z}(t, A)\, P_{s,x}(u, dz) \quad (10.8.1)$$

for all $s < u < t$, $x \in \mathbb{R}^n$, and $A \in \mathcal{B}(\mathbb{R}^n)$, the Borel field of $\mathbb{R}^n$.

Example 10.8.1. Let $B_i(t)$, $i = 1, 2, \ldots, n$, be independent Brownian motions. The stochastic process $B(t) = (B_1(t), B_2(t), \ldots, B_n(t))$ is called an $\mathbb{R}^n$-valued Brownian motion. It is a Markov process with transition probabilities

$$P_{s,x}(t, A) = \left(2\pi(t - s)\right)^{-n/2} \int_A e^{-|y-x|^2/2(t-s)} \, dy,$$

where $s < t$, $x \in \mathbb{R}^n$, and $A \in \mathcal{B}(\mathbb{R}^n)$. The norm of $x = (x_1, x_2, \ldots, x_n)$ is given by $|x| = (x_1^2 + x_2^2 + \cdots + x_n^2)^{1/2}$.

Example 10.8.2. Consider the n-dimensional analogue of Equation 7.4.7 with $\alpha = \beta = 1$, namely, let

$$P_{s,x}(t, A) = \left\{\pi(1 - e^{-2(t-s)})\right\}^{-n/2} \int_A \exp\left[-\frac{|y - e^{-(t-s)}x|^2}{1 - e^{-2(t-s)}}\right] dy,$$

In view of the fact that a one-dimensional Ornstein–Uhlenbeck process is a Markov process, we see that the above collection $\{P_{s,x}(t, \cdot)\}$ satisfies the Chapman–Kolmogorov equation in Equation (10.8.1). The resulting Markov process X_t is called an $\mathbb{R}^n$-valued Ornstein–Uhlenbeck process.

A special class of $\mathbb{R}^n$-valued Markov processes, called diffusion processes, is specified in the next definition.

Definition 10.8.3. *An $\mathbb{R}^n$-valued Markov process X_t, $a \leq t \leq b$, is called a diffusion process if its transition probabilities $\{P_{s,x}(t, \cdot)\}$ satisfy the following three conditions for any $t \in [a, b]$, $x \in \mathbb{R}^n$, and $c > 0$:*

(1) $\displaystyle \lim_{\varepsilon \downarrow 0} \frac{1}{\varepsilon} \int_{|y-x| \geq c} P_{t,x}(t + \varepsilon, dy) = 0.$

(2) $\displaystyle \lim_{\varepsilon \downarrow 0} \frac{1}{\varepsilon} \int_{|y-x| < c} (y_i - x_i) P_{t,x}(t + \varepsilon, dy) = \rho_i(t, x)$ *exists.*

(3) $\displaystyle \lim_{\varepsilon \downarrow 0} \frac{1}{\varepsilon} \int_{|y-x| < c} (y_i - x_i)(y_j - x_j) P_{t,x}(t + \varepsilon, dy) = Q_{ij}(t, x)$ *exists.*

Remark 10.8.4. Condition (1) implies that the stochastic process X_t cannot have instantaneous jumps. Moreover, observe that for any $c_2 > c_1 > 0$,

$$\frac{1}{\varepsilon} \int_{c_1 \leq |y-x| < c_2} |y_i - x_i| \, P_{t,x}(t + \varepsilon, dy) \leq c_2 \frac{1}{\varepsilon} \int_{|y-x| \geq c_1} P_{t,x}(t + \varepsilon, dy),$$

which tends to 0 as $\varepsilon \to 0$ by condition (1). It follows that the limit in condition (2) is independent of the constant c. Similarly, the limit in condition (3) is also independent of the constant c. Hence $\rho_i(t, x)$ and $Q_{ij}(t, x)$ are given by

$$\rho_i(t, x) = \lim_{\varepsilon \downarrow 0} \frac{1}{\varepsilon} \int_{\mathbb{R}^n} (y_i - x_i) P_{t,x}(t + \varepsilon, dy),$$

$$Q_{ij}(t, x) = \lim_{\varepsilon \downarrow 0} \frac{1}{\varepsilon} \int_{\mathbb{R}^n} (y_i - x_i)(y_j - x_j) P_{t,x}(t + \varepsilon, dy).$$

Definition 10.8.5. *The vector* $\rho(t,x) = (\rho_1(t,x), \rho_2(t,x), \ldots, \rho_n(t,x))$ *and the matrix* $Q(t,x) = [Q_{ij}(t,x)]_{ij}$ *are called the* drift *and* diffusion coefficient, *respectively, of the diffusion process* X_t.

A point $x \in \mathbb{R}^n$ will be denoted by either one of the notations

$$x = (x_1, \ldots, x_n), \quad x = \begin{bmatrix} x_1 \\ \vdots \\ x_n \end{bmatrix}.$$

The transpose of x is a row vector $x^T = [x_1, x_2, \ldots, x_n]$. Then the drift $\rho(t,x)$ and diffusion coefficient $Q(t,x)$ are given by

$$\rho(t,x) = \lim_{\varepsilon \downarrow 0} \frac{1}{\varepsilon} E\{X_{t+\varepsilon} - x \mid X_t = x\}, \tag{10.8.2}$$

$$Q(t,x) = \lim_{\varepsilon \downarrow 0} \frac{1}{\varepsilon} E\{(X_{t+\varepsilon} - x)(X_{t+\varepsilon} - x)^T \mid X_t = x\}. \tag{10.8.3}$$

Example 10.8.6. An $\mathbb{R}^n$-valued Brownian motion $B(t)$ is a diffusion process. To check condition (1), note that for any $c > 0$,

$$\frac{1}{\varepsilon} P\big(|B(t+\varepsilon) - x| \geq c \mid B(t) = x\big) = \frac{1}{\varepsilon} P\big(|B(t+\varepsilon) - B(t)| \geq c\big)$$

$$\leq \frac{1}{\varepsilon c^4} E\big(|B(\varepsilon)|^4\big)$$

$$= \frac{1}{c^4} n(n+2)\varepsilon \to 0, \quad \text{as } \varepsilon \to 0.$$

Here we have used the fact that $E\big(|B(\varepsilon)|^4\big) = n(n+2)\varepsilon^2$, which can be checked by direct computation. Hence condition (1) is satisfied. Then we use Equations (10.8.2) and (10.8.3) to find the diffusion coefficient and drift,

$$Q(t,x) = I, \quad \rho(t,x) = 0,$$

where I is the identity matrix of size $n \times n$ and 0 is the zero vector in $\mathbb{R}^n$.

Example 10.8.7. The Ornstein–Uhlenbeck process X_t in Example 10.8.2 is a diffusion process with the diffusion coefficient and drift

$$Q(t,x) = I, \quad \rho(t,x) = -x,$$

which can be verified using the transition probabilities $P_{s,x}(t,\cdot)$ of X_t given in Example 10.8.2.

Example 10.8.8. Consider the random telegraph process X_t with parameter $\lambda > 0$ defined in Example 10.5.15. Take $x = 1$ and $c = 1$ to check whether condition (1) in Definition 10.8.3 is satisfied:

$$\frac{1}{\varepsilon}\int_{|y-1|\geq 1} P_{t,1}(t+\varepsilon, dy) = \frac{1}{\varepsilon}P_{t,1}(t+\varepsilon, \{-1\}) = \frac{1}{2\varepsilon}(1 - e^{-\lambda\varepsilon}),$$

which converges to $\frac{1}{2}\lambda \neq 0$, as $\varepsilon \to 0$. Hence condition (1) is not satisfied. This shows that the random telegraph process is not a diffusion process.

A special class of diffusion processes is given by the solutions of stochastic differential equations. In order to avoid complicated notation, we first consider the one-dimensional stochastic integral equation

$$X_t = \xi + \int_a^t \sigma(s, X_s)\, dB(t) + \int_a^t f(s, X_s)\, ds, \quad a \leq t \leq b, \qquad (10.8.4)$$

where $\sigma(t, x)$ and $f(t, x)$ satisfy the Lipschitz and linear growth conditions. By Theorem 10.6.1, the solution X_t of this equation is a Markov process.

Let $\{P_{s,x}(t, \cdot)\}$ be the transition probabilities of the solution X_t. Then for any $t \in [a, b]$, $x \in \mathbb{R}^n$, and $c > 0$, we have

$$\int_{|y-x|\geq c} P_{t,x}(t+\varepsilon, dy) = P\big(|X_{t+\varepsilon} - x| \geq c \mid X_t = x\big)$$
$$= P\big(|X_{t+\varepsilon}^{t,x} - x| \geq c\big)$$
$$= P\big(|X_{t+\varepsilon}^{t,x} - x|^4 \geq c^4\big),$$

where $X_t^{s,x}$ is the solution of Equation (10.8.4) with initial condition $X_s = x$. First apply the above Chebyshev inequality and then use the estimate in Equation (10.7.10) to show that for any $0 \leq t \leq 1$,

$$\int_{|y-x|\geq c} P_{t,x}(t+\varepsilon, dy) \leq \frac{1}{c^4}E\big(|X_{t+\varepsilon}^{t,x} - x|^4\big) \leq \frac{1}{c^4} L\, \varepsilon^2\, e^{C_1\varepsilon},$$

where $L = C_2(1 + 27x^4 + C_1)$ with C_1 and C_2 being given in Theorem 10.7.2 for $a = t$ and $b = t + 1$. Obviously, this inequality implies that

$$\lim_{\varepsilon\downarrow 0} \frac{1}{\varepsilon}\int_{|y-x|\geq c} P_{t,x}(t+\varepsilon, dy) = 0.$$

Thus the Markov process X_t satisfies condition (1) in Definition 10.8.3. If in addition $\sigma(t, x)$ and $f(t, x)$ are continuous functions of t and x, then we will see from the next theorem that X_t is a diffusion process.

Let $B(t)$ be an $\mathbb{R}^m$-valued Brownian motion. Consider the $\mathbb{R}^n$-valued SDE in Equation (10.4.2), namely, the SIE

$$X_t = \xi + \int_a^t \sigma(s, X_s)\, dB(s) + \int_a^t f(s, X_s)\, ds, \quad a \leq t \leq b, \qquad (10.8.5)$$

where the matrix-valued function $\sigma(t, x)$ and vector-valued function $f(t, x)$ satisfy the Lipschitz and linear growth conditions in Theorem 10.4.1. As we remarked at the end of Section 10.6, the solution X_t of Equation (10.8.5) is a Markov process.

Theorem 10.8.9. *Let $\sigma(t, x)$ and $f(t, x)$ be functions specified in Theorem 10.4.1. Assume that $\sigma(t, x)$ and $f(t, x)$ are continuous on $[a, b] \times \mathbb{R}^n$. Then the solution X_t of Equation (10.8.5) is a diffusion process with diffusion coefficient $Q(t, x)$ and drift $\rho(t, x)$ given by*

$$Q(t, x) = \sigma(t, x)\,\sigma(t, x)^T, \quad \rho(t, x) = f(t, x), \tag{10.8.6}$$

where A^T denotes the transpose of a matrix A.

Proof. When $n = m = 1$, we have already proved in the above discussion that X_t satisfies condition (1) in Definition 10.8.3. For the multidimensional case, we have the equality

$$E\left(\left|\int_a^t \sigma(s, X_s)\,dB(t)\right|^2\right) = \int_a^t E\|\sigma(s, X_s)\|^2\,ds, \tag{10.8.7}$$

where $\|\sigma\|$ is the Hilbert–Schmidt norm of σ defined by Equation (10.4.3). Thus in view of Theorem 10.4.1, we can apply the same arguments as those for the case $n = m = 1$ to show that multidimensional X_t also satisfies condition (1) in Definition 10.8.3.

To find the drift of X_t, note that for any $\varepsilon > 0$,

$$X_{t+\varepsilon} = x + \int_t^{t+\varepsilon} \sigma(s, X_s)\,dB(s) + \int_t^{t+\varepsilon} f(s, X_s)\,ds.$$

It follows from the estimates in Theorem 10.7.2 that the stochastic processes $\sigma(t, X_t)$ and $f(t, X_t)$, $a \le t \le b$, belong to the space $L^2_{\mathrm{ad}}([a, b] \times \Omega)$. Hence we can take the expectation to get

$$E\{X_{t+\varepsilon} - x \mid X_t = x\} = E(X_{t+\varepsilon} - x) = \int_t^{t+\varepsilon} E(f(s, X_s))\,ds.$$

Thus by Equation (10.8.2), the drift is given by

$$\rho(t, x) = \lim_{\varepsilon \downarrow 0} \frac{1}{\varepsilon} E\{X_{t+\varepsilon} - x \mid X_t = x\} = \lim_{\varepsilon \downarrow 0} \frac{1}{\varepsilon} \int_t^{t+\varepsilon} E(f(s, X_s))\,ds.$$

Since $f(t, x)$ is continuous by assumption, the expectation $E(f(t, X_t))$ is a continuous function of t. Hence by the fundamental theorem of calculus, we see that $\rho(t, x) = f(t, x)$.

Next, we will use Equation (10.8.3) to find the diffusion coefficient of X_t. Apply the Itô product formula to derive

$$(X_{t+\varepsilon} - x)(X_{t+\varepsilon} - x)^T$$
$$= \int_t^{t+\varepsilon} \sigma(s, X_s)\,dB(s)\,(X_s - x)^T + \int_t^{t+\varepsilon} f(s, X_s)(X_s - x)^T\,ds$$
$$+ \int_t^{t+\varepsilon} (X_s - x)\,dB(s)^T\,\sigma(s, X_s)^T + \int_t^{t+\varepsilon} (X_s - x)f(s, X_s)^T\,ds$$
$$+ \int_t^{t+\varepsilon} \sigma(s, X_s)\,dB(s)\,dB(s)^T\,\sigma(s, X_s)^T.$$

Note that the first and third integrals have expectation 0. Moreover, by the Itô Table (2) in Section 7.5, we have $dB(t)\, dB(t)^T = I\, dt$ (I is the $n \times n$ identity matrix). Thus upon taking the expectation, we get

$$E\big((X_{t+\varepsilon} - x)(X_{t+\varepsilon} - x)^T\big)$$
$$= \int_t^{t+\varepsilon} f(s, X_s)(X_s - x)^T\, ds + \int_t^{t+\varepsilon} (X_s - x) f(s, X_s)^T\, ds$$
$$+ \int_t^{t+\varepsilon} \sigma(s, X_s)\sigma(s, X_s)^T\, ds.$$

Under the condition $X_t = x$, by the fundamental theorem of calculus,

$$\lim_{\varepsilon \downarrow 0} \frac{1}{\varepsilon} \int_t^{t+\varepsilon} f(s, X_s)(X_s - x)^T\, ds = \lim_{\varepsilon \downarrow 0} \frac{1}{\varepsilon} \int_t^{t+\varepsilon} (X_s - x) f(s, X_s)^T\, ds = 0.$$

Therefore, by Equation (10.8.3) the diffusion coefficient of X_t is given by

$$Q(t, x) = \lim_{\varepsilon \downarrow 0} \frac{1}{\varepsilon} E\big\{(X_{t+\varepsilon} - x)(X_{t+\varepsilon} - x)^T \,\big|\, X_t = x\big\}$$
$$= \lim_{\varepsilon \downarrow 0} \frac{1}{\varepsilon} \int_t^{t+\varepsilon} \sigma(s, X_s)\sigma(s, X_s)^T\, ds$$
$$= \sigma(t, x)\sigma(t, x)^T.$$

Thus we have completed the proof of this theorem.   □

By this theorem, the solutions of the first five examples in Section 10.1 are diffusion processes. Their diffusion coefficients and drifts can be obtained using Equation (10.8.6). The solution of Example 10.1.7 is also a diffusion process if $h(t)$ is a continuous function. We give one more example.

Example 10.8.10. The functions $\sigma(x) = \sin x$ and $f(x) = \cos x$ satisfy the conditions of Theorem 10.8.9. Hence the solution X_t of the SDE

$$dX_t = \sin X_t\, dB(t) + \cos X_t\, dt, \quad X_0 = 1,$$

is a diffusion process with the diffusion coefficient and drift given by

$$Q(x) = \sin^2 x, \quad \rho(x) = \cos x.$$

Note that by Theorem 10.6.2 the diffusion process X_t is stationary.

10.9 Semigroups and the Kolmogorov Equations

In this section we briefly describe semigroups and the Kolmogorov equations for diffusion processes without going into the technical details.

Suppose X_t is a diffusion process. Its drift $\rho(t, x)$ and diffusion coefficient $Q(t, x)$ are given by Equations (10.8.2) and (10.8.3), respectively. Then we can ask the following question.

Question 10.9.1. Given a matrix-valued function $Q(t,x)$ and a vector-valued function $\rho(t,x)$, does there exist a diffusion process X_t that has diffusion coefficient $Q(t,x)$ and drift $\rho(t,x)$? If so, how does one find the diffusion process?

Obviously, we must impose some conditions on the function $Q(t,x)$. For example, the matrix $Q(t,x)$ must be symmetric in view of Equation (10.8.3). In fact, $Q(t,x)$ is a positive matrix, namely,

$$v^T Q(t,x)\, v \geq 0, \quad \forall v \in \mathbb{R}^n,$$

where v^T is the transpose of v. This fact can be seen as follows:

$$v^T Q(t,x)\, v = \lim_{\varepsilon \downarrow 0} \frac{1}{\varepsilon} E\{v^T (X_{t+\varepsilon} - x)(X_{t+\varepsilon} - x)^T\, v \mid X_t = x\}$$

$$= \lim_{\varepsilon \downarrow 0} \frac{1}{\varepsilon} E\{|(X_{t+\varepsilon} - x)^T\, v|^2 \mid X_t = x\}$$

$$\geq 0,$$

where $|\cdot|$ is the Euclidean norm on $\mathbb{R}^n$.

There are three approaches to answer this question. We first outline these approaches for comparison.

1. Semigroup approach (Hille–Yosida theory)

$$X_t,\, t \geq 0 \quad \longleftrightarrow \quad \{P_t(x,\cdot)\} \quad \longleftrightarrow \quad \{T_t\} \quad \longleftrightarrow \quad \mathcal{A}$$

2. Partial differential equation approach (Kolmogorov equations)

$$X_t,\, a \leq t \leq b \quad \longleftrightarrow \quad \{P_{s,x}(t,\cdot)\} \quad \longleftrightarrow \quad \mathcal{A}$$

3. Stochastic differential equation approach (Itô theory)

$$X_t,\, a \leq t \leq b \quad \longleftrightarrow \quad \mathcal{A}$$

Here the operator $\mathcal{A}$ is to be defined below in terms of Q and ρ. But first we want to point out that the Itô theory gives the shortest and most direct way to construct diffusion processes from the operator $\mathcal{A}$. In fact, this was the original motivation for K. Itô to introduce his theory of stochastic integration.

(1) Semigroup Approach

For simplicity, consider only stationary diffusion processes. Let $X_t,\, t \geq 0$, be such a diffusion process taking values in $\mathbb{R}^n$. Then its diffusion coefficient $Q(t,x)$ and drift $\rho(t,x)$ depend only on x and are given by

$$Q(x) = \lim_{t \downarrow 0} \frac{1}{t} E\{(X_t - x)(X_t - x)^T \mid X_0 = x\}, \qquad (10.9.1)$$

$$\rho(x) = \lim_{t \downarrow 0} \frac{1}{t} E\{X_t - x \mid X_0 = x\}. \qquad (10.9.2)$$

Note that for each $x \in \mathbb{R}^n$, $Q(x)$ is an $n \times n$ matrix and $\rho(x)$ is a vector in $\mathbb{R}^n$ denoted by either $\rho(x) = (\rho_1(x), \rho_2(x), \ldots, \rho_n(x))$ or a column vector.

In order to continue further, we need to introduce the concept of semigroup and its infinitesimal generator.

Definition 10.9.2. *A family $\{T_t; t \geq 0\}$ of linear operators on a Banach space $\mathbb{B}$ with norm $\| \cdot \|$ is called a (C_0)-contraction semigroup if it satisfies the following conditions:*

1. $\lim_{t \downarrow 0} T_t h = h$ *for all* $h \in \mathbb{B}$ *(strong continuity);*
2. $\|T_t h\| \leq \|h\|$ *for all* $t \geq 0$ *and* $h \in \mathbb{B}$ *(contraction);*
3. $T_0 = I$ *and* $T_{s+t} = T_s T_t$ *for all* $s, t \geq 0$ *(semigroup).*

Definition 10.9.3. *Suppose $\{T_t; t \geq 0\}$ is a (C_0)-contraction semigroup on a Banach space $\mathbb{B}$. Its* infinitesimal generator *is defined to be the operator*

$$\mathcal{A}h = \lim_{t \downarrow 0} \frac{1}{t}(T_t h - h)$$

with the domain consisting of all vectors h such that the limit exists in $\mathbb{B}$.

The infinitesimal generator $\mathcal{A}$ is a densely defined operator, namely, its domain is dense in $\mathbb{B}$. For the proof, see page 237 of the book by Yosida [89]. In general, an infinitesimal generator $\mathcal{A}$ is not defined on the whole space $\mathbb{B}$ and is an unbounded operator.

Example 10.9.4. Consider the Banach space $C_b[0, \infty)$ of bounded uniformly continuous functions on $[0, \infty)$ with the sup-norm $\|f\|_\infty = \sup_{x \in [0, \infty)} |f(x)|$. For $f \in C_b[0, \infty)$, define

$$(T_t f)(x) = f(x + t) \quad \text{(translation by t).}$$

Then $\{T_t; t \geq 0\}$ is a (C_0)-contraction semigroup on $C_b[0, \infty)$. Its infinitesimal generator $\mathcal{A}$ is given by $\mathcal{A}f = f'(x)$, namely, $\mathcal{A} = d/dx$, the differentiation operator. The domain of $\mathcal{A}$ consists of all functions f in $C_b[0, \infty)$ such that $f' \in C_b[0, \infty)$. Note that $\mathcal{A}$ is an unbounded operator.

Now, we go back to the discussion of stationary diffusion processes. Such a process X_t has diffusion coefficient $Q(x)$ and drift $\rho(x)$ given by Equations (10.9.1) and (10.9.2), respectively. Assume that the transition probabilities $\{P_t(x, \cdot)\}$ satisfy the following conditions:

(A1) For any $t > 0$ and $c > 0$, $\lim_{|x| \to \infty} P_t(x, \{y; |y| < c\}) = 0$;
(A2) For any $c > 0$, $\lim_{t \downarrow 0} P_t(x, \{y; |y - x| \geq c\}) = 0$ uniformly on $x \in \mathbb{R}^n$;
(A3) For any bounded continuous function f and $t > 0$, the function

$$(T_t f)(x) = \int_{\mathbb{R}^n} f(y) \, P_t(x, dy), \quad x \in \mathbb{R}^n, \tag{10.9.3}$$

is also continuous.

Let $C_0(\mathbb{R}^n)$ denote the space of continuous functions f on $\mathbb{R}^n$ vanishing at infinity, i.e., $\lim_{|x|\to\infty} f(x) = 0$. The space $C_0(\mathbb{R}^n)$ is a Banach space with the sup-norm $\|f\|_\infty = \sup_{x\in\mathbb{R}^n} |f(x)|$.

For each $t > 0$ and $f \in C_0(\mathbb{R}^n)$, define a function $T_t f$ on $\mathbb{R}^n$ by Equation (10.9.3). By assumption (A3), the function $T_t f$ is continuous. We show that $T_t f$ vanishes at infinity. Given any $\varepsilon > 0$, since f vanishes at infinity, there exists $c > 0$ such that $|f(y)| < \varepsilon/2$ for all $|y| \geq c$. Therefore,

$$|(T_t f)(x)| \leq \int_{|y|\geq c} |f(y)| P_t(x, dy) + \int_{|y|<c} |f(y)| P_t(x, dy)$$

$$\leq \varepsilon/2 + \|f\|_\infty P_t(x, \{y; |y| < c\}).$$

We may assume that $\|f\|_\infty > 0$. By assumption (A1), there exists $K > 0$ such that $P_t(x, \{y; |y| < c\}) < \varepsilon/(2\|f\|_\infty)$ for all $|x| > K$. Then

$$|(T_t f)(x)| \leq \varepsilon/2 + \varepsilon/2 = \varepsilon, \quad \forall |x| > K.$$

This shows that $T_t f$ vanishes at infinity. Thus $T_t f \in C_0(\mathbb{R}^n)$. Hence T_t is a linear operator from $C_0(\mathbb{R}^n)$ into itself.

Let $T_0 = I$. We show that the family $\{T_t; t \geq 0\}$ is a (C_0)-contraction semigroup on $C_0(\mathbb{R}^n)$. It is easily verified that the equality $T_{s+t} = T_s T_t$ for all $s, t \geq 0$ is equivalent to the Chapman–Kolmogorov equation in Equation (10.5.10). Hence condition 3 in Definition 10.9.2 holds. Condition 2 can be easily checked as follows:

$$\|T_t f\|_\infty = \sup_{x\in\mathbb{R}^n} |(T_t f)(x)| \leq \sup_{x\in\mathbb{R}^n} \int_{\mathbb{R}^n} |f(y)| P_t(x, dy) \leq \|f\|_\infty.$$

To check condition 1, we need to point out that every function f in $C_0(\mathbb{R}^n)$ is uniformly continuous. Hence for any $\varepsilon > 0$, there exists $\delta > 0$ such that

$$|x - y| < \delta \implies |f(x) - f(y)| < \varepsilon/2.$$

Therefore, we have

$$|T_t f(x) - f(x)|$$

$$= \left| \int_{\mathbb{R}^n} (f(y) - f(x)) P_t(x, dy) \right|$$

$$\leq \int_{|y-x|<\delta} |f(y) - f(x)| P_t(x, dy) + \int_{|y-x|\geq\delta} |f(y) - f(x)| P_t(x, dy)$$

$$\leq \varepsilon/2 + 2\|f\|_\infty P_t(x, \{y; |y - x| \geq \delta\}). \tag{10.9.4}$$

Again we may assume that $\|f\|_\infty > 0$. Then by assumption (A2), there exists $t_0 > 0$ such that for any $0 < t < t_0$,

$$P_t(x, \{y; |y - x| \geq \delta\}) \leq \varepsilon/(4\|f\|_\infty), \quad \forall x \in \mathbb{R}^n. \tag{10.9.5}$$

Equations (10.9.4) and (10.9.5) imply that $\|T_t f - f\|_\infty < \varepsilon$ for all $0 < t < t_0$. Hence condition 1 in Definition 10.9.2 holds. This finishes the proof that $\{T_t; t \geq 0\}$, defined by Equation (10.9.3), is a (C_0)-contraction semigroup on the Banach space $C_0(\mathbb{R}^n)$.

Next, we give an informal derivation for the infinitesimal generator of $\{T_t; t \geq 0\}$. Let ϕ be a twice continuously differentiable function on $\mathbb{R}^n$. Then it has the Taylor expansion

$$\phi(y) = \phi(x) + (y - x) \cdot \nabla\phi(x) + \frac{1}{2}\mathrm{tr}\,(y - x)(y - x)^T D^2\phi(x) + \cdots,$$

where the first " $\cdot$ " denotes the dot product on $\mathbb{R}^n$, $\nabla\phi$ is the gradient of ϕ, $\mathrm{tr}A$ denotes the trace of a matrix A, and $D^2\phi$ is the second derivative of ϕ. Put $y = X_t$ and then take the expectation to get

$$\begin{aligned}
(T_t\phi)(x) &- \phi(x) \\
&= E[\phi(X_t) \mid X_0 = x] - \phi(x) \\
&= E(X_t - x) \cdot \nabla\phi(x) + \frac{1}{2}\mathrm{tr}\,E\big((X_t - x)(X_t - x)^T\big)D^2\phi(x) + \cdots.
\end{aligned}$$

Hence by Equations (10.9.1) and (10.9.2), we have

$$\lim_{t \downarrow 0} \frac{1}{t}\big\{(T_t\phi)(x) - \phi(x)\big\} = \rho(x) \cdot \nabla\phi(x) + \frac{1}{2}\mathrm{tr}\,Q(x)D^2\phi(x),$$

which is the infinitesimal generator of $\{T_t; t \geq 0\}$. From the above discussion, we have shown the next theorem.

Theorem 10.9.5. *Assume that a diffusion process X_t satisfies the conditions in $(A1), (A2)$, and $(A3)$. Then the family $\{T_t; t \geq 0\}$ of linear operators defined by Equation (10.9.3) is a (C_0)-contraction semigroup on the Banach space $C_0(\mathbb{R}^n)$ with infinitesimal generator $\mathcal{A}$ given by*

$$(\mathcal{A}\phi)(x) = \frac{1}{2}\mathrm{tr}\,Q(x)D^2\phi(x) + \rho(x) \cdot \nabla\phi(x), \tag{10.9.6}$$

or equivalently, in terms of partial derivatives of ϕ,

$$(\mathcal{A}\phi)(x) = \frac{1}{2}\sum_{i,j=1}^n Q_{ij}(x)\frac{\partial^2\phi}{\partial x_i \partial x_j} + \sum_{i=1}^n \rho_i(x)\frac{\partial\phi}{\partial x_i},$$

where $Q_{ij}(x)$ are the entries of $Q(x)$ and $\rho_i(x)$ are the components of $\rho(x)$.

In general, if a stationary diffusion process X_t satisfies certain conditions, then the associated family $\{T_t; t \geq 0\}$ of linear operators is a semigroup on some space of functions defined on $\mathbb{R}^n$. Moreover, its infinitesimal generator $\mathcal{A}$ is determined by the diffusion coefficient $Q(x)$ and drift $\rho(x)$ of X_t as in Equation (10.9.6).

Example 10.9.6. The infinitesimal generator of a Brownian motion $B(t)$ on $\mathbb{R}^n$ is given by

$$\mathcal{A}\phi(x) = \frac{1}{2}\Delta\phi(x),$$

where $\Delta\phi$ is the Laplacian of ϕ. On the other hand, the infinitesimal generator of an Ornstein–Uhlenbeck process on $\mathbb{R}^n$ (see Example 10.8.2) is given by

$$\mathcal{A}\phi(x) = \frac{1}{2}\Delta\phi(x) - x \cdot \nabla\phi(x).$$

Now consider the reverse direction, namely, given an operator $\mathcal{A}$ in the form of Equation (10.9.6), how can we construct $\{T_t\}$, $\{P_t(x, \cdot)\}$, and X_t so that $\mathcal{A}$ is the infinitesimal generator of X_t? Recall that this is the question raised in the beginning of this section.

Observe that we can use the semigroup property to show that

$$\frac{d}{dt}T_t = \lim_{\varepsilon \to 0} \frac{T_{t+\varepsilon} - T_t}{\varepsilon} = T_t \lim_{\varepsilon \downarrow 0} \frac{T_\varepsilon - I}{\varepsilon} = T_t \mathcal{A} = \mathcal{A}T_t, \quad t > 0.$$

Thus it looks like $T_t = e^{t\mathcal{A}}$. This is indeed the case when $\mathcal{A}$ is a bounded operator and $e^{t\mathcal{A}}$ is the bounded operator defined by the power series

$$e^{t\mathcal{A}} = \sum_{n=0}^{\infty} \frac{t^n}{n!}\mathcal{A}^n.$$

However, when $\mathcal{A}$ is an unbounded operator, we cannot use the above series to define $e^{t\mathcal{A}}$. Hence the operator T_t can be thought of as a substitute for $e^{t\mathcal{A}}$. The existence of $\{T_t\}$ is given by the next theorem.

Theorem 10.9.7. (Hille–Yosida theorem) *Let $\mathcal{A}$ be a densely defined linear operator from a Banach space into itself. Then $\mathcal{A}$ is the infinitesimal generator of a (C_0)-contraction semigroup if and only if $(I - n^{-1}\mathcal{A})^{-1}$ exists such that the operator norm $\|(I - n^{-1}\mathcal{A})^{-1}\| \leq 1$ for any natural number n.*

Suppose we are given a differential operator $\mathcal{A}$ of the form

$$(\mathcal{A}\phi)(x) = \frac{1}{2}\mathrm{tr}\, Q(x)D^2\phi(x) + \rho(x) \cdot \nabla\phi(x),$$

where the matrix $Q(x)$ is symmetric and $v^T Q(x)v > 0$ for any nonzero $v \in \mathbb{R}^n$. We can take a Banach space $\mathbb{B}$ to be either $C_b(\mathbb{R}^n)$ of bounded uniformly continuous functions on $\mathbb{R}^n$ or $C_0(\mathbb{R}^n)$ of continuous functions vanishing at infinity. We need to make sure that $\mathcal{A}$ is densely defined on $\mathbb{B}$. Then we proceed to find $\{T_t\}$, $\{P_t(x, \cdot)\}$, and X_t as follows:

1. Check whether the operator $\mathcal{A}$ satisfies the condition in the Hille–Yosida theorem. If so, then there is a (C_0)-contraction semigroup $\{T_t\,;\, t \geq 0\}$ whose infinitesimal generator is $\mathcal{A}$.

2. Check whether the semigroup $\{T_t; t \geq 0\}$ has the properties (i) $T_t 1 = 1$ and (ii) $T_t f \geq 0$ for any $f \geq 0$ in $\mathbb{B}$. If so, then we can apply the Riesz representation theorem to obtain a family $\{P_t(x, \cdot); t \geq 0, x \in \mathbb{R}^n\}$ of probability measures on $\mathbb{R}^n$ such that $(T_t f)(x) = \int_{\mathbb{R}^n} f(y) P_t(x, dy)$ for any $t \geq 0$, $x \in \mathbb{R}^n$, and $f \in \mathbb{B}$.

3. The family $\{P_t(x, \cdot); t \geq 0, x \in \mathbb{R}^n\}$ satisfies the Chapman–Kolmogorov equation because of the semigroup property of $\{T_t; t \geq 0\}$. Then we can apply the n-dimensional analogue of Theorem 10.5.14 to obtain a Markov process X_t. Check that X_t is a diffusion process.

This concludes the first approach via the Hille–Yosida theory of semigroups for constructing diffusion processes from the infinitesimal generators. For further information, see volume 1 of Dynkin's books [12] and the book by Itô [39].

(2) Partial Differential Equation Approach

Suppose X_t, $a \leq t \leq b$, is a diffusion process. We will informally derive partial differential equations that are satisfied by transition probabilities of X_t. To avoid complicated notation, we consider the one-dimensional case. Let $\{P_{s,x}(t, \cdot); a \leq s < t \leq b, x \in \mathbb{R}\}$ be the transition probabilities of a diffusion process X_t, $a \leq t \leq b$, with diffusion coefficient $Q(t, x)$ and drift $\rho(t, x)$.

Let $F_{s,x}(t, y)$ denote the distribution function of X_t given $X_s = x$, i.e.,

$$F_{s,x}(t, y) = P_{s,x}(t, (-\infty, y]).$$

Then the Chapman–Kolmogorov equation takes the form

$$F_{s,x}(t, y) = \int_{\mathbb{R}} F_{u,z}(t, y) \, dF_{s,x}(u, z). \tag{10.9.7}$$

Let $t_0 \in (a, b]$ and $y_0 \in \mathbb{R}$ be fixed and consider $F_{s,x}(t_0, y_0)$ as a function of $s \in [a, t_0)$ and $x \in \mathbb{R}$. By Equation (10.9.7) with $t = t_0, y = y_0$, and $u = s + \varepsilon$ for small $\varepsilon > 0$, we have

$$F_{s,x}(t_0, y_0) = \int_{\mathbb{R}} F_{s+\varepsilon,z}(t_0, y_0) \, dF_{s,x}(s + \varepsilon, z).$$

The integrand has the following Taylor expansion:

$$F_{s+\varepsilon,z}(t_0, y_0) \approx F_{s+\varepsilon,x}(t_0, y_0) + \left(\frac{\partial}{\partial x} F_{s+\varepsilon,x}(t_0, y_0) \right)(z - x)$$
$$+ \frac{1}{2} \left(\frac{\partial^2}{\partial x^2} F_{s+\varepsilon,x}(t_0, y_0) \right)(z - x)^2.$$

Hence we have the approximation

$$F_{s,x}(t_0, y_0) \approx F_{s+\varepsilon,x}(t_0, y_0) + \left(\frac{\partial}{\partial x} F_{s+\varepsilon,x}(t_0, y_0) \right) \varepsilon \, \rho(s, x)$$
$$+ \frac{1}{2} \left(\frac{\partial^2}{\partial x^2} F_{s+\varepsilon,x}(t_0, y_0) \right) \varepsilon \, Q(s, x).$$

Bring the first term in the right-hand side to the left-hand side, then divide both sides by ε and let $\varepsilon \to 0$ to get

$$-\frac{\partial}{\partial s}F_{s,x}(t_0, y_0) = \rho(s, x)\frac{\partial}{\partial x}F_{s,x}(t_0, y_0) + \frac{1}{2}Q(s, x)\frac{\partial^2}{\partial x^2}F_{s,x}(t_0, y_0). \quad (10.9.8)$$

The terminal condition for this equation is

$$\lim_{s\uparrow t_0} F_{s,x}(t_0, y_0) = \begin{cases} 1, & \text{if } x < y_0; \\ 0, & \text{if } x > y_0. \end{cases} \quad (10.9.9)$$

Equation (10.9.8) is called the *Kolmogorov backward equation*. The term refers to the fact that the time variable s moves backward from t_0.

Theorem 10.9.8. *Suppose $Q(t, x)$ and $\rho(t, x)$ are continuous functions of t and x and satisfy the Lipschitz and linear growth conditions in x. In addition, assume that there exists a constant $c > 0$ such that*

$$Q(t, x) \geq c, \quad \forall a \leq t \leq b, \, x \in \mathbb{R}. \quad (10.9.10)$$

Then the Kolmogorov backward equation in Equation (10.9.8) has a unique solution satisfying the condition in Equation (10.9.9). Moreover, there exists a continuous Markov process X_t, $a \leq t \leq b$, with transition probabilities given by $P_{s,x}(t_0, dy_0) = dF_{s,x}(t_0, y_0)$.

Remark 10.9.9. In the multidimensional case, the inequality for $Q(t, x)$ in Equation (10.9.10) takes the form

$$v^T Q(t, x)v \geq c|v|^2, \quad \forall v \in \mathbb{R}^n.$$

This theorem and Theorem 10.9.10 below can be easily formulated for the multidimensional case.

When the diffusion process X_t is stationary, $F_{s,x}(s+t, y)$ does not depend on s. Hence we have a well-defined function

$$F_t(x, y) = F_{s,x}(s+t, y), \quad t > 0.$$

Since $F_{s,x}(s+t, y)$ does not depend on s, $\frac{d}{ds}F_{s,x}(s+t, y) = 0$. This implies the second equality below:

$$\frac{\partial}{\partial t}F_t(x, y_0) = \frac{\partial}{\partial(s+t)}F_{s,x}(s+t, y_0) = \left(-\frac{\partial}{\partial s}F_{s,x}(u, y_0)\right)\Big|_{u=s+t}.$$

Then use Equation (10.9.8) to get

$$\frac{\partial}{\partial t}F_t(x, y_0) = \rho(x)\frac{\partial}{\partial x}F_t(x, y_0) + \frac{1}{2}Q(x)\frac{\partial^2}{\partial x^2}F_t(x, y_0), \quad (10.9.11)$$

which is the Kolmogorov backward equation for stationary diffusion processes. The initial condition is given by

$$\lim_{t \downarrow 0} F_t(x, y_0) = \begin{cases} 1, & \text{if } x < y_0; \\ 0, & \text{if } x > y_0. \end{cases}$$

For a reasonably regular function $f(x)$ on $\mathbb{R}$, define

$$u(t, x) = \int_{\mathbb{R}} f(y) \, P_t(x, dy) = \int_{\mathbb{R}} f(y) \, dF_t(x, y).$$

It follows from Equation (10.9.11) that u satisfies the following equation:

$$\frac{\partial u}{\partial t} = \mathcal{A}u, \quad u(0, x) = f(x),$$

where $\mathcal{A} = \rho(x) \frac{d}{dx} + \frac{1}{2} Q(x) \frac{d^2}{dx^2}$ is the infinitesimal generator of X_t.

Next, we derive another partial differential equation for Markov processes. This time we assume that the transition probabilities of a Markov process X_t have density functions

$$P_{s,x}(t, dy) = p_{s,x}(t, y) \, dy.$$

Then the Chapman–Kolmogorov equation takes the form

$$p_{s,x}(t, y) = \int_{\mathbb{R}} p_{u,z}(t, y) p_{s,x}(u, z) \, dz. \tag{10.9.12}$$

Let $s_0 \in [a, b]$ and $x_0 \in \mathbb{R}$ be fixed and consider $p_{s_0,x_0}(t, y)$ as a function of $t \in (s_0, b]$ and $x \in \mathbb{R}$. For a reasonably good function ξ, define

$$\theta(t) = \int_{\mathbb{R}} \xi(y) p_{s_0,x_0}(t, y) \, dy. \tag{10.9.13}$$

Let $\varepsilon > 0$ be a small number. By Equation (10.9.12),

$$p_{s_0,x_0}(t + \varepsilon, y) = \int_{\mathbb{R}} p_{t,z}(t + \varepsilon, y) p_{s_0,x_0}(t, z) \, dz.$$

Use this equation for $\theta(t+\varepsilon)$, change the order of integration, and then expand ξ to derive the estimate

$$\theta(t + \varepsilon) = \int_{\mathbb{R}} \xi(y) \left[\int_{\mathbb{R}} p_{t,z}(t + \varepsilon, y) p_{s_0,x_0}(t, z) \, dz \right] dy$$

$$= \int_{\mathbb{R}} \left[\int_{\mathbb{R}} \xi(y) p_{t,z}(t + \varepsilon, y) \, dy \right] p_{s_0,x_0}(t, z) \, dz$$

$$\approx \int_{\mathbb{R}} \left[\int_{\mathbb{R}} \left\{ \xi(z) + \xi'(z)(y - z) + \frac{1}{2} \xi''(z)(y - z)^2 \right\} \right.$$

$$\left. \times p_{t,z}(t + \varepsilon, y) \, dy \right] p_{s_0,x_0}(t, z) \, dz.$$

Note that the integral with respect to the variable y produces the drift and diffusion coefficient. Hence we have

$$\theta(t+\varepsilon) \approx \int_{\mathbb{R}} \Big(\xi(z) + \xi'(z)\,\varepsilon\,\rho(t,z) + \frac{1}{2}\xi''(z)\,\varepsilon\,Q(t,z) \Big) p_{s_0,x_0}(t,z)\,dz$$

$$= \theta(t) + \varepsilon \int_{\mathbb{R}} \Big(\xi'(z)\rho(t,z) + \frac{1}{2}\xi''(z)Q(t,z) \Big) p_{s_0,x_0}(t,z)\,dz.$$

Therefore,

$$\theta'(t) = \int_{\mathbb{R}} \Big(\xi'(y)\rho(t,y) + \frac{1}{2}\xi''(y)Q(t,y) \Big) p_{s_0,x_0}(t,y)\,dy.$$

Then use the integration by parts formula to get

$$\theta'(t) = \int_{\mathbb{R}} \xi(y) \bigg\{ -\frac{\partial}{\partial y}\Big(\rho(t,y)p_{s_0,x_0}(t,y) \Big) + \frac{1}{2}\frac{\partial^2}{\partial y^2}\Big(Q(t,y)p_{s_0,x_0}(t,y) \Big) \bigg\}\,dy.$$

On the other hand, we can find $\theta'(t)$ from Equation (10.9.13) by bringing the differentiation inside the integral,

$$\theta'(t) = \int_{\mathbb{R}} \xi(y) \Big(\frac{\partial}{\partial t}p_{s_0,x_0}(t,y) \Big)\,dy.$$

Hence the last two integrals are equal for all reasonably good functions ξ. This can happen only when the other factors in the integrands are equal, i.e.,

$$\frac{\partial}{\partial t}p_{s_0,x_0}(t,y) = -\frac{\partial}{\partial y}\Big(\rho(t,y)p_{s_0,x_0}(t,y) \Big) + \frac{1}{2}\frac{\partial^2}{\partial y^2}\Big(Q(t,y)p_{s_0,x_0}(t,y) \Big).$$

$$(10.9.14)$$

The initial condition for this equation is

$$\lim_{t\downarrow s_0} p_{s_0,x_0}(t,y) = \delta_{x_0}(y), \qquad (10.9.15)$$

where δ_{x_0} is the Dirac delta function at x_0.

Equation (10.9.14) is called the *Kolmogorov forward equation*. The term refers to the fact that the time variable t moves forward from s_0. The equation is also called the *Fokker–Planck equation*.

Theorem 10.9.10. *Let $Q(t,x)$ and $\rho(t,x)$ satisfy the conditions in Theorem 10.9.8. In addition, assume that the partial derivatives $\partial\rho/\partial x$, $\partial Q/\partial x$, and $\partial^2 Q/\partial x^2$ satisfy the Lipschitz and linear growth conditions in x. Then the Kolmogorov forward equation in Equation (10.9.14) has a unique solution satisfying the condition in Equation (10.9.15).*

The solution $p_{s_0,x_0}(t,y)$ is often referred to as the *fundamental solution* of the Kolmogorov forward equation. Observe that the backward equation is for the transition distribution functions, while the forward equation is for the transition density functions. Thus naturally the assumption in Theorem 10.9.10 is stronger than that of Theorem 10.9.8.

(3) Stochastic Differential Equation Approach

Suppose the given diffusion coefficient $Q(t, x)$ and the drift $\rho(t, x)$ satisfy the conditions in Theorem 10.9.8. Define

$$\sigma(t, x) = \sqrt{Q(t, x)}.$$

The function $\sigma(t, x)$ satisfies the Lipschitz condition because

$$|\sigma(t, x) - \sigma(t, y)| = \frac{|Q(t, x) - Q(t, y)|}{\sqrt{Q(t, x)} + \sqrt{Q(t, y)}} \leq \frac{1}{2\sqrt{c}} |Q(t, x) - Q(t, y)|.$$

Moreover, it is obvious that $\sigma(t, x)$ satisfies the linear growth conditions.

Theorem 10.9.11. *Assume that $Q(t, x)$ and $\rho(t, x)$ satisfy the conditions in Theorem 10.9.8. Let $\sigma(t, x) = \sqrt{Q(t, x)}$. Then the unique continuous solution X_t of the stochastic integral equation*

$$X_t = \xi + \int_a^t \sigma(s, X_s) \, dB(s) + \int_a^t \rho(s, X_x) \, ds \qquad (10.9.16)$$

is a diffusion process with diffusion coefficient $Q(t, x)$ and drift $\rho(t, x)$. The transition distribution function $F_{s,x}(t, y)$ of X_t, with fixed t and y, is the unique solution of the Kolmogorov backward equation. Furthermore, if the additional condition in Theorem 10.9.10 is also assumed, then the transition probabilities of X_t have density functions $p_{s,x}(t, y)$ and, with fixed s and x, $p_{s,x}(t, y)$ is the unique solution of the Kolmogorov forward equation.

Remark 10.9.12. In the multidimensional case, $\sigma(t, x)$ is taken to be a matrix such that

$$\sigma(t, x)\sigma(t, x)^T = Q(t, x).$$

Note that if $\sigma(t, x)$ is such a matrix, then $\sigma(t, x)U$ is also such a matrix for any matrix U such that $UU^T = I$. Thus the choice of $\sigma(t, x)$ to be used in the multidimensional case of Equation (10.9.16) is not unique. However, since $UB(t)$ is also a Brownian motion, the transition probabilities of X_t do not depend on the choice of $\sigma(t, x)$.

We give below a simple example to compare the above three different approaches for constructing diffusion processes.

Example 10.9.13. Consider the problem of constructing a diffusion process X_t having the infinitesimal generator

$$(\mathcal{A}\phi)(x) = \phi''(x) - x\phi'(x).$$

The corresponding diffusion coefficient $Q(x)$ and drift $\rho(x)$ are given by

$$Q(x) = 2, \quad \rho(x) = -x.$$

If we want to use the semigroup approach, then by Theorem 10.9.7 we need to check that the inverse $(I - n^{-1}\mathcal{A})^{-1}$ exists such that $\|(I - n^{-1}\mathcal{A})^{-1}\| \leq 1$ for any natural number n. The verification of this fact is not so easy.

If we want to use the partial differential equation approach, then we can use either Theorem 10.9.8 or 10.9.10. Suppose we decide to use Theorem 10.9.10. Then we need to solve the Kolmogorov forward equation

$$\frac{\partial}{\partial t} p_t(x_0, y) = \frac{\partial}{\partial y} [y p_t(x_0, y)] + \frac{\partial^2}{\partial y^2} [p_t(x_0, y)] \qquad (10.9.17)$$

with the initial condition

$$\lim_{t \downarrow 0} p_t(x_0, y) = \delta_{x_0}(y). \qquad (10.9.18)$$

It is reasonable to expect that the limit $p(y) = \lim_{t \to \infty} p_t(x_0, y)$ exists and is independent of the initial position x_0. The function $p(y)$ is the density function of the so-called invariant measure of the diffusion process X_t. Since $\frac{\partial}{\partial t} p(y) = 0$, Equation (10.9.17) for $p(y)$ becomes

$$p(y) + y p'(y) + p''(y) = 0,$$

which can be easily solved to be

$$p(y) = \frac{1}{\sqrt{2\pi}} e^{-y^2/2}, \qquad -\infty < y < \infty.$$

Now we can use $p(y)$ to solve Equation (10.9.17). Again it is reasonable to expect that $p_t(x_0, y)$ is of the form

$$p_t(x_0, y) = \frac{1}{\sqrt{2\pi\theta(t)}} e^{-(y - \lambda(t)x_0)^2/2\theta(t)}, \qquad (10.9.19)$$

where the functions $\theta(t)$ and $\lambda(t)$ are to be determined. It is straightforward to check that

$$\frac{\partial}{\partial t} p_t(x_0, y) = p_t(x_0, y) \left\{ -\frac{\theta'(t)}{2\theta(t)} + \frac{\theta'(t)(y - \lambda(t)x_0)^2}{2\theta(t)^2} \right.$$
$$\left. + \frac{\lambda'(t)x_0(y - \lambda(t)x_0)}{\theta(t)} \right\}. \qquad (10.9.20)$$

On the other hand, we can easily find that

$$\frac{\partial}{\partial y} [y p_t(x_0, y)] + \frac{\partial^2}{\partial y^2} [p_t(x_0, y)]$$
$$= p_t(x_0, y) \left\{ 1 - \frac{y(y - \lambda(t)x_0)}{\theta(t)} + \frac{(y - \lambda(t)x_0)^2}{\theta(t)^2} - \frac{1}{\theta(t)} \right\}. \qquad (10.9.21)$$

Put Equations (10.9.20) and (10.9.21) into Equation (10.9.17) to get

$$-\frac{\theta'(t)}{2\theta(t)} + \frac{\theta'(t)(y - \lambda(t)x_0)^2}{2\theta(t)^2} + \frac{\lambda'(t)x_0(y - \lambda(t)x_0)}{\theta(t)}$$

$$= 1 - \frac{y(y - \lambda(t)x_0)}{\theta(t)} + \frac{(y - \lambda(t)x_0)^2}{\theta(t)^2} - \frac{1}{\theta(t)}.$$

By comparing the coefficients of y^2 and y and the constant terms in both sides of the equation, we see that this equation holds if and only if $\theta(t)$ and $\lambda(t)$ satisfy the equations

$$\theta'(t) = -2\theta(t) + 2, \quad -\theta'(t)\lambda(t) + 2\theta(t)\lambda'(t) = -2\lambda(t).$$

The initial condition in Equation (10.9.18) yields the initial conditions

$$\theta(0) = 0, \quad \lambda(0) = 1.$$

Then it is easy to derive the solutions

$$\theta(t) = 1 - e^{-2t}, \quad \lambda(t) = e^{-t}.$$

Put $\theta(t)$ and $\lambda(t)$ into Equation (10.9.19) to get the solution

$$p_t(x_0, y) = \frac{1}{\sqrt{2\pi(1 - e^{-2t})}} \exp\left[-\frac{(y - e^{-t}x_0)^2}{2(1 - e^{-2t})}\right] \tag{10.9.22}$$

for the Kolmogorov forward equation in Equation (10.9.17). Note that $p_t(x_0, y)$ coincides with the density function given by Equation (7.4.7) with $s = 0$ and $x = x_0$. Thus the resulting stochastic process X_t is an Ornstein–Uhlenbeck process given in Example 7.4.5.

Finally, we use the stochastic differential equation approach to construct a diffusion process X_t having diffusion coefficient $Q(x) = 2$ and drift $\rho(x) = -x$. By Theorem 10.9.11, $\sigma(x) = \sqrt{2}$ and we need to solve the following stochastic integral equation:

$$X_t = x_0 + \int_0^t \sqrt{2}\, dB(s) - \int_0^t X_s\, ds.$$

This equation can be written as a stochastic differential equation,

$$dX_t = \sqrt{2}\, dB(t) - X_t\, dt, \quad X_0 = x_0,$$

which we solved in Example 7.4.5 using Itô's formula. From Equation (7.4.6), we have the solution

$$X_t = e^{-t}x_0 + \sqrt{2}\int_0^t e^{-(t-u)}\, dB(u).$$

Thus for each $t > 0$, X_t is Gaussian with mean $e^{-t}x_0$ and variance

$$2 \int_0^t e^{-2(t-u)} \, du = 1 - e^{-2t}.$$

Hence the density function of X_t is the same as the one in Equation (10.9.22).

From this example, it is clear that the Itô theory provides a very powerful tool to construct diffusion processes.

Exercises

1. Let $X_{n+1}(t)$ be defined as in the proof of Theorem 10.3.5. Prove that

$$E\big(|X_t^{(n+1)}|^2\big) \le C_1 e^{C_2(t-a)} + \frac{C_2^n (t-a)^n}{n!} E(\xi^2),$$

   where $C_1 = 3E(\xi^2) + 3C(1+b-a)(b-a)$ and $C_2 = 3C(1+b-a)$.

2. Show that the *pinned Brownian motion*

$$X_t = (t-1) \int_0^t \frac{1}{s-1} \, dB(s), \quad 0 \le t < 1,$$

   is the solution of the stochastic differential equation

$$dX_t = dB(t) + \frac{1}{t-1} X_t \, dt, \quad X_0 = 0.$$

3. Prove that Equation (10.5.1) is equivalent to the equality

$$E(X \, e^{i\lambda_1 Y_1 + \cdots + i\lambda_n Y_n}) = E(\theta(Y_1, \ldots, Y_n) \, e^{i\lambda_1 Y_1 + \cdots + i\lambda_n Y_n})$$

   for all $\lambda_1, \ldots, \lambda_n \in \mathbb{R}$. Hence this equality also characterizes the function $\theta(y_1, \ldots, y_n)$.

4. Suppose X and $\mathcal{G}$ are independent and $Y_1, Y_2, \ldots, Y_n$ are $\mathcal{G}$-measurable. Let $\psi(X, Y_1, Y_2, \ldots, Y_n)$ have expectation. Prove that

$$E\{\psi(X, Y_1, Y_2, \ldots, Y_n) \,|\, \mathcal{G}\}$$
$$= E\big(\psi(X, y_1, y_2, \ldots, y_n)\big)\big|_{y_1 = Y_1, y_2 = Y_2, \ldots, y_n = Y_n}.$$

   In particular, let X be independent of the random variables $Y_1, Y_2, \ldots, Y_n$. Suppose $\psi(X, Y_1, Y_2, \ldots, Y_n)$ has expectation. Then

$$E\{\psi(X, Y_1, Y_2, \ldots, Y_n) \,|\, Y_1, Y_2, \ldots, Y_n\}$$
$$= E\big(\psi(X, y_1, y_2, \ldots, y_n)\big)\big|_{y_1 = Y_1, y_2 = Y_2, \ldots, y_n = Y_n}.$$

5. Let $B(t)$ be a Brownian motion. Check whether the stochastic process $X_t = B(t)^2$ is a Markov process.

6. Let $N(t)$ be a Poisson process and $\widetilde{N}(t)$ the compensated Poisson process. Check whether $N(t)$ and $\widetilde{N}(t)$ are Markov processes.

7. Let $B(t)$ be a Brownian motion. Derive the following conditional density function of $B(1)$ given $B(2)$:

$$f_{B(1)\,|\,B(2)}(x\,|\,y) = \frac{1}{\sqrt{\pi}} e^{-x^2 + xy - \frac{1}{4}y^2};$$

and show that $E\{B(1)\,|\,B(2)\} = \frac{1}{2}B(2)$.

8. Let Z be a random variable independent of $Y_1, Y_2, \ldots, Y_n$. Show that

$$P(Y_n + Z \le x\,|\,Y_1, Y_2, \ldots, Y_n) = P(Y_n + Z \le x\,|\,Y_n).$$

9. Let $f \in L^2[a,b]$ and $X_t = \int_a^t f(s)\,dB(s)$, $a \le t \le b$. For $s < t$, define

$$P_{s,x}(t, A) = P(X_t \in A\,|\,X_s = x).$$

Show that the collection $\{P_{s,x}(t, \cdot); a \le s < t \le b, x \in \mathbb{R}\}$ satisfies the Chapman–Kolmogorov equation. Hence X_t is a Markov process.

10. Suppose X_t is a Markov process. Show that the Chapman–Kolmogorov equation can be stated in terms of conditional expectation as

$$E\{1_{\{X_t \in A\}}\,|\,X_s = x\} = E\left[E\{1_{\{X_t \in A\}}\,|\,X_u\}\,|\,X_s = x\right]$$

for any $s < u < t$.

11. Prove that the collection $\{P_{s,x}(t, \cdot)\}$ of probability measures in Example 10.5.15 satisfies the Chapman–Kolmogorov equation.

12. Find the expectation and variance of X_t in Equation (10.6.10) for fixed t.

13. Let X_t be the Ornstein–Uhlenbeck process in Example 10.8.2. Show that X_t is a diffusion process with the diffusion coefficient $\rho(t, x) = I$ and the drift $f(t, x) = -x$.

14. Let $B(t)$ be an n-dimensional Brownian motion. Prove that $E(|B(t)|^4) = n(n+2)t^2$.

15. Prove the equality in Equation (10.8.7).

16. Check that the differentiation operator d/dx is an unbounded operator on the Banach space $C_b[0, \infty)$ defined in Example 10.9.4.

17. Let $B(t)$ be a Brownian motion on $\mathbb{R}^n$ and let U be an $n \times n$ matrix such that $UU^T = I$. Show that $UB(t)$ is also a Brownian motion on $\mathbb{R}^n$.

18. Solve the stochastic differential equation $dX_t = X_t\,dB(t) + \lambda X_t\,dt$ with the initial condition $X_0 = x$ for any real number λ. Find the transition density of the diffusion process X_t.

11

Some Applications and Additional Topics

In this last chapter we will give several concrete examples to show just a few applications of stochastic integration in applied fields. We will also discuss some additional topics related to applications.

11.1 Linear Stochastic Differential Equations

In this section we will derive explicit solutions of linear stochastic differential equations to be used in Sections 11.2 and 11.3.

Consider a first-order linear ordinary differential equation

$$\frac{dx_t}{dt} = f(t)x_t + g(t), \quad a \le t \le b, \quad x_a = x, \tag{11.1.1}$$

where $f(t)$ is a continuous function. To solve this differential equation, bring the term $f(t)x_t$ to the left-hand side and then multiply both sides by the integrating factor

$$h(t) = e^{-\int_a^t f(s)\,ds} \tag{11.1.2}$$

to obtain the following equation:

$$h(t)\left(\frac{dx_t}{dt} - f(t)x_t\right) = h(t)g(t). \tag{11.1.3}$$

Now note that

$$\frac{d}{dt}\big(h(t)x_t\big) = h(t)\left(\frac{dx_t}{dt} - f(t)x_t\right). \tag{11.1.4}$$

From Equations (11.1.3) and (11.1.4) we see that

$$\frac{d}{dt}\big(h(t)x_t\big) = h(t)g(t),$$

which has the solution

$$h(t)x_t = x + \int_a^t h(s)g(s)\, ds.$$

Therefore, the solution x_t of Equation (11.1.1) is given by

$$x_t = xh(t)^{-1} + \int_a^t h(t)^{-1}h(s)g(s)\, ds$$

$$= xe^{\int_a^t f(s)\, ds} + \int_a^t g(s)\, e^{\int_s^t f(u)\, du}\, ds.$$

Now, by a *linear stochastic differential equation*, we mean a stochastic differential equation of the form

$$dX_t = \{\phi(t)X_t + \theta(t)\}\, dB(t) + \{f(t)X_t + g(t)\}\, dt, \quad X_a = x, \qquad (11.1.5)$$

which is the symbolic expression of the linear stochastic integral equation

$$X_t = x + \int_a^t \{\phi(s)X_s + \theta(s)\}\, dB(s) + \int_a^t \{f(s)X_s + g(s)\}\, ds$$

for $a \le t \le b$. Several special cases of this equation have already appeared in Examples 10.1.1, 10.1.5, and 10.1.7.

In view of the integrating factor in Equation (11.1.2) and the exponential processes in Section 8.7, we can guess that an integrating factor for Equation (11.1.5) is given by H_t as follows:

$$H_t = e^{-Y_t}, \quad Y_t = \int_a^t f(s)\, ds + \int_a^t \phi(s)\, dB(s) - \frac{1}{2}\int_a^t \phi(s)^2\, ds.$$

We need to find $d(H_t X_t)$ just as in the ordinary differential equation case. By the Itô product formula, we have

$$d(H_t X_t) = H_t\, dX_t + X_t\, dH_t + (dH_t)(dX_t). \qquad (11.1.6)$$

Apply Itô's formula to find dH_t as follows:

$$dH_t = -H_t\, dY_t + \frac{1}{2}H_t\, (dY_t)^2$$

$$= H_t\left(-f(t)\, dt - \phi(t)\, dB(t) + \frac{1}{2}\phi(t)^2\, dt\right) + \frac{1}{2}H_t\phi(t)^2\, dt$$

$$= H_t\{-f(t)\, dt - \phi(t)\, dB(t) + \phi(t)^2\, dt\}. \qquad (11.1.7)$$

It follows from Equations (11.1.5) and (11.1.7) that

$$(dH_t)(dX_t) = -H_t\phi(t)\{\phi(t)X_t + \theta(t)\}\, dt. \qquad (11.1.8)$$

Put Equations (11.1.7) and (11.1.8) into Equation (11.1.6) to get

$$d(H_t X_t) = H_t\{dX_t - f(t)X_t\, dt - \phi(t)X_t\, dB(t) - \theta(t)\phi(t)\, dt\}. \qquad (11.1.9)$$

Observe that the quantity inside $\{\cdots\}$ is not quite what we would obtain by moving the terms involving X_t in Equation (11.1.5) to the left-hand side. It has the extra term $-\theta(t)\phi(t)\, dt$. But we can take care of this extra term rather easily, namely, Equations (11.1.5) and (11.1.9) imply that

$$d(H_t X_t) = H_t\{\theta(t)\, dB(t) + g(t)\, dt - \theta(t)\phi(t)\, dt\},$$

which yields that

$$H_t X_t = x + \int_a^t H_s \theta(s)\, dB(s) + \int_a^t H_s\{g(s) - \theta(s)\phi(s)\}\, ds.$$

Upon dividing both sides by H_t we get the solution X_t of Equation (11.1.5). We state this result as the next theorem.

Theorem 11.1.1. *The solution of the linear stochastic differential equation*

$$dX_t = \{\phi(t)X_t + \theta(t)\}\, dB(t) + \{f(t)X_t + g(t)\}\, dt, \quad X_a = x,$$

is given by

$$X_t = xe^{Y_t} + \int_a^t e^{Y_t - Y_s}\, \theta(s)\, dB(s) + \int_a^t e^{Y_t - Y_s}\{g(s) - \theta(s)\phi(s)\}\, ds,$$

where $Y_t = \int_a^t \phi(s)\, dB(s) + \int_a^t \{f(s) - \tfrac{1}{2}\phi(s)^2\}\, ds.$

In general, suppose Z_t is an Itô process and consider a linear stochastic differential equation

$$dX_t = \{\phi(t)X_t + \theta(t)\}\, dZ_t + \{f(t)X_t + g(t)\}\, dt, \quad X_a = x. \qquad (11.1.10)$$

To solve this equation, we can of course rewrite this equation in the form stated in the above theorem and apply the formula in Theorem 11.1.1 to get the solution, which is expressed in terms of the Brownian motion $B(t)$. However, in applied problems, it is often desirable to express the solution in terms of the Itô process Z_t.

To find a formula for the solution of Equation (11.1.10) in terms of Z_t, we can simply go through each step in the above derivation to modify the computation. The solution X_t is given by

$$X_t = xe^{Y_t} + \int_a^t e^{Y_t - Y_s}\theta(s)\, dZ_s + \int_a^t e^{Y_t - Y_s} g(s)\, ds - \int_a^t e^{Y_t - Y_s}\theta(s)\phi(s)(dZ_s)^2,$$

$$(11.1.11)$$

where $Y_t = \int_a^t \phi(s)\, dZ_s + \int_a^t f(s)\, ds - \frac{1}{2}\int_a^t \phi(s)^2 (dZ_s)^2$ and $(dZ_s)^2$ is computed according to the Itô Table (1) in Section 7.4.

Example 11.1.2. Let $dZ_t = \alpha \, dB(t) - \beta Z_t \, dt$ be an Ornstein–Uhlenbeck process and consider the stochastic differential equation

$$dX_t = (X_t + 1) \, dZ_t + X_t \, dt, \quad X_0 = x.$$

We can apply the formula in Equation (11.1.11) to find the solution

$$X_t = x e^{Z_t + (1 - \frac{\alpha^2}{2})t} + \int_0^t e^{Z_t - Z_s + (1 - \frac{1}{2}\alpha^2)(t-s)} \, dZ_s$$

$$- \alpha^2 \int_0^t e^{Z_t - Z_s + (1 - \frac{1}{2}\alpha^2)(t-s)} \, ds.$$

11.2 Application to Finance

In this section we will follow the book by Øksendal [66] to briefly explain the Black–Scholes model [2] for option pricing.

Let $B_1(t), B_2(t), \ldots, B_m(t)$ be m independent Brownian motions defined on a probability space $(\Omega, \mathcal{F}, P)$. Let $\{\mathcal{F}_t; t \geq 0\}$ be the filtration given by $\mathcal{F}_t = \sigma\{B_j(s); 1 \leq j \leq m, s \leq t\}$.

(1) Nonexistence of an Arbitrage

A *market* is an $\mathbb{R}^{n+1}$-valued Itô process $X_t = (X_t^{(0)}, X_t^{(1)}, \ldots, X_t^{(n)})$ for $0 \leq t \leq T$ and the components are specified by

$$dX_t^{(0)} = \gamma(t) X_t^{(0)} \, dt, \quad X_0^{(0)} = 1; \tag{11.2.1}$$

$$dX_t^{(i)} = \sum_{j=1}^m \sigma_{ij}(t) \, dB_j(t) + f_i(t) \, dt, \quad 1 \leq i \leq n, \tag{11.2.2}$$

with adapted stochastic processes $\gamma(t), \sigma_{ij}(t)$, and $f_i(t)$ satisfying conditions to be specified later. Here $X_t^{(0)}$ refers to the unit price of the safe investment and $X_t^{(i)}, 1 \leq i \leq n$, refers to the unit price of the ith risky investment. In practical problems, the stochastic processes $X_t^{(i)}, 1 \leq i \leq n$, are solutions of stochastic differential equations. Since the solutions are Itô processes, they can be written in the form of Equation (11.2.2). By Theorem 11.1.1, the solution of Equation (11.2.1) is given by

$$X_t^{(0)} = e^{\int_0^t \gamma(s) \, ds}, \quad t \geq 0. \tag{11.2.3}$$

Recall that a vector in $\mathbb{R}^n$ can be written as either $(x_1, x_2, \ldots, x_n)$ or an $n \times 1$ column vector with the corresponding components. Define

$$\widehat{X}_t = (X_t^{(1)}, X_t^{(2)}, \ldots, X_t^{(n)}), \tag{11.2.4}$$

which is the risky part of the investment. Let $\sigma(t)$ be the $n \times m$ matrix with entries $\sigma_{ij}(t)$ and let $f(t)$ be the column vector with components $f_i(t)$. Then Equation (11.2.2) can be rewritten as a matrix equation

$$d\widehat{X}_t = \sigma(t)\,dB(t) + f(t)\,dt, \qquad (11.2.5)$$

where $B(t) = (B_1(t), B_2(t), \ldots, B_m(t))$. In view of Equations (11.2.3) and (11.2.5), a market X_t is specified by the stochastic processes $\gamma(t)$, $\sigma(t)$, $f(t)$, which we indicate by the relationship

$$X_t \longleftrightarrow \{\gamma(t),\, \sigma(t),\, f(t)\}.$$

A *portfolio* in the market X_t is an $\{\mathcal{F}_t\}$-adapted stochastic process

$$p(t) = (p_0(t), p_1(t), \ldots, p_n(t)),$$

where $p_i(t)$ represents the number of units of the ith investment at time t. The *value* of a portfolio $p(t)$ is given by

$$V_p(t) = \sum_{i=0}^{n} p_i(t) X_t^{(i)} = p(t) \cdot X_t,$$

where " $\cdot$ " is the dot product on $\mathbb{R}^{n+1}$.

A portfolio $p(t)$ is said to be *self-financing* if its value $V_p(t)$ satisfies the following equality

$$V_p(t) = V_p(0) + \int_0^t p(s) \cdot dX_s,$$

which can be written in a stochastic differential form as

$$dV_p(t) = p(t) \cdot dX_t. \qquad (11.2.6)$$

This equation means that no money is brought in or taken out from the system at any time, a fact suggesting the term *self-financing* for the portfolio.

Note that if we single out $p_0(t)$, then Equation (11.2.6) can be written as

$$d\big(p_0(t)X_t^{(0)}\big) + \sum_{i=1}^{n} d\big(p_i(t)X_t^{(i)}\big) = p_0(t)\,dX_t^{(0)} + \sum_{i=1}^{n} p_i(t)\,dX_t^{(i)}.$$

This shows that if $p_i(t), i = 1, 2, \ldots, n$, are given, then we can always find $p_0(t)$ such that $p(t) = (p_0(t), p_1(t), \ldots, p_n(t))$ is self-financing.

A self-financing portfolio $p(t)$ is called *admissible* if its value $V_p(t)$ is lower bounded for almost all (t, ω) in $[0, T] \times \Omega$, i.e., there exists a constant $C \geq 0$ such that

$$V_p(t, \omega) \geq -C, \quad \text{for almost all } (t, \omega) \in [0, T] \times \Omega.$$

The lower boundedness of a portfolio indicates the fact that there is a debt limit that the creditors can tolerate. Mathematically, it is a condition to ensure that a local martingale is a supermartingale.

Definition 11.2.1. *An admissible portfolio $p(t)$ is called an arbitrage in a market X_t, $0 \leq t \leq T$, if its value $V_p(t)$ satisfies the conditions*

$$V_p(0) = 0, \quad V_p(T) \geq 0, \quad P\{V_p(T) > 0\} > 0.$$

The next theorem gives a sufficient condition for the nonexistence of an arbitrage in a market. We will sketch the proof below.

Theorem 11.2.2. *Assume that there exists an $m \times 1$ column-vector-valued $\{\mathcal{F}_t\}$-adapted stochastic process $h(t)$ satisfying the following conditions:*

(a) *$\sigma(t,\omega)h(t,\omega) = \gamma(t,\omega)\widehat{X}_t(\omega) - f(t,\omega)$ for almost all $(t,\omega) \in [0,T] \times \Omega$;*

(b) *$E \exp\left(\frac{1}{2} \int_0^T |h(t)|^2 \, dt\right) < \infty.$*

Then the market X_t, $0 \leq t \leq T$, has no arbitrage.

Remark 11.2.3. The matrix $\sigma(t,\omega)$ defines a linear transformation from $\mathbb{R}^m$ into $\mathbb{R}^n$ by the multiplication $x \mapsto \sigma(t,\omega)x$. Hence condition (a) means that the vector $\gamma(t,\omega)\widehat{X}_t(\omega) - f(t,\omega)$ belongs to the range of $\sigma(t,\omega)$ for almost all $(t,\omega) \in [0,T] \times \Omega$. As for condition (b), it can be replaced by the weaker condition that

$$E \exp\left[\int_0^t h(s) \cdot dB(s) - \frac{1}{2} \int_0^t |h(s)|^2 \, ds\right] = 1, \quad \forall t \in [0,T], \qquad (11.2.7)$$

where "$\cdot$" is the dot product in $\mathbb{R}^m$ (cf. Lemma 8.7.3 and the remark following the lemma). On the other hand, it has been proved by Karatzas [48] that if a market has no arbitrage, then there exists an $m \times 1$ column-vector-valued $\{\mathcal{F}_t\}$-adapted stochastic process $h(t)$ such that condition (a) holds.

Here is the main idea in the proof of the above theorem. First of all, it is easy to check that a portfolio $p(t)$ is an arbitrage for a market X_t if and only if it is an arbitrage for the normalized market

$$\widetilde{X}_t = (X_t^{(0)})^{-1} X_t, \quad 0 \leq t \leq T, \qquad (11.2.8)$$

where $X_t^{(0)}$ is the first component of X_t and is given by Equation (11.2.3). Hence without loss of generality we may assume that $X_t^{(0)} \equiv 1$, or equivalently, $\gamma \equiv 0$. Next, use the given $h(t)$ to define a stochastic process

$$B_h(t) = B(t) - \int_0^t h(s) \, ds, \quad 0 \leq t \leq T.$$

Then by condition (b) or even the weaker condition in Equation (11.2.7), we can apply the Girsanov theorem to see that $B_h(t)$ is a Brownian motion with respect to the probability measure

$$dQ = e^{\int_0^T h(t) \cdot dB(t) - \frac{1}{2} \int_0^T |h(t)|^2 \, dt} \, dP. \qquad (11.2.9)$$

In fact, the Girsanov theorem we have just used above is the multidimensional version of Theorem 8.9.4. Then by Equation (11.2.5) and condition (a),

$$
\begin{aligned}
d\widehat{X}_t &= \sigma(t)\{dB_h(t) + h(t)\,dt\} + f(t)\,dt \\
&= \sigma(t)\,dB_h(t) + \{\sigma(t)h(t) + f(t)\}\,dt \\
&= \sigma(t)\,dB_h(t) + \gamma(t)\widehat{X}_t\,dt \\
&= \sigma(t)\,dB_h(t).
\end{aligned}
\tag{11.2.10}
$$

Now, suppose $p(t)$ is an arbitrage in the normalized market X_t. Then its value V_p is given by

$$
V_p(t) = \int_0^t p(s) \cdot dX_s = \int_0^t \widehat{p}(s) \cdot d\widehat{X}_s = \int_0^t \widehat{p}(s) \cdot \big(\sigma(s)\,dB_h(s)\big),
$$

where $\widehat{p}(s) = (p_1(s), p_2(s), \dots, p_n(s))$. The first equality is due to the fact that $V_p(0) = 0$ and the self-financing property, the second equality holds because $X_t^{(0)} \equiv 1$, while the third equality follows from Equation (11.2.10). Thus by the multidimensional version of Theorem 5.5.2, the stochastic process $V_p(t)$ is a local martingale with respect to Q. On the other hand, by the admissibility of $p(t)$, $V_p(t)$ is lower bounded for almost all $t \in [0, T]$ and $\omega \in \Omega$. It follows that $V_p(t)$ is a supermartingale. Therefore, we have

$$
E_Q\big(V_p(T)\big) = E_Q\big(E_Q\{V_p(T)\,|\,\mathcal{F}_0\}\big) \leq E_Q\big(V_p(0)\big) = 0.
\tag{11.2.11}
$$

Note that the probability measure Q defined by Equation (11.2.9) is equivalent to P. Hence the conditions for $p(t)$ being an arbitrage are equivalent to

$$
V_p(0) = 0, \quad V_p(T) \geq 0, \quad Q\text{--a.s.}, \quad \text{and} \quad Q\{V_p(T) > 0\} > 0,
$$

which imply that $E_Q\big(V_p(T)\big) > 0$. Obviously, this is a contradiction with Equation (11.2.11). Therefore, there is no arbitrage in the market X_t.

(2) Completeness of a Market

A lower-bounded $\mathcal{F}_T$-measurable random variable Φ is called a T-claim. A T-claim Φ is said to be *attainable* in a market X_t, $0 \leq t \leq T$, if there exist a real number r and an admissible portfolio $p(t)$ such that

$$
\Phi = V_p(T) = r + \int_0^T p(t) \cdot dX_t.
\tag{11.2.12}
$$

Such a portfolio $p(t)$ is called a *hedging portfolio* for the T-claim Φ. Let $\widetilde{V}_p(t)$ be the value of the portfolio $p(t)$ in the normalized market $\widetilde{X}_t$ defined by Equation (11.2.8). By the self-financing property,

$$
\widetilde{V}_p(t) = r + \int_0^t p(s) \cdot d\widetilde{X}_s = r + \int_0^t (X_s^{(0)})^{-1}\widehat{p}(s) \cdot d\widehat{X}_s.
\tag{11.2.13}
$$

Suppose the market X_t satisfies the conditions in Theorem 11.2.2. Then we can put Equation (11.2.10) into Equation (11.2.13) to obtain

$$\widetilde{V}_p(t) = r + \int_0^t (X_s^{(0)})^{-1} \widehat{p}(s) \cdot \left(\sigma(s) \, dB_h(s) \right). \tag{11.2.14}$$

Obviously, this equality shows that $\widetilde{V}_p(t)$ is a local martingale with respect to Q defined by Equation (11.2.9). For reasons of integrability, the portfolio $p(t)$ in Equation (11.2.12) is often assumed to have the property that the associated stochastic process $\widetilde{V}_p(t)$ in Equation (11.2.14) is actually a martingale with respect to Q.

Definition 11.2.4. *A market is said to be* complete *if every bounded T-claim is attainable.*

Theorem 11.2.5. *Let X_t, $0 \le t \le T$, be a market satisfying the assumption in Theorem 11.2.2. In addition, assume that*

$$\sigma\{B_h(s)\,;\, 0 \le s \le t\} = \sigma\{B(s)\,;\, 0 \le s \le t\}, \quad \forall 0 \le t \le T. \tag{11.2.15}$$

Then the market X_t, $0 \le t \le T$, is complete if and only if the matrix $\sigma(t, \omega)$ has a left inverse for almost all (t, ω) in $[0, T] \times \Omega$, namely, there exists an $\{\mathcal{F}_t\}$-adapted $m \times n$ matrix-valued stochastic process $L(t, \omega)$ such that

$$L(t, \omega) \sigma(t, \omega) = I_m, \quad for \ almost \ all \ (t, \omega) \in [0, T] \times \Omega,$$

where I_m is the $m \times m$ identity matrix.

We will sketch the proof of the sufficiency part of the theorem. At the same time, we will show how to find a hedging portfolio for a T-claim.

Let Φ be a bounded T-claim. We need to find $r \in \mathbb{R}$ and an admissible portfolio $p(t)$ such that Equation (11.2.12) holds. By Equation (11.2.14), we can easily see that the $\widehat{p}(t)$ part of Equation (11.2.12) is equivalent to

$$\left(X_T^{(0)}\right)^{-1} \Phi = \left(X_T^{(0)}\right)^{-1} V_p(T) = \widetilde{V}_p(T)$$

$$= r + \int_0^T \left(X_t^{(0)}\right)^{-1} \widehat{p}(t) \cdot \left(\sigma(t) \, dB_h(t) \right). \tag{11.2.16}$$

Hence we will first find r and $\widehat{p}(t)$ such that Equation (11.2.16) holds. Then we find $p_0(t)$ to get an admissible portfolio $p(t)$ satisfying Equation (11.2.12).

Notice that the random variable $\left(X_T^{(0)}\right)^{-1} \Phi$ is bounded. Moreover, it is $\mathcal{F}_T^{B_h}$-measurable since the filtration given by $B_h(t)$ is assumed to be the same as the underlined filtration $\{\mathcal{F}_t\}$ given by $B(t)$. Hence $\left(X_T^{(0)}\right)^{-1} \Phi$ belongs to $L_{B_h}^2(\Omega)$. Apply Theorem 9.8.1 to $\left(X_T^{(0)}\right)^{-1} \Phi - E\{\left(X_T^{(0)}\right)^{-1} \Phi\}$ to obtain a stochastic process $\theta(t)$ in $L_{\text{ad}}^2([0, T] \times \Omega)$ such that

$$\left(X_T^{(0)}\right)^{-1}\Phi = E\left\{\left(X_T^{(0)}\right)^{-1}\Phi\right\} + \int_0^T \theta(t) \cdot dB_h(t). \tag{11.2.17}$$

By comparing Equations (11.2.16) and (11.2.17), we see that

$$r = E\left\{\left(X_T^{(0)}\right)^{-1}\Phi\right\}$$

and $\widehat{p}(t)$ is a solution of the following equation:

$$\left(X_t^{(0)}\right)^{-1} \widehat{p}(t) \cdot \left(\sigma(t)v\right) = \theta(t) \cdot v, \quad \forall v \in \mathbb{R}^m. \tag{11.2.18}$$

Note that $u \cdot v = u^* v$ for $u, v \in \mathbb{R}^m$. Here we use A^*, rather than A^T as in Chapter 10, to denote the transpose of a matrix A in order to avoid confusion with the terminal time T in this section. Then Equation (11.2.18) is equivalent to the matrix equation $\left(X_t^{(0)}\right)^{-1} \widehat{p}(t)^* \sigma(t) = \theta(t)^*$, or equivalently,

$$\sigma(t)^* \widehat{p}(t) = X_t^{(0)} \theta(t). \tag{11.2.19}$$

By assumption, $L(t)\sigma(t) = I$. Hence $\sigma(t)^* L(t)^* = I$. Thus if we put

$$\widehat{p}(t) = X_t^{(0)} L(t)^* \theta(t), \tag{11.2.20}$$

then we have

$$\sigma(t)^* \widehat{p}(t) = \sigma(t)^* \left(X_t^{(0)} L(t)^* \theta(t)\right) = X_t^{(0)} \theta(t).$$

This proves that $\widehat{p}(t)$ given by Equation (11.2.20) is a solution of Equation (11.2.19). Finally, we need to find $p_0(t)$ such that $p(t) = (p_0(t), \widehat{p}(t))$ is a hedging portfolio for the given T-claim Φ. Since $p(t)$ is self-financing, its value $V_p(t)$ must satisfy the equation

$$dV_p(t) = p(t) \cdot dX_t,$$

which is the same as the following equation:

$$d\left(p_0(t)X_t^{(0)}\right) + d\left(\widehat{p}(t) \cdot \widehat{X}_t\right) = \gamma(t) p_0(t)X_t^{(0)} \, dt + \widehat{p}(t) \cdot d\widehat{X}_t.$$

Put $Y_t = p_0(t)X_t^{(0)}$. Then the above equation can be written as

$$dY_t = \gamma(t) Y_t \, dt + \widehat{p}(t) \cdot d\widehat{X}_t - d\left(\widehat{p}(t) \cdot \widehat{X}_t\right). \tag{11.2.21}$$

Notice that we have already derived $\widehat{p}(t)$ in Equation (11.2.20) and $\widehat{X}_t$ is the given market. Hence Equation (11.2.21) is a linear stochastic differential equation for Y_t. Then we can use Theorem 11.1.1 to find the explicit formula for the solution Y_t. Thus we get the solution

$$p_0(t) = \left(X_t^{(0)}\right)^{-1} Y_t = e^{-\int_0^t \gamma(s)\, ds} Y_t.$$

The portfolio we have derived can be shown to be a hedging portfolio for the given T-claim Φ. Hence the market X_t is complete.

Obviously, the above discussion also gives a procedure on how to find a hedging portfolio for a given T-claim Φ.

(3) Option Pricing

Let Φ be a T-claim. An *option* on Φ is a guarantee to be paid the amount $\Phi(\omega)$ at time $t = T$. From the viewpoint of a buyer, the maximal price for purchasing such an option at time $t = 0$ is

$$\wp_b(\Phi) = \sup \left\{ x \, ; \, \exists \, p(t) \text{ such that } -x + \int_0^T p(t) \cdot dX_t + \Phi \geq 0 \text{ a.s.} \right\}.$$

Since Φ is a T-claim, it is lower bounded. Suppose $-C$ is a lower bound for Φ, i.e., $\Phi \geq -C$ almost surely. Then the number $x = -C$ belongs to the above set $\{x; \cdots\}$ by taking $p(t) = 0$. It follows that the above set $\{x; \cdots\}$ is nonempty and essinf $\Phi \leq \wp_b(\Phi)$.

On the other hand, from the viewpoint of a seller, the minimum price for selling such an option at time $t = 0$ is

$$\wp_s(\Phi) = \inf \left\{ y \, ; \, \exists \, p(t) \text{ such that } y + \int_0^T p(t) \cdot dX_t \geq \Phi \text{ a.s.} \right\},$$

if the set $\{y; \cdots\}$ is not empty. Otherwise, define $\wp_s(\Phi) = \infty$.

Suppose the market X_t, $0 \leq t \leq T$, satisfies the assumption in Theorem 11.2.2, i.e., there exists an $m \times 1$ column-vector-valued $\{\mathcal{F}_t\}$-adapted stochastic process $h(t)$ satisfying conditions (a) and (b) in the theorem. Then we have the following inequalities for any T-claim Φ:

$$\wp_b(\Phi) \leq E_Q\left[(X_T^{(0)})^{-1}\Phi\right] \leq \wp_s(\Phi), \tag{11.2.22}$$

where E_Q is the expectation with respect to the probability measure Q defined by Equation (11.2.9). We only prove the second inequality since the first one can be proved by similar arguments. If $\wp_s(\Phi) = \infty$, then we have nothing to prove. So assume that $\wp_s(\Phi) < \infty$. In that case, the above set $\{y; \cdots\}$ is not empty. Let y be an element in this set. Then there exists an admissible portfolio $p(t)$ such that

$$V_p(T) = y + \int_0^T p(t) \cdot dX_t \geq \Phi, \quad \text{almost surely.}$$

Multiply both sides by $(X_T^{(0)})^{-1}$ and use Equation (11.2.14) to show that

$$(X_T^{(0)})^{-1}V_p(T) = y + \int_0^T (X_s^{(0)})^{-1}\widehat{p}(s) \cdot \left(\sigma(s)\, dB_h(s)\right) \geq (X_T^{(0)})^{-1}\Phi.$$

Then take the expectation with respect to the probability measure Q to get

$$y \geq E_Q\left[(X_T^{(0)})^{-1}\Phi\right],$$

which is true for any y in the set $\{y; \cdots\}$. This implies the second inequality in Equation (11.2.22).

In general, strict inequalities in Equation (11.2.22) can occur. Obviously, the price of a T-claim can be well-defined only when $\wp_b(\Phi) = \wp_s(\Phi)$. This is the rationale for the next definition.

Definition 11.2.6. *The price of a T-claim Φ is said to exist if $\wp_b(\Phi) = \wp_s(\Phi)$. The common value, denoted by $\wp(\Phi)$, is called the* price *of Φ at time $t = 0$.*

Theorem 11.2.7. *Let X_t, $0 \le t \le T$, be a complete market and suppose there exists an $m \times 1$ column-vector-valued $\{\mathcal{F}_t\}$-adapted stochastic process $h(t)$ satisfying conditions* (a) *and* (b) *in Theorem 11.2.2. Then for any T-claim Φ with $E_Q\{(X_T^{(0)})^{-1}\Phi\} < \infty$, the price of Φ at time $t = 0$ is given by*

$$\wp(\Phi) = E_Q\{(X_T^{(0)})^{-1}\Phi\}, \tag{11.2.23}$$

where Q is the probability measure $dQ = e^{\int_0^T h(t) \cdot dB(t) - \frac{1}{2}\int_0^T |h(t)|^2\, dt}\, dP$.

Equation (11.2.23) follows from Equation (11.2.22) and the inequalities

$$\wp_s(\Phi) \le E_Q[(X_T^{(0)})^{-1}\Phi] \le \wp_b(\Phi). \tag{11.2.24}$$

To prove the first inequality, let Φ be a T-claim. For each $n \ge 1$, define

$$\Phi_n(\omega) = \begin{cases} \Phi(\omega), & \text{if } \Phi(\omega) \le n; \\ n, & \text{if } \Phi(\omega) > n. \end{cases}$$

Since Φ is a T-claim, it is lower bounded. Hence for each n, Φ_n is a bounded T-claim. Then by the completeness of the market, there exist $y_n \in \mathbb{R}$ and a portfolio $p_n(t)$ such that

$$y_n + \int_0^T p_n(t) \cdot dX_t = \Phi_n, \tag{11.2.25}$$

which, by Equation (11.2.14), is equivalent to

$$y_n + \int_0^T (X_s^{(0)})^{-1}\widehat{p}(s) \cdot (\sigma(s)\, dB_h(s)) = (X_T^{(0)})^{-1}\Phi_n.$$

Take the expectation with respect to Q to get

$$y_n = E_Q[(X_T^{(0)})^{-1}\Phi_n]. \tag{11.2.26}$$

On the other hand, by the definition of $\wp_s$, Equation (11.2.25) implies that

$$\wp_s(\Phi_n) \le y_n. \tag{11.2.27}$$

Now observe that the definition of $\wp_s$ implies that $\wp_s(\Phi) \le \wp_s(\Theta)$ for any T-claims $\Theta \le \Phi$ almost surely. Since $\Phi_n \le \Phi$, we have

$$\wp_s(\varPhi) \le \wp_s(\varPhi_n).$$ (11.2.28)

It follows from Equations (11.2.26), (11.2.27), and (11.2.28) that

$$\wp_s(\varPhi) \le E_Q\big[(X_T^{(0)})^{-1}\varPhi_n\big].$$

Let $n \to \infty$ and use the monotone convergence theorem to conclude that

$$\wp_s(\varPhi) \le E_Q\big[(X_T^{(0)})^{-1}\varPhi\big],$$

which proves the first inequality in Equation (11.2.24). The second inequality in this equation can be proved by similar arguments.

(4) Black–Scholes Model

We now explain a simplified version of the Black–Scholes model due to Black and Scholes [2] and Merton [60]. Suppose a market $X_t = (X_t^{(0)}, X_t^{(1)})$ is given by

$$\begin{cases} dX_t^{(0)} = \gamma(t)X_t^{(0)}\, dt, \quad X_0^{(0)} = 1, \\ dX_t^{(1)} = \alpha(t)X_t^{(1)}\, dB(t) + J(t)X_t^{(1)}\, dt. \end{cases}$$ (11.2.29)

By Theorem 11.1.1, the solutions are given by

$$X_t^{(0)} = e^{\int_0^t \gamma(s)\, ds},$$ (11.2.30)

$$X_t^{(1)} = x_1 \exp\left[\int_0^t \alpha(s)\, dB(s) + \int_0^t \left(J(s) - \frac{1}{2}\alpha(s)^2\right) ds\right].$$ (11.2.31)

Hence the market X_t, $0 \le t \le T$, in Equation (11.2.29) is specified by

$$\gamma(t), \quad \sigma(t) = \alpha(t)X_t^{(1)}, \quad f(t) = J(t)X_t^{(1)},$$

with $X_t^{(1)}$ being given by Equation (11.2.31). The equation in condition (a) of Theorem 11.2.2 becomes

$$\alpha(t)X_t^{(1)}h(t) = \gamma(t)X_t^{(1)} - J(t)X_t^{(1)},$$

which has the obvious solution for $h(t)$

$$h(t) = \frac{\gamma(t) - J(t)}{\alpha(t)}.$$ (11.2.32)

Thus we need to assume the following condition in order for condition (b) of Theorem 11.2.2 to hold:

$$E \exp\left[\frac{1}{2}\int_0^T \left(\frac{\gamma(t) - J(t)}{\alpha(t)}\right)^2 dt\right] < \infty.$$ (11.2.33)

Under this assumption, we can apply Theorem 11.2.2 to assert that the market has no arbitrage.

For the completeness of the market, we need to assume the condition in Equation (11.2.15). On the other hand, the condition in Equation (11.2.33) implies implicitly that $\alpha(t) \neq 0$. Hence $\sigma(t)^{-1}$ exists. Then by Theorem 11.2.5, the market is complete.

Assuming the conditions in Equations (11.2.15) and (11.2.33), we can use Theorem 11.2.7 to compute the price $\wp(\Phi)$ of a T-claim Φ. In order to obtain an explicit formula for computing $\wp(\Phi)$, we need to impose conditions on $\{\gamma(t), \alpha(t), J(t)\}$, which specifies the market, and the form of Φ.

So, we impose the condition that $\gamma(t)$ and $\alpha(t)$ are deterministic functions. Moreover, we consider a T-claim Φ of the form

$$\Phi = \Gamma(X_T^{(1)}).$$

Use Theorem 11.2.7 and Equation (11.2.30) to find the price $\wp(\Phi)$,

$$\wp(\Phi) = E_Q\{(X_T^{(0)})^{-1}\Phi\} = e^{-\int_0^T \gamma(t)\,dt} E_Q\{F(X_T^{(1)})\}. \qquad (11.2.34)$$

By Equation (11.2.31), we have

$$E_Q\{F(X_T^{(1)})\} = E_Q\left\{F\left(x_1\, e^{\int_0^T \alpha(t)\,dB(t) + \int_0^T \left(J(t) - \frac{1}{2}\alpha(t)^2\right) dt}\right)\right\}. \qquad (11.2.35)$$

Now we use $B_h(t)$ with h being given by Equation (11.2.32) to rewrite the exponent in the above equation as

$$\int_0^T \alpha(t)\,dB(t) + \int_0^T \left(J(t) - \tfrac{1}{2}\alpha(t)^2\right) dt$$

$$= \int_0^T \alpha(t)\left(dB_h(t) + h(t)\,dt\right) + \int_0^T \left(J(t) - \tfrac{1}{2}\alpha(t)^2\right) dt$$

$$= \int_0^T \alpha(t)\,dB_h(t) + \int_0^T \left[\gamma(t) - \tfrac{1}{2}\alpha(t)^2\right] dt. \qquad (11.2.36)$$

Observe that the stochastic process $J(t)$ disappears from the last equation. Moreover, the last integral yields a constant since the functions $\gamma(t)$ and $\alpha(t)$ are assumed to be deterministic functions.

From Equations (11.2.34), (11.2.35), and (11.2.36) we get

$$\wp(\Phi) = e^{-\int_0^T \gamma(t)\,dt} E_Q\left\{F\left(x_1\, e^{\int_0^T \alpha(t)\,dB_h(t) + \int_0^T [\gamma(t) - \frac{1}{2}\alpha(t)^2]\,dt}\right)\right\}.$$

Note that $B_h(t)$ is a Brownian motion with respect to Q, and $\alpha(t)$ is assumed to be a deterministic function. Hence $\int_0^T \alpha(t)\,dB_h(t)$ is a Wiener integral and so it has a normal distribution with mean 0 and variance $\|\alpha\|^2 = \int_0^T \alpha(t)^2\,dt$ with respect to Q. From the last equation, we immediately obtain the formula for $\wp(\Phi)$ given in the next theorem.

Theorem 11.2.8. *Let X_t, $0 \leq t \leq T$, be a market specified by Equation (11.2.29) with $\gamma(t)$ and $\alpha(t)$ being deterministic functions in $L^1[0,T]$ and $L^2[0,T]$, respectively. Assume that the condition in Equation (11.2.33) holds. Then the price $\wp(\Phi)$ of a T-claim $\Phi = F(X_T^{(1)})$ at time $t = 0$ is given by*

$$\wp(\Phi) = e^{-\int_0^T \gamma(t)\,dt} \frac{1}{\sqrt{2\pi c}} \int_{\mathbb{R}} F\left(x_1 e^{y + \int_0^T [\gamma(t) - \frac{1}{2}\alpha(t)^2]\,dt}\right) e^{-\frac{1}{2c}y^2}\,dy,$$

where c is the constant $c = \int_0^T \alpha(t)^2\,dt$.

It is worthwhile to point out that the price $\wp(\Phi)$ does not depend on the randomness part $J(t)$ of the market.

Next, we will derive a hedging portfolio for the T-claim $\Phi = F(X_T^{(1)})$. Assume that the functions $\gamma(t) = \gamma$ and $\alpha(t) = \alpha$ are constants. Then the market is given by

$$X_t^{(0)} = e^{\gamma t}, \quad X_t^{(1)} = x_1 e^{\alpha B_h(t) + (\gamma - \frac{1}{2}\alpha^2)t}. \tag{11.2.37}$$

By Equation (11.2.19), $p_1(t)$ of a hedging portfolio $p(t) = (p_0(t), p_1(t))$ is a solution of the equation

$$e^{-\gamma t} p_1(t)\,\alpha X_t^{(1)} = \theta(t), \tag{11.2.38}$$

where $\theta(t)$ is a stochastic process satisfying Equation (11.2.17), i.e.,

$$e^{-\gamma T} F(X_T^{(1)}) = E\{e^{-\gamma T} F(X_T^{(1)})\} + \int_0^T \theta(t)\,dB_h(t).$$

By Theorem 9.8.1, $\theta(t)$ is given by

$$\theta(t) = E_Q\left\{ \frac{\delta}{\delta t}\left(e^{-\gamma T} F(X_T^{(1)})\right) \Big| \mathcal{F}_t^{B_h} \right\}, \tag{11.2.39}$$

where $\delta/\delta t$ is the variational derivative defined by Equation (9.7.3).

Assume that F is a C^1-function. Then we can use the explicit form of $X_T^{(1)}$ from Equation (11.2.37) to find that

$$\frac{\delta}{\delta t} F(X_T^{(1)}) = \alpha F'\left(x_1 e^{\alpha B_h(T) + (\gamma - \frac{1}{2}\alpha^2)T}\right) x_1 e^{\alpha B_h(T) + (\gamma - \frac{1}{2}\alpha^2)T}.$$

Write the exponent as $\alpha\{B_h(T) - B_h(t)\} + \alpha B_h(t) + (\gamma - \frac{1}{2}\alpha^2)T$ and then use Exercise 4 in Chapter 10 with $X = \alpha\{B_h(T) - B_h(t)\}$, $Y_1 = \alpha B_h(t)$, and $\mathcal{G} = \mathcal{F}_t^{B_h}$ to derive the following equality:

$$E_Q\left\{ \frac{\delta}{\delta t} F(X_T^{(1)}) \Big| \mathcal{F}_t^{B_h} \right\} = \frac{\alpha}{\sqrt{2\pi(T-t)}}$$

$$\times \int_{\mathbb{R}} F'\left(x_1 e^{\alpha y + \alpha B_h(t) + (\gamma - \frac{1}{2}\alpha^2)T}\right) x_1 e^{\alpha y + \alpha B_h(t) + (\gamma - \frac{1}{2}\alpha^2)T} e^{-\frac{1}{2(T-t)}y^2}\,dy,$$

which can be simplified using $X_t^{(1)}$ from Equation (11.2.37),

$$E_Q\left\{\frac{\delta}{\delta t}F(X_T^{(1)})\,\Big|\,\mathcal{F}_t^{B_n}\right\} = \frac{\alpha}{\sqrt{2\pi(T-t)}}X_t^{(1)}$$
$$\times \int_{\mathbb{R}} F'\big(X_t^{(1)}\,e^{\alpha y+(\gamma-\frac{1}{2}\alpha^2)(T-t)}\big)\,e^{\alpha y+(\gamma-\frac{1}{2}\alpha^2)(T-t)}\,e^{-\frac{1}{2(T-t)}y^2}\,dy.$$

Put this equality into Equation (11.2.39) to find $\theta(t)$:

$$\theta(t) = e^{-\gamma T}X_t^{(1)}\,e^{(\gamma-\frac{1}{2}\alpha^2)(T-t)}\frac{\alpha}{\sqrt{2\pi(T-t)}}$$
$$\times \int_{\mathbb{R}} F'\big(X_t^{(1)}\,e^{\alpha y+(\gamma-\frac{1}{2}\alpha^2)(T-t)}\big)\,e^{\alpha y-\frac{1}{2(T-t)}y^2}\,dy. \qquad (11.2.40)$$

From Equations (11.2.38) and (11.2.40), we get the value of $p_1(t)$. Once we have $p_1(t)$, then we can derive $p_0(t)$ for a portfolio $p(t) = (p_0(t), p_1(t))$ using Equation (11.2.6), which in the present case reduces to

$$dp_0(t) = -e^{-\gamma t}X_t^{(1)}\,dp_1(t) - \alpha\,e^{-\gamma t}X_t^{(1)}\,(dp_1(t))(dB(t)).$$

We summarize the above discussion as the next theorem.

Theorem 11.2.9. *Let X_t, $0 \le t \le T$, be a market specified by Equation (11.2.29) with $\gamma(t) = \gamma$ and $\alpha(t) = \alpha$ being constant functions. Assume that the condition in Equation (11.2.33) holds. Then for a C^1-function F, a hedging portfolio $p(t) = (p_0(t), p_1(t))$ for the T-claim $\Phi = F(X_T^{(1)})$ is given by*

$$p_1(t) = C\int_{\mathbb{R}} F'\big(X_t^{(1)}\,e^{\alpha y+(\gamma-\frac{1}{2}\alpha^2)(T-t)}\big)\,e^{\alpha y-\frac{1}{2(T-t)}y^2}\,dy,$$
$$p_0(t) = -\int_0^t X_s^{(1)}e^{-\gamma s}\,dp_1(s) - \alpha\int_0^t X_s^{(1)}e^{-\gamma s}\,(dp_1(s))(dB(s)),$$

where $C = e^{-\frac{1}{2}\alpha^2(T-t)}\dfrac{1}{\sqrt{2\pi(T-t)}}$.

Example 11.2.10. Take a simple Black–Scholes model specified by constant functions $\gamma(t) = \gamma$, $\alpha(t) = \alpha$, and an adapted stochastic process $J(t)$. Then the market $X_t = (X_t^{(0)}, X_t^{(1)})$ is given by

$$X_t^{(0)} = e^{\gamma t}, \quad X_t^{(1)} = x_1\,e^{\alpha B(t)+\int_0^t J(s)\,ds-\frac{1}{2}\alpha^2 t}.$$

Let $F(x) = x^n$, n even, and consider the corresponding T-claim $\Phi = \big(X_T^{(1)}\big)^n$. We can use Theorem 11.2.8 to derive the price of Φ at time $t = 0$,

$$\wp(\Phi) = x_1^n e^{(n-1)\gamma T+\frac{1}{2}n(n-1)\alpha^2 T}.$$

Note that when $(n-1)\gamma + \frac{1}{2}n(n-1)\alpha^2 > 0$, $\wp(\Phi)$ increases to ∞ as $T \to \infty$. This reflects the fact that as time passes, predictions become less reliable.

We can use Theorem 11.2.9 to find a hedging portfolio $p(t) = (p_0(t), p_1(t))$ for the T-claim $\Phi = \left(X_T^{(1)}\right)^n$,

$$p_0(t) = -n(n-1)\alpha x_1^n \int_0^t e^{n\alpha B_h(s) - \frac{1}{2}n\alpha^2 s + (n-1)\gamma s} \, dB_h(s),$$

$$p_1(t) = nx_1^n \, e^{(n-1)\alpha B_h(t) + (n-1)\gamma T + \frac{1}{2}n(n-1)\alpha^2 T \frac{1}{2}(n^2-1)\alpha^2 t},$$

where $h(t) = \frac{\gamma - J(t)}{\alpha}$.

11.3 Application to Filtering Theory

In this section we explain very briefly the Kalman–Bucy linear filtering. For more information, see the books by Kallianpur [45] and by Øksendal [66].

The state ξ_t of a system (input process) at time t is specified by a linear stochastic differential equation

$$d\xi_t = \alpha(t) \, dB(t) + \beta(t)\xi_t \, dt, \quad t \ge 0, \quad \xi_0 \text{ at } t = 0, \tag{11.3.1}$$

where $\alpha(t)$ and $\beta(t)$ are deterministic functions, $B(t)$ is a Brownian motion, and the initial distribution ξ_0 is assumed to be independent of the Brownian motion $B(t)$.

On the other hand, the observation Z_t (output process) of the system at time t is given by the stochastic process

$$dZ_t = f(t) \, dW(t) + g(t) \, \xi_t \, dt, \quad t \ge 0, \quad Z_0 = 0, \tag{11.3.2}$$

where $f(t)$ and $g(t)$ are deterministic functions and $W(t)$ is a Brownian motion assumed to be independent of $B(t)$ and ξ_0. Note that the input ξ_t appears in Equation (11.3.2) and is being filtered out by the observable output Z_t.

Here is the filtering problem: *Based on the observed values Z_s, $0 \le s \le t$, what is the best estimator $\widehat{\xi}_t$ of the state ξ_t of the system at time t?*

There are several ways to define the meaning of being the best estimator. The one that is the easiest to handle mathematically is the least mean square error, namely, we seek $\widehat{\xi}_t$ to minimize the mean square error

$$R_t \equiv E\left\{(\xi_t - \widehat{\xi}_t)^2\right\} \le E\left\{(\xi_t - Y)^2\right\}$$

for any square integrable random variable Y that is measurable with respect to the σ-field

$$\mathcal{F}_t^Z \equiv \sigma\{Z_s \, ; \, s \le t\}.$$

Note that such an estimator $\widehat{\xi}_t$ is the orthogonal projection of ξ_t onto the Hilbert space $L^2(\mathcal{F}_t^Z)$ consisting of square integrable random variables that are $\mathcal{F}_t^Z$-measurable. This estimator turns out to coincide with the conditional expectation of ξ_t given the σ-field $\mathcal{F}_t^Z$, namely,

$$\widehat{\xi}_t = E\{\xi_t \mid \mathcal{F}_t^Z\}. \tag{11.3.3}$$

Therefore, in the sense of the least mean square error, the conditional expectation in Equation (11.3.3) is the best estimator for the state ξ_t based on the observation Z_s, $0 \le s \le t$.

Next comes the real question: *How to compute $E\{\xi_t \mid \mathcal{F}_t^Z\}$.*

Consider a very simple example with the following functions specifying Equations (11.3.1) and (11.3.2):

$$\alpha(t) = \beta(t) = 0, \quad f(t) = g(t) = 1, \quad \xi_0 \sim N(0,1),$$

where $N(0,1)$ denotes the standard normal distribution. In this case, $\xi_t = \xi_0$ is a constant stochastic process and $Z_t = W(t) + t\,\xi_0$. By assumption $W(t)$ and ξ_0 are independent. One can carry out a rather lengthy computation to find the conditional expectation

$$E\{\xi_t \mid Z_t\} = \frac{1}{1+t} Z_t.$$

On the other hand, using the next theorem, it is rather easy to derive the equality $E\{\xi_t \mid \mathcal{F}_t^Z\} = \frac{1}{1+t} Z_t$, which implies the above equality.

Theorem 11.3.1. (Kalman–Bucy [46]) *Let the state ξ_t of a system be given by Equation (11.3.1) with deterministic functions $\alpha(t)$ and $\beta(t)$. Assume that the initial distribution ξ_0 is independent of the Brownian motion $B(t)$ and has mean μ_0 and variance σ_0^2.*

Suppose the observation Z_t of the system is given by Equation (11.3.2) with deterministic functions $f(t)$ and $g(t)$, and a Brownian motion $W(t)$. Assume that $W(t)$ is independent of $B(t)$ and ξ_0.

Then the conditional expectation $\widehat{\xi}_t = E\{\xi_t \mid \mathcal{F}_t^Z\}$ is the solution of the linear stochastic differential equation

$$d\widehat{\xi}_t = \frac{g(t)R_t}{f(t)^2}\, dZ_t + \left(\beta(t) - \frac{g(t)^2 R_t}{f(t)^2}\right)\widehat{\xi}_t\, dt, \quad \widehat{\xi}_0 = \mu_0, \tag{11.3.4}$$

where R_t is the solution of the Riccati equation

$$\frac{dR_t}{dt} = \alpha(t)^2 + 2\beta(t)R_t - \frac{g(t)^2}{f(t)^2} R_t^2, \quad R_0 = \sigma_0^2. \tag{11.3.5}$$

Moreover, R_t equals the mean square error, i.e., $R_t = E\{(\xi_t - \widehat{\xi}_t)^2\}$.

For a rather nice proof of this theorem and some examples, see the book by Øksendal [66]. Here we give two simple examples.

Example 11.3.2. Consider a linear filtering problem with the state ξ_t of the system and the observation Z_t specified by

$$d\xi_t = 0, \quad E\xi_0 = \mu_0, \ \mathrm{Var}(\xi_0) = \sigma_0^2,$$

$$dZ_t = c\,dW(t) + \xi_t\,dt, \quad Z_0 = 0,$$

where $c \neq 0$. Hence we have the corresponding functions

$$\alpha(t) \equiv 0, \quad \beta(t) \equiv 0, \quad f(t) \equiv c, \quad g(t) \equiv 1.$$

Note that $\xi_t = \xi_0$ for all t, so that ξ_t is a constant stochastic process. First we need to solve the Riccati equation

$$\frac{dR_t}{dt} = -\frac{1}{c^2}R_t^2, \quad R_0 = \sigma_0^2.$$

The solution is easily found to be given by

$$R_t = \frac{c^2\sigma_0^2}{c^2 + \sigma_0^2 t}.$$

Put R_t into Equation (11.3.4) to get the following linear equation:

$$d\widehat{\xi_t} = \frac{\sigma_0^2}{c^2 + \sigma_0^2 t}\,dZ_t - \frac{\sigma_0^2}{c^2 + \sigma_0^2 t}\xi_t\,dt, \quad \widehat{\xi_0} = \mu_0.$$

Then we use the formula in Equation (11.1.11) to get the solution

$$\widehat{\xi_t} = \frac{\mu_0}{c^2 + \sigma_0^2 t} + \frac{\sigma_0^2}{c^2 + \sigma_0^2 t}Z_t.$$

Example 11.3.3. Let the state ξ_t of a system and the observation Z_t in a linear filtering problem be specified by

$$d\xi_t = dB(t) - \frac{1+t}{2}\xi_t\,dt, \quad \mu_0 = 0, \ \sigma_0 = 1,$$

$$dZ_t = \frac{1}{1+t}\,dW(t) + \frac{1}{1+t}\xi_t\,dt, \quad Z_0 = 0.$$

Hence we have the corresponding functions

$$\alpha(t) \equiv 1, \quad \beta(t) = -\frac{1+t}{2}, \quad f(t) = g(t) = \frac{1}{1+t}.$$

Put these functions into Equation (11.3.5) to get the Riccati equation

$$\frac{dR_t}{dt} = 1 - (1+t)R_t - R_t^2, \quad R_0 = 1.$$

The solution of this equation is given by

$$R_t = \frac{1}{1+t}.$$

Then put this function R_t, together with the above $\beta(t)$, $f(t)$, and $g(t)$, into Equation (11.3.4) to get the linear stochastic differential equation

$$d\widehat{\xi}_t = dZ_t - \left(\frac{1+t}{2} + \frac{1}{1+t}\right)\widehat{\xi}_t\, dt, \quad \widehat{\xi}_0 = 0.$$

Finally we use the formula in Equation (11.1.11) to obtain the solution

$$\widehat{\xi}_t = \frac{1}{1+t}\, e^{-\frac{1}{2}(t+\frac{1}{2}t^2)} \int_0^t (1+s)\, e^{\frac{1}{2}(s+\frac{1}{2}s^2)}\, dZ_s.$$

11.4 Feynman–Kac Formula

Consider a nonrelativistic particle of mass m moving in $\mathbb{R}$ under the influence of a conservative force given by a potential V. In quantum mechanics, the state of the particle at time t is specified as a function $u(t, x)$, $t \geq 0$, $x \in \mathbb{R}$, satisfying the Schrödinger equation

$$i\hbar \frac{\partial u}{\partial t} = -\frac{\hbar^2}{2m} \Delta u + V(x)u, \quad u(0, x) = \psi(x), \qquad (11.4.1)$$

where $\hbar$ is the Planck constant and $\int_{\mathbb{R}} |\psi(x)|^2\, dx = 1$. In 1948, Feynman [16] gave an informal expression for the solution $u(t, x)$ as follows:

$$u(t, x) = \mathcal{N} \int_{\mathcal{C}_x} \exp\left[\frac{i}{\hbar} \int_0^t \left(\frac{m}{2}\dot{y}(s)^2 - V(y(s))\right) ds\right] \psi(y(t))\, \mathcal{D}_t^\infty[y],$$

where $\mathcal{N}$ is a symbolic renormalization constant, $\mathcal{C}_x$ is the space of continuous functions $y(t)$ on $[0, \infty)$ with $y(0) = x$, and $\mathcal{D}_t^\infty[y]$ is a symbolic expression for the nonexistent infinite-dimensional Lebesgue measure. Since then, this informal expression has been called a *Feynman integral*. However, it is not really an integral, namely, there exists no measure that gives this integral. Consequently, there has been a vast number of research papers in the literature with the aim to give a mathematical meaning to the Feynman integral. Among the many approaches, white noise theory has been quite successful; see the books [27] and [56].

On the other hand, suppose we adopt the imaginary time, i.e., replace t in Equation (11.4.1) with $-it$. Moreover, put $\hbar = m = 1$ for simplicity. Then the Schrödinger equation becomes

$$\frac{\partial u}{\partial t} = \frac{1}{2} \Delta u - V(x)u, \quad u(0, x) = \psi(x).$$

The corresponding $u(t, x)$ was shown by Kac [44] in 1949 to take the form

$$u(t, x) = E^x \left\{ \psi(B(t))\, e^{-\int_0^t V(B(s))\, ds} \right\}, \qquad (11.4.2)$$

where E^x is the conditional expectation given that the Brownian motion starts at x, or equivalently, the expectation when we use $B(t) + x$ for a Brownian motion $B(t)$ with $B(0) = 0$. The representation $u(t, x)$ in Equation (11.4.2) is known as the *Feynman–Kac formula*. Below we sketch a heuristic derivation of this formula for Itô processes.

Let $B(t)$ be an $\mathbb{R}^n$-valued Brownian motion. Consider an $\mathbb{R}^n$-valued stochastic differential equation

$$dX_t = \sigma(X_t)\, dB(t) + f(X_t)\, dt, \tag{11.4.3}$$

where the matrix-valued function $\sigma(x)$ and the column-vector-valued function $f(x)$ satisfy the Lipschitz condition. Then by the multidimensional version of Theorem 10.6.2, the unique continuous solution X_t is a stationary Markov process. Moreover, by Theorems 10.8.9 and 10.9.5, X_t is a diffusion process with infinitesimal generator given by

$$(\mathcal{A}\phi)(x) = \frac{1}{2}\operatorname{tr}\sigma(x)\sigma(x)^T D^2\phi(x) + f(x)\cdot\nabla\phi(x). \tag{11.4.4}$$

Now, let $\psi(x)$ be a C^2-function on $\mathbb{R}^n$ with compact support, namely, it vanishes outside a bounded set. Let $V(x)$ be a lower-bounded continuous function on $\mathbb{R}^n$. Define a function $u(t, x)$ by

$$u(t, x) = E^x\left\{\psi(X_t)\, e^{-\int_0^t V(X_s)\, ds}\right\}, \tag{11.4.5}$$

where E^x is the conditional expectation given that $X_0 = x$.

It follows from the assumptions on ψ and V that $u(t, x)$ is a real-valued function on $[0, \infty) \times \mathbb{R}^n$. Moreover, both $\psi(X_t)$ and $\exp[-\int_0^t V(X_s)\, ds]$ are Itô processes. Hence the product $\psi(X_t)\exp[-\int_0^t V(X_s)\, ds]$ is an Itô process. This implies that $u(t, x)$ is differentiable in the t variable.

Let $t \geq 0$ be fixed and consider the function $g(x) = u(t, x)$. The function $\mathcal{A}g$ is defined to be the limit

$$(\mathcal{A}g)(x) = \lim_{\varepsilon\downarrow 0} \frac{E^x\left[g(X_\varepsilon)\right] - g(x)}{\varepsilon}. \tag{11.4.6}$$

In order to find out the limit, first note that

$$E^x\left[g(X_\varepsilon)\right] = E^x\left[E^{X_\varepsilon}\left\{\psi(X_t)\, e^{-\int_0^t V(X_s)\, ds}\right\}\right].$$

Then we use a crucial argument, i.e., the one we used in the proof of Theorem 10.6.2, to show that

$$E^{X_\varepsilon}\left\{\psi(X_t)\, e^{-\int_0^t V(X_s)\, ds}\right\} = E^x\left\{\psi(X_{t+\varepsilon})\, e^{-\int_0^t V(X_{s+\varepsilon})\, ds}\,\Big|\, \mathcal{F}_\varepsilon\right\}.$$

Therefore,

$$E^x\big[g(X_\varepsilon)\big] = E^x\Big\{\psi(X_{t+\varepsilon})\, e^{-\int_0^t V(X_{s+\varepsilon})\, ds}\Big\}$$

$$= E^x\Big\{\psi(X_{t+\varepsilon})\, e^{-\int_0^{t+\varepsilon} V(X_s)\, ds}\, e^{\int_0^\varepsilon V(X_s)\, ds}\Big\}$$

$$= u(t+\varepsilon, x) + E^x\Big\{\psi(X_{t+\varepsilon})\, e^{-\int_0^{t+\varepsilon} V(X_s)\, ds}\Big(e^{\int_0^\varepsilon V(X_s)\, ds} - 1\Big)\Big\}.$$

Recall that $g(x) = u(t, x)$ with t fixed. Thus we have

$$E^x\big[g(X_\varepsilon)\big] - g(x) = u(t+\varepsilon, x) - u(t, x)$$

$$+ E^x\Big\{\psi(X_{t+\varepsilon})\, e^{-\int_0^{t+\varepsilon} V(X_s)\, ds}\Big(e^{\int_0^\varepsilon V(X_s)\, ds} - 1\Big)\Big\}. \qquad (11.4.7)$$

But as remarked above, the function $u(t, x)$ is differentiable in t. Hence

$$\lim_{\varepsilon\downarrow 0} \frac{1}{\varepsilon}\big(u(t+\varepsilon, x) - u(t, x)\big) = \frac{\partial u}{\partial t}(t, x). \qquad (11.4.8)$$

On the other hand, by the continuity of V,

$$\lim_{\varepsilon\downarrow 0} \frac{1}{\varepsilon}\Big(e^{\int_0^\varepsilon V(X_s)\, ds} - 1\Big) = V(x),$$

which, together with the continuity of ψ, implies that

$$\lim_{\varepsilon\downarrow 0} \frac{1}{\varepsilon} E^x\Big\{\psi(X_{t+\varepsilon})\, e^{-\int_0^{t+\varepsilon} V(X_s)\, ds}\Big(e^{\int_0^\varepsilon V(X_s)\, ds} - 1\Big)\Big\}$$

$$= V(x)E^x\Big\{\psi(X_t)\, e^{-\int_0^t V(X_s)\, ds}\Big\}$$

$$= V(x)u(t, x). \qquad (11.4.9)$$

Putting Equations from (11.4.6) to (11.4.9) together, we get

$$\mathcal{A}u(t, x) = (\mathcal{A}g)(x) = \frac{\partial u}{\partial t}(t, x) + V(x)u(t, x).$$

Moreover, it is obvious that $u(0, x) = \psi(x)$.

We summarize the above heuristic derivation as the next theorem.

Theorem 11.4.1. (Feynman–Kac formula) *Let X_t be the continuous solution of the stochastic differential equation in Equation (11.4.3) and let $\mathcal{A}$ be its infinitesimal generator. Suppose $\psi(x)$ is a C^2-function on $\mathbb{R}^n$ with compact support and $V(x)$ is a lower-bounded continuous function on $\mathbb{R}^n$. Then*

$$u(t, x) = E^x\Big\{\psi(X_t)\, e^{-\int_0^t V(X_s)\, ds}\Big\}, \quad t \ge 0, \ x \in \mathbb{R}^n,$$

is a solution of the partial differential equation

$$\frac{\partial u}{\partial t} = \mathcal{A}u - Vu, \quad t \ge 0, \quad u(0) = \psi. \qquad (11.4.10)$$

Remark 11.4.2. A solution of Equation (11.4.10) is unique in the space of $C^{1,2}$-functions $v(t, x)$ such that v is bounded on $K \times \mathbb{R}^n$ for any compact set $K \subset [0, \infty)$. For the proof of this fact, see the books [49] and [66].

A good application of this theorem is the derivation of a formula, known as the arc-sine law of Lévy, that answers the following question.

Question 11.4.3. Let L_t^+ be the amount of time that a Brownian motion is positive during $[0, t]$. What is the distribution function of L_t^+?

Let sgn^+ denote the positive part of the signum function, namely,

$$\mathrm{sgn}^+(x) = \begin{cases} 1, & \text{if } x > 0; \\ 0, & \text{if } x \le 0. \end{cases}$$

Then the random variable L_t^+ can be expressed as the following integral:

$$L_t^+ = \int_0^t \mathrm{sgn}^+(B(s)) \, ds. \qquad (11.4.11)$$

In order to find out the distribution function of L_t^+, consider the case of one-dimensional Brownian motion in the Feynman–Kac formula with $\psi \equiv 1$ and $V = a \, \mathrm{sgn}^+$, $a \ge 0$. Then $\mathcal{A} = \frac{d^2}{dx^2}$ and we have the function

$$u(t, x) = E^x \left\{ e^{-a \int_0^t \mathrm{sgn}^+(B(s)) \, ds} \right\}. \qquad (11.4.12)$$

By Theorem 11.4.1 and its remark, $u(t, x)$ is the solution of the following partial differential equation:

$$\frac{\partial u}{\partial t} = \frac{1}{2} u'' - (a \, \mathrm{sgn}^+) \, u, \quad u(0, x) = 1, \qquad (11.4.13)$$

where u'' denotes the second derivative of u with respect to x.

Let $U(\zeta, x)$ be the Laplace transform of $u(t, x)$ in the t variable, i.e.,

$$U(\zeta, x) = (\mathcal{L}u(\cdot, x))(\zeta) = \int_0^\infty e^{-\zeta t} u(t, x) \, dt, \quad \zeta > 0. \qquad (11.4.14)$$

Use the property $(\mathcal{L}\frac{\partial u}{\partial t})(\zeta) = \zeta U - u(0)$ and the condition $u(0) = 1$ to get the Laplace transform of Equation (11.4.13),

$$\zeta U - 1 = \frac{1}{2} U'' - (a \, \mathrm{sgn}^+) U,$$

which, for each fixed ζ, is a second-order differential equation in the x variable and can be rewritten as follows:

$$\begin{cases} U'' - 2(a + \zeta)U = -2, & \text{if } x > 0; \\ U'' - 2\zeta U = -2, & \text{if } x < 0. \end{cases} \qquad (11.4.15)$$

These two equations have particular solutions $\frac{1}{a+\zeta}$ and $\frac{1}{\zeta}$, respectively. The first equation has characteristic equation $r^2 - 2(a + \zeta) = 0$, which has two roots $r = \pm\sqrt{2(a + \zeta)}$. The positive one should be discarded because the corresponding exponential function goes to ∞ as $x \to \infty$. On the other hand, the characteristic equation of the second equation has two roots $r = \pm\sqrt{2\zeta}$ and the negative one should be discarded. Hence the solution of Equation (11.4.15) is given by

$$U(\zeta, x) = \begin{cases} Ae^{-x\sqrt{2(a+\zeta)}} + \dfrac{1}{a + \zeta}, & \text{if } x > 0; \\[2ex] Be^{x\sqrt{2\zeta}} + \dfrac{1}{\zeta}, & \text{if } x < 0. \end{cases}$$

The continuity of U and U' at $x = 0$ implies that

$$A = \frac{\sqrt{a + \zeta} - \sqrt{\zeta}}{\sqrt{\zeta}\,(a + \zeta)}, \quad B = \frac{\sqrt{\zeta} - \sqrt{a + \zeta}}{\zeta\sqrt{a + \zeta}}.$$

Hence we have the value of $U(\zeta, x)$ at $x = 0$,

$$U(\zeta, 0) = \frac{1}{\sqrt{\zeta(a + \zeta)}}, \quad \zeta > 0.$$

Now we have the following facts for the Laplace transform:

$$\frac{1}{\sqrt{t}} \longmapsto \sqrt{\pi}\frac{1}{\sqrt{\zeta}}, \quad e^{-at}\frac{1}{\sqrt{t}} \longmapsto \sqrt{\pi}\frac{1}{\sqrt{a + \zeta}}, \quad f * g \longmapsto (\mathcal{L}f)(\mathcal{L}g),$$

where $f * g$ is the convolution of f and g defined by

$$f * g(t) = \int_0^t f(t - \tau)g(\tau)\,d\tau, \quad t > 0.$$

Hence the inverse Laplace transform of $U(\zeta, 0)$ is given by the convolution

$$\left(\mathcal{L}^{-1}U(\cdot, 0)\right)(t) = \frac{1}{\pi}\int_0^t \frac{1}{\sqrt{t - \tau}}\,e^{-a\tau}\frac{1}{\sqrt{\tau}}\,d\tau.$$

On the other hand, recall that $U(\zeta, x)$ is the Laplace transform of $u(t, x)$. Therefore, we conclude that

$$u(t, 0) = \frac{1}{\pi}\int_0^t e^{-a\tau}\frac{1}{\sqrt{\tau(t - \tau)}}\,d\tau. \tag{11.4.16}$$

It follows from Equations (11.4.11), (11.4.12), and (11.4.16) that

$$E^0\{e^{-aL_t^+}\} = \frac{1}{\pi}\int_0^t e^{-a\tau}\frac{1}{\sqrt{\tau(t - \tau)}}\,d\tau, \quad \forall\, a \geq 0, \, 0 < \tau < t.$$

This equality implies that the random variable L_t^+ has a density function

$$f(\tau) = \begin{cases} \dfrac{1}{\pi} \dfrac{1}{\sqrt{\tau(t-\tau)}}, & 0 < \tau < t; \\ 0, & \text{elsewhere.} \end{cases}$$

Hence for $0 < \tau < t$, we have

$$P\big(L_t^+ \leq \tau\big) = \int_0^\tau \frac{1}{\pi} \frac{1}{\sqrt{x(t-x)}}\, dx.$$

Make a change of variables $x = ty^2$ to get

$$P\big(L_t^+ \leq \tau\big) = \frac{2}{\pi} \int_0^{\sqrt{\tau/t}} \frac{1}{\sqrt{1-y^2}}\, dy = \frac{2}{\pi} \sin^{-1} \sqrt{\frac{\tau}{t}}.$$

Thus we have proved the following theorem.

Theorem 11.4.4. (Arc-sine law of Lévy) *Let $L_t^+ = \int_0^t \operatorname{sgn}^+(B(s))\, ds$ be the amount of time that a Brownian motion is positive during $[0,t]$. Then*

$$P\big(L_t^+ \leq \tau\big) = \frac{2}{\pi} \sin^{-1} \sqrt{\frac{\tau}{t}}, \quad 0 \leq \tau \leq t.$$

11.5 Approximation of Stochastic Integrals

In this section we will describe a relationship between the Leibniz–Newton calculus and the Itô calculus due to Wong and Zakai [87] [88].

Let $B(t)$ be a Brownian motion. Suppose $h(t,x)$ is a continuous function on $[a,b] \times \mathbb{R}$. Then the stochastic integral $\int_a^b h(t, B(t))\, dB(t)$ is defined in Chapter 5. For the numerical analysis of this integral, we need to replace the Brownian motion $B(t)$ by a sequence $\{\Phi_n(t)\}$ of stochastic processes that are reasonably smooth so that we can use the ordinary calculus. For simplicity, we fix an approximating sequence $\{\Phi_n(t)\}$ satisfying the following conditions:

(a) For each n and almost all ω, $\Phi_n(\cdot, \omega)$ is a continuous function of bounded variation on $[a,b]$;

(b) For each $t \in [a,b]$, $\Phi_n(t) \to B(t)$ almost surely as $n \to \infty$;

(c) For almost all ω, the sequence $\{\Phi_n(\cdot, \omega)\}$ is uniformly bounded, i.e.,

$$\sup_{n \geq 1} \sup_{a \leq t \leq b} |\Phi_n(t, \omega)| < \infty.$$

For example, let $\{\Delta_n\}$ be an increasing sequence of partitions of $[a,b]$ such that $\|\Delta_n\| \to 0$ as $n \to \infty$ and let $\Phi_n(t, \omega)$ be the polygonal curve with consecutive vertices the points on $B(t, \omega)$ given by the partition points of Δ_n. Then the sequence $\{\Phi_n(t)\}$ satisfies the above conditions.

When we replace $B(t)$ by $\Phi_n(t)$, then the Itô integral $\int_a^b h(t, B(t)) \, dB(t)$ becomes $\int_a^b h(t, \Phi_n(t)) \, d\Phi_n(t)$ which, by condition (a), is a Riemann–Stieltjes integral. Obviously, it is natural to ask the following question.

Question 11.5.1. Does the sequence $\int_a^b h(t, \Phi_n(t)) \, d\Phi_n(t)$, $n \geq 1$, converge to the Itô integral $\int_a^b h(t, B(t)) \, dB(t)$ almost surely as $n \to \infty$?

Example 11.5.2. Take the function $h(t, x) = x$. Then we have

$$\int_a^b \Phi_n(t) \, d\Phi_n(t) = \frac{1}{2} \left(\Phi_n(b)^2 - \Phi_n(a)^2 \right) \; \to \; \frac{1}{2} \left(B(b)^2 - B(a)^2 \right).$$

On the other hand, we have the Itô and Stratonovich integrals

$$\int_a^b B(t) \, dB(t) = \frac{1}{2} \left(B(b)^2 - B(a)^2 - (b - a) \right),$$

$$\int_a^b B(t) \circ dB(t) = \frac{1}{2} \left(B(b)^2 - B(a)^2 \right).$$

Hence we see that the Riemann–Stieltjes integrals $\int_a^b \Phi_n(t) \, d\Phi_n(t)$ converge to the Stratonovich integral $\int_a^b B(t) \circ dB(t)$ rather than the Itô integral $\int_a^b B(t) \, dB(t)$.

Now we will try to get an answer to the above question. Let $h(t, x)$ be a continuously differentiable function in both variables t and x. Then we have a Riemann–Stieltjes integral $\int_a^b h(t, \Phi_n(t)) \, d\Phi_n(t)$. Define a function

$$F(t, x) = \int_0^x h(t, y) \, dy, \quad t \in [a, b], \; x \in \mathbb{R}.$$

By the chain rule in the Leibniz–Newton calculus, we have

$$dF(t, \Phi_n(t)) = \frac{\partial F}{\partial t}(t, \Phi_n(t)) \, dt + h(t, \Phi_n(t)) \, d\Phi_n(t).$$

Therefore,

$$\int_a^b h(t, \Phi_n(t)) \, d\Phi_n(t)$$

$$= F(b, \Phi_n(b)) - F(a, \Phi_n(a)) - \int_a^b \frac{\partial F}{\partial t}(t, \Phi_n(t)) \, dt$$

$$\longrightarrow F(b, B(b)) - F(a, B(a)) - \int_a^b \frac{\partial F}{\partial t}(t, B(t)) \, dt. \qquad (11.5.1)$$

On the other hand, by Itô's formula, we have

$$dF(t, B(t)) = \frac{\partial F}{\partial t}(t, B(t)) \, dt + h(t, B(t)) \, dB(t) + \frac{1}{2}\frac{\partial h}{\partial x}(t, B(t)) \, dt,$$

which yields the following equality:

$$F(b, B(b)) - F(a, B(a)) - \int_a^b \frac{\partial F}{\partial t}(t, B(t)) \, dt$$

$$= \int_a^b h(t, B(t)) \, dB(t) + \frac{1}{2}\int_a^b \frac{\partial h}{\partial x}(t, B(t)) \, dt. \qquad (11.5.2)$$

Notice that the right-hand side of this equation equals the Stratonovich integral $\int_a^b h(t, B(t)) \circ dB(t)$. From Equations (11.5.1) and (11.5.2), we have the next theorem.

Theorem 11.5.3. *Let $\{\Phi_n(t)\}$ be an approximating sequence of a Brownian motion $B(t)$ satisfying conditions $(a), (b)$, and (c). Let $h(t, x)$ be a continuous function having continuous partial derivative $\frac{\partial h}{\partial x}$. Then*

$$\lim_{n \to \infty} \int_a^b h(t, \Phi_n(t)) \, d\Phi_n(t) = \int_a^b h(t, B(t)) \, dB(t) + \frac{1}{2}\int_a^b \frac{\partial h}{\partial x}(t, B(t)) \, dt$$

$$= \int_a^b h(t, B(t)) \circ dB(t).$$

Next, consider the approximation of a stochastic differential equation. Let $\sigma(t, x)$ and $f(t, x)$ be two functions satisfying the Lipschitz and linear growth conditions. On the one hand, the stochastic differential equation

$$dX_t = \sigma(t, X_t) \, dB(t) + f(t, X_t) \, dt, \quad X_a = \xi, \qquad (11.5.3)$$

has a unique continuous solution X_t. On the other hand, for each n and ω, let $X_t^{(n)}(\omega)$ be the unique solution of the ordinary differential equation

$$dY_t = \sigma(t, Y_t) \, d\Phi_n(t, \omega) + f(t, Y_t) \, dt, \quad Y_a = \xi(\omega). \qquad (11.5.4)$$

Question 11.5.4. Does the solution $X_t^{(n)}$ of Equation (11.5.4) converge to the solution X_t of Equation (11.5.3)?

Example 11.5.5. The ordinary differential equation

$$dY_t = Y_t \, d\Phi_n(t), \quad Y_0 = 1,$$

has the solution $X_t^{(n)} = e^{\Phi_n(t)}$, which converges to $X_t = e^{B(t)}$. In view of Example 10.1.5, $X_t^{(n)}$ converges to the solution of $dX_t = X_t \circ dB(t)$, $X_0 = 1$, rather than that of $dX_t = X_t \, dB(t)$, $X_0 = 1$.

Now we will find an answer to the above question. Assume that $\sigma(t, x)$ does not vanish and is continuously differentiable. Define a function

$$G(t, x) = \int_0^x \frac{1}{\sigma(t, y)} \, dy, \quad t \in [a, b], \ x \in \mathbb{R}.$$

Then by the chain rule in the Leibniz–Newton calculus, we have

$$dG(t, X_t^{(n)}) = \frac{\partial G}{\partial t}(t, X_t^{(n)}) \, dt + d\Phi_n(t) + \frac{f(t, X_t^{(n)})}{\sigma(t, X_t^{(n)})} \, dt.$$

Therefore,

$$G(t, X_t^{(n)}) = G(a, X_a^{(n)}) + \Phi_n(t) - \Phi_n(a)$$
$$+ \int_a^t \left[\frac{\partial G}{\partial t}(s, X_s^{(n)}) + \frac{f(s, X_s^{(n)})}{\sigma(s, X_s^{(n)})} \right] ds.$$

It is reasonable to expect that for each t, the sequence $X_t^{(n)}$ converges almost surely to, say, X_t as $n \to \infty$. Hence X_t satisfies the equation

$$G(t, X_t) = G(a, X_a) + B(t) - B(a)$$
$$+ \int_a^t \left[\frac{\partial G}{\partial t}(s, X_s) + \frac{f(s, X_s)}{\sigma(s, X_s)} \right] ds. \qquad (11.5.5)$$

On the other hand, since X_t is an Itô process, it must be of the form

$$X_t = X_a + \int_a^t \theta(s) \, dB(s) + \int_a^t g(s) \, ds, \qquad (11.5.6)$$

where θ and g are to be determined. By Itô's formula,

$$dG(t, X_t) = \frac{\partial G}{\partial t}(t, X_t) \, dt + \frac{1}{\sigma(t, X_t)} \theta(t) \, dB(t)$$
$$+ \frac{1}{\sigma(t, X_t)} g(t) \, dt - \frac{1}{2} \frac{1}{\sigma(t, X_t)^2} \frac{\partial \sigma}{\partial x}(t, X_t) \, \xi(t)^2 \, dt.$$

Hence we have the equality

$$G(t, X_t) = G(a, X_a) + \int_a^t \frac{1}{\sigma(s, X_s)} \theta(s) \, dB(s) + \int_a^t \left[\frac{\partial G}{\partial t}(s, X_s) \right.$$
$$\left. + \frac{1}{\sigma(s, X_s)} g(s) - \frac{1}{2} \frac{1}{\sigma(s, X_s)^2} \frac{\partial \sigma}{\partial x}(t, X_t) \, \theta(s)^2 \right] ds. \qquad (11.5.7)$$

By comparing Equations (11.5.5) and (11.5.7), we see that

$$\theta(s) = \sigma(s, X_s), \quad g(s) = f(s, X_s) + \frac{1}{2} \sigma(s, X_s) \frac{\partial \sigma}{\partial x}(s, X_s).$$

Therefore, Equation (11.5.6) becomes

$$X_t = X_a + \int_a^t \sigma(s, X_s)\, dB(s) + \int_a^t \left[f(s, X_s) + \frac{1}{2}\sigma(s, X_s)\frac{\partial \sigma}{\partial x}(s, X_s) \right] ds,$$

which can be written in the Stratonovich form as

$$X_t = X_a + \int_a^t \sigma(s, X_s) \circ dB(s) + \int_a^t f(s, X_s)\, ds.$$

Thus we have derived the next theorem.

Theorem 11.5.6. *Let* $\{\Phi_n(t)\}$ *be an approximating sequence of a Brownian motion* $B(t)$ *satisfying conditions* (a), (b), *and* (c). *Assume that*

(1) $\sigma(t, x)$ *and* $f(t, x)$ *satisfy the Lipschitz and linear growth conditions;*
(2) $\frac{\partial \sigma}{\partial x}$ *exists and* $\sigma\frac{\partial \sigma}{\partial x}$ *satisfies the Lipschitz and linear growth conditions.*

Let $X_t^{(n)}$ *be the solution of the ordinary differential equation*

$$dX_t^{(n)} = \sigma(t, X_t^{(n)})\, d\Phi_n(t) + f(t, X_t^{(n)})\, dt, \quad X_a^{(n)} = \xi.$$

Then $X_t^{(n)} \to X_t$ *almost surely as* $n \to \infty$ *for each* $t \in [a, b]$ *and* X_t *is the unique continuous solution of the stochastic differential equation*

$$dX_t = \sigma(t, X_t)\, dB(t) + \left(f(t, X_t) + \frac{1}{2}\sigma(t, X_t)\frac{\partial \sigma}{\partial x}(t, X_t) \right) dt, \quad X_a = \xi,$$

or equivalently,

$$dX_t = \sigma(t, X_t) \circ dB(t) + f(t, X_t)\, dt, \quad X_a = \xi.$$

11.6 White Noise and Electric Circuits

A *wide-sense stationary Gaussian process* X_t is a Gaussian process determined by a constant m, the *mean*, and a function $C(t)$, the *covariance function*, as follows:

(1) $EX_t = m$ for all $t \in \mathbb{R}$;
(2) $E(X_t - m)(X_s - m) = C(t - s)$ for all $s, t \in \mathbb{R}$.

We will assume that $C(t)$ is a continuous function on $\mathbb{R}$. Note that for any $z_1, z_2, \ldots, z_n \in \mathbb{C}$ and $t_1, t_2, \ldots, t_n \in \mathbb{R}$,

$$\sum_{j,k=1}^n z_j C(t_j - t_k)\overline{z_k} = E\left\{ \left| \sum_{j=1}^n (X_{t_j} - m) \right|^2 \right\} \geq 0,$$

which shows that the function $C(t)$ is positive definite. Hence by the Bochner theorem [5], there exists a unique measure μ on $\mathbb{R}$ such that

$$C(t) = \int_{\mathbb{R}} e^{it\lambda} \, d\mu(\lambda), \quad t \in \mathbb{R}.$$

Note that $\mu(\mathbb{R}) = C(0) \geq 0$. If μ has a density function f, then

$$C(t) = \int_{\mathbb{R}} e^{it\lambda} f(\lambda) \, d\lambda, \quad t \in \mathbb{R}.$$

The function $f(\lambda)$ is called the *spectral density function* of X_t. If $C(t)$ is an integrable function, then $f(\lambda)$ is given by

$$f(\lambda) = \int_{\mathbb{R}} e^{-2\pi i \lambda t} C(t) \, dt, \quad \lambda \in \mathbb{R}. \tag{11.6.1}$$

Note that $f(0) - \int_{\mathbb{R}} C(t) \, dt$.

Example 11.6.1. Let $B_1(t)$ and $B_2(t)$ be two independent Brownian motions with $t \geq 0$. Define a stochastic process $B(t)$ by

$$B(t) = \begin{cases} B_1(t), & \text{if } t \geq 0; \\ B_2(-t), & \text{if } t < 0. \end{cases}$$

Then $B(t), t \in \mathbb{R}$, is a Brownian motion. Since $E[B(t)B(s)] = t \wedge s$ for $t, s \geq 0$, $B(t)$ is not a wide-sense stationary Gaussian process. However, for each fixed $\varepsilon > 0$, the stochastic process

$$X_t^{(\varepsilon)} = \frac{1}{\varepsilon}\big(B(t+\varepsilon) - B(t)\big), \quad t \in \mathbb{R},$$

is a wide-sense stationary Gaussian process. To see this fact, first note that $X_t^{(\varepsilon)}$ is a Gaussian process and $EX_t^{(\varepsilon)} = 0$ for all $t \in \mathbb{R}$. For $0 \leq s \leq t$, it is easy to check that

$$E\big(X_s^{(\varepsilon)} X_t^{(\varepsilon)}\big) = \frac{1}{\varepsilon^2}\Big(s + \varepsilon - \min\{t, s+\varepsilon\}\Big). \tag{11.6.2}$$

Observe that

$$\min\{t, s+\varepsilon\} = s + \min\{t - s, \varepsilon\} = s + \varepsilon \min\Big\{\frac{t-s}{\varepsilon}, 1\Big\}$$

$$= s + \varepsilon\Big(1 - \max\Big\{1 - \frac{t-s}{\varepsilon}, 0\Big\}\Big)$$

$$= s + \varepsilon - \varepsilon \max\Big\{1 - \frac{t-s}{\varepsilon}, 0\Big\}. \tag{11.6.3}$$

Put Equation (11.6.3) into Equation (11.6.2) to get

$$E\big(X_s^{(\varepsilon)} X_t^{(\varepsilon)}\big) = \frac{1}{\varepsilon} \max\Big\{1 - \frac{t-s}{\varepsilon}, 0\Big\}, \quad \forall 0 \leq s \leq t.$$

Obviously, this equality also holds for any $s \leq t \leq 0$. For the case $s \leq 0 \leq t$, it is a little bit tedious to derive $E\big(X_s^{(\varepsilon)} X_t^{(\varepsilon)}\big)$, but we have the following equality for the covariance function of $X_t^{(\varepsilon)}$:

$$C_\varepsilon(t-s) = E\big(X_s^{(\varepsilon)} X_t^{(\varepsilon)}\big) = \frac{1}{\varepsilon} \max\Big\{1 - \frac{|t-s|}{\varepsilon}, 0\Big\}, \quad \forall\, s, t \in \mathbb{R}.$$

This shows that $X_t^{(\varepsilon)}$ is a wide-sense stationary Gaussian process. Then we can use Equation (11.6.1) to find the spectral density function of $X_t^{(\varepsilon)}$,

$$f_\varepsilon(\lambda) = \Big(\frac{\sin \pi\lambda\varepsilon}{\pi\lambda\varepsilon}\Big)^2, \quad \lambda \in \mathbb{R}, \tag{11.6.4}$$

where it is understood that $f_\varepsilon(0) = 1$.

The above example leads to the following question.

Question 11.6.2. What is the limiting process of $X_\varepsilon(t) = \frac{1}{\varepsilon}\big(B(t+\varepsilon) - B(t)\big)$ as $\varepsilon \downarrow 0$?

Symbolically, we have $\lim_{\varepsilon \downarrow 0} X_\varepsilon(t) = \dot{B}(t)$. But $\dot{B}(t)$ is not an ordinary stochastic process because Brownian paths are nowhere differentiable. Let us observe more closely the density function $f_\varepsilon(\lambda)$ in Equation (11.6.4). For a very small number ε, the graph of $f_\varepsilon(\lambda)$ is very close to 1 for a large range of λ and oscillates to 0 outside this range. Moreover, we have

$$\lim_{\varepsilon \downarrow 0} f_\varepsilon(\lambda) = 1, \quad \forall\, \lambda \in \mathbb{R}. \tag{11.6.5}$$

The limiting function, being identically equal to 1, is not the spectral density function of an ordinary wide-sense stationary Gaussian process. This is the motivation for the concept of white noise in the next definition.

Definition 11.6.3. *A* white noise *is defined to be a generalized wide-sense stationary Gaussian process* Z_t *with mean* $EZ_t = 0$ *and covariance function* $E(Z_t Z_s) = \delta_0(t-s)$. *Here* δ_0 *is the Dirac delta function at* 0.

The white noise Z_t is termed a "generalized" stochastic process because its covariance function $\delta_0(t)$ is a generalized function. By Equation (11.6.1), the spectral density function of Z_t is given by

$$f(\lambda) = \int_{\mathbb{R}} e^{-2\pi i\lambda t} \delta_0(t)\, dt = 1, \quad \forall\, \lambda \in \mathbb{R},$$

which means that Z_t represents a sound with equal intensity at all frequencies. This is the reason why the noise is called "white," in analogy to full-spectrum light being referred to as white in color. From Example 11.6.1, we see that the derivative $\dot{B}(t)$ is a white noise. In fact, white noise is often defined as the informal derivative of a Brownian motion.

A white noise Z_t has the properties that Z_t is independent at different times t and has identical distribution with $EZ_t = 0$ and $\text{var}(Z_t) = \infty$. Because of its distinct properties, white noise is often used as an idealization of a random noise that is independent at different times and has a very large fluctuation at any time.

A deterministic system under the influence of such a random noise gives rise to the following random equation:

$$\frac{dX_t}{dt} = f(t, X_t) + \sigma(t, X_t)Z_t, \quad X_a = \xi. \tag{11.6.6}$$

This equation is just an informal expression. In order to make sense out of this equation, let us rewrite it in the stochastic differential form

$$dX_t = f(t, X_t)\, dt + \sigma(t, X_t)Z_t\, dt,$$

which means the following stochastic integral equation:

$$X_t = \xi + \int_a^t f(s, X_s)\, ds + \int_a^t \sigma(s, X_s)Z_s\, ds.$$

How do we interpret the second integral? Take Z_t to be $\dot{B}(t)$ and write $Z_s\, ds$ as $dB(s)$. But then there are two ways to interpret the integral with respect to $dB(s)$, namely, as an Itô integral and as a Stratonovich integral. Thus the random equation in Equation (11.6.6) has two interpretations as follows:

$$dX_t = \sigma(t, X_t)\, dB(t) + f(t, X_t)\, dt, \tag{11.6.7}$$

$$dX_t = \sigma(t, X_t) \circ dB(t) + f(t, X_t)\, dt. \tag{11.6.8}$$

Note that Equation (11.6.8) can be rewritten as

$$dX_t = \sigma(t, X_t)\, dB(t) + \left\{ f(t, X_t) + \frac{1}{2}\sigma(t, X_t)\frac{\partial \sigma}{\partial x}(t, X_t) \right\} dt.$$

Hence if $\sigma(t, x)$ does not depend on x, then Equations (11.6.7) and (11.6.8) coincide and the two interpretations of Equation (11.6.6) are the same.

Consider an electric circuit, in which the charge Q_t at time t at a fixed point satisfies the equation

$$I\frac{d^2}{dt^2}Q_t + R\frac{d}{dt}Q_t + \frac{1}{C}Q_t = f(t) + g(t)\,Z_t, \tag{11.6.9}$$

where I is the inductance, R is the resistance, C is the capacitance, $f(t)$ is a deterministic function representing the potential source, $g(t)$ is an adapted stochastic process, and Z_t is a white noise. In fact, a harmonic oscillator with external force $f(t)$ under the white noise influence also takes the form of Equation (11.6.9).

We can convert Equation (11.6.9) into a first-order matrix differential equation. Let $X_t = (X_t^{(1)}, X_t^{(2)})$ with $X_t^{(1)} = Q_t$ and $X_t^{(2)} = \frac{d}{dt}Q_t$. Then Equation (11.6.9) can be rewritten as a linear matrix stochastic differential equation, which can be solved by the two-dimensional version of Theorem 11.1.1. From the solution X_t, we automatically obtain the solution Q_t of Equation (11.6.9).

On the other hand, we can leave Equation (11.6.9) as it is with Z_t being replaced by the informal derivative $\dot{B}(t)$ of a Brownian motion $B(t)$. First find the fundamental solutions of the associated homogeneous equation, which does not involve the white noise. Then we use the method of the variation of parameter as in ordinary differential equations to derive a particular solution of Equation (11.6.9). Finally, we can obtain the general solution of Equation (11.6.9) from the fundamental solutions and the particular solution. Below we give several simple examples to illustrate this technique.

Example 11.6.4. Consider the second-order differential equation

$$\frac{d^2}{dt^2}Q_t + 3\frac{d}{dt}Q_t + 2Q_t = \dot{B}(t). \qquad (11.6.10)$$

Since the characteristic equation $r^2 + 3r + 2 = 0$ has two roots $r = -1, -2$, we obtain the fundamental solutions $\{e^{-t}, e^{-2t}\}$. To find a particular solution of Equation (11.6.10), let

$$Q_t = u_1 e^{-t} + u_2 e^{-2t}, \qquad (11.6.11)$$

where u_1 and u_2 are to be determined. Then we have

$$\frac{d}{dt}Q_t = u_1' e^{-t} - u_1 e^{-t} + u_2' e^{-2t} - 2u_2 e^{-2t}.$$

Impose the condition on u_1 and u_2 such that

$$u_1' e^{-t} + u_2' e^{-2t} = 0. \qquad (11.6.12)$$

Then $\frac{d}{dt}Q_t$ is given by

$$\frac{d}{dt}Q_t = -u_1 e^{-t} - 2u_2 e^{-2t}. \qquad (11.6.13)$$

Differentiate both sides to get

$$\frac{d^2}{dt^2}Q_t = -u_1' e^{-t} + u_1 e^{-t} - 2u_2' e^{-2t} + 4u_2 e^{-2t}. \qquad (11.6.14)$$

Put Equations (11.6.11), (11.6.13), and (11.6.14) into Equation (11.6.10) to obtain the following equation:

$$u_1' e^{-t} + 2u_2' e^{-2t} = -\dot{B}(t). \qquad (11.6.15)$$

Now we can solve Equations (11.6.12) and (11.6.15) to get $u_1' = e^t \dot{B}(t)$ and $u_2' = -e^{2t} \dot{B}(t)$. Therefore,

$$u_1 = \int_0^t e^s \, dB(s), \quad u_2 = -\int_0^t e^{2s} \, dB(s).$$

Put u_1 and u_2 into Equation (11.6.11) to get a particular solution and then we have the general solution of Equation (11.6.10),

$$Q_t = c_1 \, e^{-t} + c_2 \, e^{-2t} + \int_0^t \left(e^{-(t-s)} - e^{-2(t-s)} \right) dB(s),$$

where the integral is a Wiener integral.

Example 11.6.5. Consider the second-order differential equation

$$\frac{d^2}{dt^2} Q_t + 3 \frac{d}{dt} Q_t + 2Q_t = B(t) \dot{B}(t). \qquad (11.6.16)$$

We will interpret this white noise equation in the Itô sense. This equation has the same fundamental solutions $\{e^{-t}, e^{-2t}\}$ as those in Example 11.6.4. To find a particular solution, let

$$Q_t = u_1 e^{-t} + u_2 e^{-2t}. \qquad (11.6.17)$$

By the same computation as that in Example 11.6.4, we have the following equations for u_1' and u_2':

$$\begin{cases} u_1' e^{-t} + u_2' e^{-2t} = 0, \\ u_1' e^{-t} + 2u_2' e^{-2t} = -B(t) \dot{B}(t). \end{cases}$$

Hence u_1 and u_2 are given by

$$u_1 = \int_0^t e^s B(s) \dot{B}(s) \, ds = \int_0^t e^s B(s) \, dB(s),$$

$$u_2 = -\int_0^t e^{2s} B(s) \dot{B}(s) \, ds = -\int_0^t e^{2s} B(s) \, dB(s),$$

where the integrals are Itô integrals. Put u_1 and u_2 into Equation (11.6.17) to get a particular solution. Then we have the general solution

$$Q_t = c_1 \, e^{-t} + c_2 \, e^{-2t} + \int_0^t \left(e^{-(t-s)} - e^{-2(t-s)} \right) B(s) \, dB(s).$$

Example 11.6.6. Consider the second-order differential equation

$$\frac{d^2}{dt^2} Q_t + 3 \frac{d}{dt} Q_t + 2Q_t = \ddot{B}(t), \qquad (11.6.18)$$

where the informal second derivative $\ddot{B}(t)$ of a Brownian motion $B(t)$ can be regarded as a higher-order white noise. Since this equation has the same fundamental solutions $\{e^{-t}, e^{-2t}\}$ as those in Example 11.6.4, we will just find a particular solution of this equation. As before, let

$$Q_t = u_1 e^{-t} + u_2 e^{-2t}. \qquad (11.6.19)$$

By the same computation as that in Example 11.6.4, we have

$$\begin{cases} u_1' e^{-t} + u_2' e^{-2t} = 0, \\ u_1' e^{-t} + 2u_2' e^{-2t} = -\ddot{B}(t). \end{cases}$$

Now, the situation to find u_1 and u_2 is very different from Examples 11.6.4 and 11.6.5 because we cannot get rid of $\dot{B}(t)$ and $\dot{B}(0)$. However, we can apply integration by parts formula to obtain

$$u_1 = e^t \dot{B}(t) - \dot{B}(0) - \int_0^t e^s \, dB(s),$$

$$u_2 = -e^{2t} \dot{B}(t) + \dot{B}(0) + 2 \int_0^t e^{2s} \, dB(s).$$

Therefore, we have the particular solution

$$Q_t = e^{-t}\Big\{ e^t \dot{B}(t) - \dot{B}(0) - \int_0^t e^s \, dB(s) \Big\}$$

$$+ e^{-2t}\Big\{ -e^{2t} \dot{B}(t) + \dot{B}(0) + 2 \int_0^t e^{2s} \, dB(s) \Big\}$$

$$= (e^{-2t} - e^{-t})\dot{B}(0) + \int_0^t \big(2e^{-2(t-s)} - e^{-(t-s)} \big) \, dB(s).$$

Notice the cancellation of the $\dot{B}(t)$ terms in the first equality. Although $\dot{B}(0)$ remains, this term can be dropped because e^{-t} and e^{-2t} are the fundamental solutions. Hence a particular solution of Equation (11.6.18) is given by

$$Q_t = \int_0^t \big(2e^{-2(t-s)} - e^{-(t-s)} \big) \, dB(s).$$

In the above examples, we have used the white noise in an informal manner to derive particular solutions. The white noise $\dot{B}(t)$ is combined with dt to form $dB(t)$ for the Itô theory of stochastic integration. In Example 11.6.6 we are very lucky to have the cancellation of the $\dot{B}(t)$ terms and even luckier to be able to drop the $\dot{B}(0)$ term (in Q_t rather than in u_1 and u_2). So, this brings up the following question:

Can we justify the above informal manipulation of white noise?

The answer is affirmative. For the mathematical theory of white noise, see the books [27], [28], [56], and [65].

Exercises

1. Let $f(t), g(t)$, and $h(t)$ be continuous functions on $[a, b]$. Show that the solution of the stochastic differential equation

$$dX_t = f(t)\,dB(t) + \big(g(t)X_t + h(t)\big)\,dt, \quad X_a = x,$$

is a Gaussian process, i.e., $\sum_{i=1}^{n} c_i X_{t_i}$ is a Gaussian random variable for any $c_i \in \mathbb{R}, t_i \in [a, b], i = 1, 2, \ldots, n$. Find the mean function $m(t) = EX_t$ and the covariance function $C(s, t) = E\{[X_s - m(s)][X_t - m(t)]\}$.

2. Let a market be given by $X_t = (1, B(t), B(t)^2)$. Find a stochastic process $p_0(t)$ such that the portfolio $p(t) = (p_0(t), B(t), t)$ is self-financing in the market X_t.

3. Suppose a market is given by $X_t = (1, t)$. Show that the portfolio $p(t) = (-tB(t) + \int_0^t B(s)\,ds, B(t))$ is self-financing, but not admissible.

4. Suppose a market is given by $X_t = (1, B(t))$. Show that the portfolio $p(t) = (-\frac{1}{2}B(t)^2 - \frac{1}{2}t, B(t))$ is admissible, but not an arbitrage.

5. Suppose a market is given by $X_t = (1, t)$. Show that the portfolio $p(t) = (-tB(t)^2 + \int_0^t B(s)^2\,ds, B(t)^2)$ is an arbitrage.

6. Suppose X_t is a local martingale on $[0, T]$ and is lower bounded for almost all $t \in [0, T]$ and $\omega \in \Omega$. Show that X_t is a supermartingale.

7. Let $B_1(t)$ and $B_2(t)$ be independent Brownian motions. Check whether a market has an arbitrage for each one of the following markets:

   (a) $X_t = (1, 2 + B_1(t), -t + B_1(t) + B_2(t))$.

   (b) $X_t = (1, 2 + B_1(t) + B_2(t), -t - B_1(t) - B_2(t))$.

   (c) $X_t = (e^t, B_1(t), B_2(t))$.

8. Let $B(t)$ be a Brownian motion. Find the conditional expectation:

   (a) $E[B(s) \mid B(t)]$ for $s < t$.

   (b) $E[\int_0^t B(s)\,ds \mid B(t)]$.

   (c) $E[\int_0^t s\,dB(s) \mid B(t)]$.

9. Let ξ and X be independent Gaussian random variables with mean 0 and variances σ_1^2 and σ_2^2, respectively. Find the conditional expectation $E[\xi \mid X + \xi]$.

10. Let the state ξ_t and the observation Z_t in a Kalman–Bucy linear filtering be given by $d\xi_t = dB(t)$ with ξ_0 being normally distributed with mean 1 and variance 2 and $dZ_t = 2\,dW(t) + \xi_t\,dt$ with $Z_0 = 0$. Find the estimator $\widehat{\xi}_t$ for ξ_t.

11. Let $\Phi_n(t)$ be the polygonal approximation of a Brownian motion $B(t)$. Show that $\{\Phi_n(t)\}$ satisfies conditions (a), (b), and (c) in Section 11.5.

12. Let $\{\Phi_n(t)\}$ be an approximating sequence of a Brownian motion $B(t)$ satisfying conditions (a), (b), and (c) in Section 11.5. Find the explicit expression of the solution $X_t^{(n)}$ of the ordinary differential equation

$$\frac{d}{dt}Y_t = Y_t\,d\Phi_n(t) + 3t, \quad Y_0 = 1.$$

Find the limit $\lim_{n\to\infty} X_t^{(n)}$.

13. Let $f(t,x)$ be a continuous function with continuous partial derivative $\frac{\partial f}{\partial x}$. Use Theorem 11.5.3 to show that

$$\int_a^b f(t,B(t))\,dB(t) = \lim_{n\to\infty}\left[\int_a^b f(t,\Phi_n(t))\,d\Phi_n(t) - \frac{1}{2}\int_a^b \frac{\partial f}{\partial x}(t,\Phi_n(t))\,dt\right].$$

14. Let $X_t^{(n)}$ be the solution of the ordinary differential equation

$$dX_t^{(n)} = \sigma(t,X_t^{(n)})\,d\Phi_n(t) + \left(f(t,X_t^{(n)}) - \frac{1}{2}\sigma(t,X_t^{(n)})\sigma'(t,X_t^{(n)})\right)dt$$

with initial condition $X_a^{(n)} = \xi$. Let X_t be the solution of the stochastic differential equation

$$dX_t = \sigma(t,X_t)\,dB(t) + f(t,X_t)\,dt, \quad X_a = \xi.$$

Use Theorem 11.5.6 to show that $X_t^{(n)}$ converges to X_t almost surely as $n \to \infty$ for each $t \in [a,b]$.

15. Let X_t be the Ornstein–Uhlenbeck process given in Theorem 7.4.7. Show that X_t is a wide-sense stationary Gaussian process. Find its spectral density function.

16. Find a particular solution for each of the following white noise equations:

(a) $\frac{d^2}{dt^2}X_t + X_t = \dot{B}(t)$.

(b) $\frac{d^2}{dt^2}X_t + X_t = B(t)\dot{B}(t)$.

(c) $\frac{d^2}{dt^2}X_t + X_t = \ddot{B}(t)$.

17. Find a particular solution of Equation (11.6.16) when the equation is interpreted in the Stratonovich sense.

References

1. Arnold, L.: *Stochastic Differential Equations: Theory and Applications.* John Wiley & Sons, 1974.
2. Black, F. and Scholes, M.: The pricing of options and corporate liabilities; *J. Political Economy* **81** (1973) 637–659.
3. Brown, R.: A brief account of microscopical observations made in the months of June, July and August, 1827, on the particles contained in the pollen of plants; and on the general existence of active molecules in organic and inorganic bodies; *Phil. Mag.* **4** (1828) 161–173.
4. Cameron R. H. and Martin, W. T.: Transformation of Wiener integrals under trnaslations; *Annals of Mathematics* **45** (1944) 386–396.
5. Chung, K. L.: *A Course in Probability Theory.* Second edition, Academic Press, 1974.
6. Chung, K. L. and Williams, R. J.: *Introduction to Stochastic Integration.* Second edition, Birkhäuser, 1990.
7. Dellacherie, C.: *Capacités et Processus Stochastiques.* Springer-Verlag, 1972.
8. Dellacherie, C. and Meyer, P. A.: *Probabilities and Potentials B. Theory of Martingales.* North Holland, 1982.
9. Doob, J. L.: *Stochastic Processes.* John Wiley & Sons, 1953.
10. Dudley, R. M.: *Real Analysis and Probability.* Wadsworth & Brooks/Cole, 1989.
11. Durrett, R.: *Stochastic Calculus.* CRC Press, 1996.
12. Dynkin, E. B.: *Markov Processes I, II.* Springer-Verlag, 1965.
13. Elliott, R. J.: *Stochastic Calculus and Applications.* Springer-Verlag, 1982.
14. Elworthy, K. D.: *Stochastic Differential Equations on Manifolds.* Cambridge University Press, 1982.
15. Emery, M.: *An Introduction to Stochastic Processes.* Universitext, Springer-Verlag, 1989.
16. Feynman, R. P.: Space-time approach to nonrelativistic quantum mechanics; *Reviews of Modern Physics* **20** (1948) 367–387.
17. Fleming, W.H. and Rishel, R.W.: *Deterministic and Stochastic Optimal Control.* Springer-Verlag, 1975.

268   References

18. Freedman, D.: *Brownian Motion and Diffusion*. Springer-Verlag, 1983.

19. Friedman, A.: *Stochastic Differential Equations and Applications*, volume I, Academic Press, 1975.

20. Friedman, A.: *Stochastic Differential Equations and Applications*, volume II, Academic Press, 1976.

21. Gard, T. C.: *Introduction to Stochastic Differential Equations*. Marcel Dekker, 1988.

22. Gihman, I. I. and Skorohod, A. V.: *Stochastic Differential Equations*. Springer-Verlag, 1974.

23. Girsanov, I. V.; On transforming a certain class of stochastic processes by absolutely continuous substitution of measures; *Theory of Probability and Applications* **5** (1960) 285–301.

24. Gross, L.: Abstract Wiener spaces; *Proc. 5th Berkeley Symp. Math. Stat. and Probab.* **2**, part 1 (1965) 31–42, University of California Press, Berkeley.

25. Hida, T.: *Brownian Motion*. Springer-Verlag, 1980.

26. Hida, T. and Hitsuda, M.: *Gaussian Processes*. Translations of Mathematics Monographs, vol. 120, Amer. Math. Soc., 1993.

27. Hida, T., Kuo, H.-H., Potthoff, J., and Streit, L.: *White Noise: An Infinite Dimensional Calculus*. Kluwer Academic Publishers, 1993.

28. Holden, H., Øksendal, B., Ubøe, J., and Zhang, T.: *Stochastic Partial Differential Equations*. Birkhäuser, 1996.

29. Hunt, G. A.: Random Fourier transforms; *Trans. Amer. Math. Soc.* **71** (1951) 38–69.

30. Ikeda, N. and Watanabe, S.: *Stochastic Differential Equations and Diffusion Processes*. Second edition, North-Holland/Kodansha, 1989.

31. Itô, K.: Stochastic integral; *Proc. Imp. Acad. Tokyo* **20** (1944) 519–524.

32. Itô, K.: On a stochastic integral equation; *Proc. Imp. Acad. Tokyo* **22** (1946) 32–35.

33. Itô, K.: *On Stochastic Differential Equations*. Memoir, Amer. Math. Soc., vol. **4**, 1951.

34. Itô, K.: Multiple Wiener integral; *J. Math. Soc. Japan* **3** (1951) 157–169.

35. Itô, K.: On a formula concerning stochastic differentials; *Nagoya Math. J.* **3** (1951) 55–65 **3** (1951) 157–169.

36. Itô, K.: Extension of stochastic integrals; *Proc. International Symp. Stochastic Differential Equations*, K. Itô (ed.) (1978) 95–109, Kinokuniya.

37. Itô, K.: *Introduction to Probability Theory*. Cambridge University Press, 1978.

38. Itô, K.: *Selected Papers*. D. W. Stroock and S. R. S. Varadhan (eds.), Springer-Verlag, 1987.

39. Itô, K.: *Stochastic Processes*. O. E. Barndorff-Nielsen and K. Sato (editors) Springer-Verlag, 2004.

40. Itô, K. and McKean, H. P.: *Diffusion Processes and Their Sample Paths*. Springer-Verlag, 1965.

41. Itô, K. and Nisio, M.: On the convergence of sums of independent Banach space valued random variables; *Osaka J. Math.* **5** (1968) 35–48.

42. Iyanaga, S. and Kawada, Y. (editors): Encyclopedic Dictionary of Mathematics, by Mathematical Society of Japan. English translation reviewed by K. O. May, MIT Press, 1977.

43. Jacod, J.: *Calcul Stochastique et Problèmes de Martingales*. Lecture Notes in Math., **714** (1979) Springer-Verlag.

44. Kac, M.: On distributions of certain Wiener functionals; *Trans. Amer. Math. Soc.* **65** (1949) 1–13.

45. Kallianpur, G.: *Stochastic Filtering Theory*. Springer-Verlag, 1980.

46. Kalman, R. E. and Bucy, R. S.: New results in linear filtering and prediction theory; *Trans. Amer. Soc. Mech. Engn., Series D, J. Basic Eng.* **83** (1961) 95–108.

47. Karatzas, I. On the pricing of American options; *Appl. Math. Optimization* **17** (1988) 37–60.

48. Karatzas, I. *Lectures on the Mathematics of Finance*. Amer. Math. Soc., 1997.

49. Karatzas, I. and Shreve, S. E.: *Brownian Motion and Stochastic Calculus*. Second edition, Springer-Verlag, 1991.

50. Kloeden, P. E. and Platen, E.: *Numerical Solution of Stochastic Differential Equations*. Springer-Verlag, 1992.

51. Knight, F. B.: *Essentials of Brownian Motion*. Amer. Math. Soc., 1981.

52. Kolmogorov, A. N.: *Grundbegriffe der Wahrscheinlichkeitsrechnung*. Ergeb. Math. **2**, 1933 (English translation: Foundations of Probability Theory. Chelsea Publ. Co., 1950).

53. Kopp, P.: *Martingales and Stochastic Integrals*. Cambridge University Press, 1984.

54. Kunita, H.: *Stochastic Flows and Stochastic Differential Equations*. Cambridge University Press, 1990.

55. Kuo, H.-H.: *Gaussian Measures in Banach Spaces*. Lecture Notes in Math. **463**, Springer-Verlag, 1975.

56. Kuo, H.-H.: *White Noise Distribution Theory*. CRC Press, 1996.

57. Lamperti, J.: *Probability*. W. A. Benjamin, Inc., 1966.

58. Lévy, P.: *Processus Stochastiques et Mouvement Brownien*. Gauthier-Villars, Paris, 1948.

59. McKean, H. P.: *Stochastic Integrals*. Academic Press, 1969.

60. Merton, R.: Theory of rational option pricing; *Bell Journal of Economics and Management Science* **4** (1973) 141–183.

61. Métivier, M.: *Semimartingales: a course on stochastic processes*. Walter de Gruyter & Co., 1982.

62. Métivier, M. and Pellaumail, J.: *Stochastic Integration*. Academic Press, 1980.

63. Mikosch, T.: *Elementary Stochastic Calculus with Finance in View*. World Scientific, 1998.

64. Novikov, A. A.: On an identity for stochastic integrals; *Theory of Probability and Its Applications* **17** (1972) 717–720.

65. Obata, N.: *White Noise Calculus and Fock Space*. Lecture Notes in Math. **1577**, Springer-Verlag, 1994.

66. Øksendal, B.: *Stochastic Differential Equations.* 5th edition, Springer, 2000.

67. Paley, R. and Wiener, N.: *Fourier Transforms in the Complex Domain.* Amer. Math. Soc. Coll. Publ. XIX, 1934.

68. Protter, P.: *Stochastic Integration and Differential Equations.* Second edition, Springer-Verlag, 2004.

69. Reed, M. and Simon, B.: *Methods of Modern Mathematical Physics I: Functional Analysis.* Academic Press, 1972.

70. Revuz, D. and Yor, M.: *Continuous Martingales and Brownian Motion.* Springer-Verlag, 1991.

71. Rogers, L.C.G. and Williams, D.: *Diffusions, Markov Processes, and Martingales.* vol. 2, John Wiley & Sons, 1987.

72. Ross, S. M.: *Introduction to Probability Models.* 4th edition, Academic Press, 1989.

73. Royden, H. L.: *Real Analysis.* 3rd Edition, Prentice Hall, 1988.

74. Rudin, W.: *Principles of Mathematical Analysis.* 3rd edition, McGraw-Hill, 1976.

75. Sengupta, A. N.: *Pricing Derivatives: The Financial Concepts Underlying the Mathematics of Pricing Derivatives.* McGraw-Hill, 2005.

76. Stratonovich, R. L.: A new representation for stochastic integrals and equations; *J. Siam Control* **4** (1966) 362–371.

77. Stroock, D. W.: *Topics in Stochastic Differential Equations.* Tata Institute of Fundamental Research, Springer-Verlag, 1981.

78. Stroock, D. W. and Varadhan, S. R. S.: *Multidimensional Diffusion Processes.* Springer-Verlag, 1979.

79. Sussman, H.: On the gap between deterministic and stochastic ordinary differential equations; *Annals of Probability* **6** (1978) 19–41.

80. Tanaka, H.: Note on continuous additive functionals of the one-dimensional Brownian path; *Z. Wahrscheinlichkeitstheorie* **1** (1963) 251–257.

81. Varadhan, S. R. S.: *Stochastic Processes.* Courant Institute of Mathematical Sciences, New York University, 1968.

82. Wiener, N.: Differential space; *J. Math. Phys.* **58** (1923) 131–174.

83. Wiener, N.: The homogeneous chaos; *Amer. J. Math.* **60** (1938) 897–936.

84. Williams, D.: *Diffusions, Markov Processes, and Martingales.* John Wiley & Sons, 1979.

85. Williams, D.: *Probability with Martingales.* Cambridge University Press, 1991.

86. Wong, E. and Hajek, B.: *Stochastic Processes in Engineering Systems.* Springer-Verlag, 1985.

87. Wong, E. and Zakai, M.: On the convergence of ordinary integrals to stochastic integrals; *Ann. Math. Stat.* **36** (1965) 1560–1564.

88. Wong, E. and Zakai, M.: On the relation between ordinary and stochastic differential equations; *Intern. J. Engr. Sci.* **3** (1965) 213–229.

89. Yosida, K.: *Functional Analysis.* Springer-Verlag, Sixth edition, 1980.

Glossary of Notation

Index

Universitext   *(continued)*

Jones/Morris/Pearson: Abstract Algebra and Famous Impossibilities
Kac/Cheung: Quantum Calculus
Kannan/Krueger: Advanced Analysis
Kelly/Matthews: The Non-Euclidean Hyperbolic Plane
Kostrikin: Introduction to Algebra
Kuo: Introduction to Stochastic Integration
Kurzweil/Stellmacher: The Theory of Finite Groups: An Introduction
Lang: Introduction to Differentiable Manifolds
Lorenz: Algebra: Volume I: Fields and Galois Theory
Luecking/Rubel: Complex Analysis: A Functional Analysis Approach
MacLane/Moerdijk: Sheaves in Geometry and Logic
Marcus: Number Fields
Martinez: An Introduction to Semiclassical and Microlocal Analysis
Matsuki: Introduction to the Mori Program
McCarthy: Introduction to Arithmetical Functions
McCrimmon: A Taste of Jordan Algebras
Meyer: Essential Mathematics for Applied Fields
Mines/Richman/Ruitenburg: A Course in Constructive Algebra
Moise: Introductory Problems Course in Analysis and Topology
Morris: Introduction to Game Theory
Poizat: A Course In Model Theory: An Introduction to Contemporary Mathematical Logic
Polster: A Geometrical Picture Book
Porter/Woods: Extensions and Absolutes of Hausdorff Spaces
Procesi: Lie Groups
Radjavi/Rosenthal: Simultaneous Triangularization
Ramsay/Richtmyer: Introduction to Hyperbolic Geometry
Reisel: Elementary Theory of Metric Spaces
Ribenboim: Classical Theory of Algebraic Numbers
Rickart: Natural Function Algebras
Rotman: Galois Theory
Rubel/Colliander: Entire and Meromorphic Functions
Runde: A Taste of Topology
Sagan: Space-Filling Curves
Samelson: Notes on Lie Algebras
Schiff: Normal Families
Shapiro: Composition Operators and Classical Function Theory
Simonnet: Measures and Probability
Smith: Power Series From a Computational Point of View
Smith/Kahanpää/Kekäläinen/Traves: An Invitation to Algebraic Geometry
Smorynski: Self-Reference and Modal Logic
Stillwell: Geometry of Surfaces
Stroock: An Introduction to the Theory of Large Deviations
Sunder: An Invitation to von Neumann Algebras
Tondeur: Foliations on Riemannian Manifolds
Toth: Finite Möbius Groups, Minimal Immersions of Spheres, and Moduli
Van Brunt: The Calculus of Variations
Weintraub: Galois Theory
Wong: Weyl Transforms
Zhang: Matrix Theory: Basic Results and Techniques
Zong: Sphere Packings
Zong: Strange Phenomena in Convex and Discrete Geometry